Advanced Computing Solutions for Healthcare

Edited by

Sivakumar Rajagopal
Department of Sensor and Biomedical Technology
School of Electronics Engineering, Vellore Institute of Technology
Vellore, Tamilnadu-632014, India

Prakasam P.
Department of Communication Engineering
School of Electronics Engineering, Vellore Institute of Technology
Vellore, Tamilnadu-632014, India

Konguvel E.
Department of Embedded Technology
School of Electronics Engineering, Vellore Institute of Technology
Vellore, Tamilnadu-632014, India

Shamala Subramaniam
Department of Communication Technology and Networks
Faculty of Computer Science and Information Technology
University Putra Malaysia, Serdang-43400, Malaysia

Ali Safaa Sadiq Al Shakarchi

Department of Computer Science
Nottingham Trent University, Nottingham, UK

&

B. Prabadevi

School of Computer Science Engineering and Information
Systems Vellore Institute of Technology, Vellore
Tamilnadu-632014, India

Advanced Computing Solutions for Healthcare

Editors: Sivakumar Rajagopal, Prakasam P., Konguvel E., Shamala Subramaniam, Ali Safaa Sadiq Al Shakarchi & B. Prabadevi

ISBN (Online): 978-981-5274-13-4

ISBN (Print): 978-981-5274-14-1

ISBN (Paperback): 978-981-5274-15-8

need for a court order if at any point you breach any terms of this License Agreement. In no event will any delay or failure by Bentham Science Publishers in enforcing your compliance with this License Agreement constitute a waiver of any of its rights.

3. You acknowledge that you have read this License Agreement, and agree to be bound by its terms and conditions. To the extent that any other terms and conditions presented on any website of Bentham Science Publishers conflict with, or are inconsistent with, the terms and conditions set out in this License Agreement, you acknowledge that the terms and conditions set out in this License Agreement shall prevail.

Bentham Science Publishers Pte. Ltd.
80 Robinson Road #02-00
Singapore 068898
Singapore
Email: subscriptions@benthamscience.net

BENTHAM SCIENCE

CONTENTS

FOREWORD I

In the ever-evolving landscape of healthcare, where the amalgamation of technology and innovation is steering the course of progress, "Advanced Computing Solutions for Healthcare" emerges as a beacon illuminating the transformative potential that lies at the intersection of advanced computing and the well-being of humanity.

Over the past decade, we have witnessed an unprecedented shift in healthcare systems, propelled by the relentless march of technology—morphing our conventional understanding into a realm dominated by computers, the internet and mobile devices. This paradigm shift has not only ushered in remarkable advancements but has also presented us with a host of challenges that demand scholarly deliberation.

I am delighted to introduce this comprehensive volume, meticulously curated by astute editors, which not only acknowledges the complexities of this technological metamorphosis but also provides an insightful platform for scholarly discourse. "Advanced Computing Solutions for Healthcare" is a compendium that navigates the intricate landscape where cutting-edge computing meets the imperatives of healthcare, thereby revolutionizing patient care, research methodologies and administrative processes.

This book ventures into the forefront of technological innovation, presenting a panorama of solutions that hold the promise of revolutionizing healthcare outcomes. From the realm of AI-driven diagnostics, predictive analytics and secure data management to the far-reaching implications of telemedicine, each chapter offers a deep dive into the potential of technology to shape a healthier future.

I extend my heartfelt congratulations to the editors for providing a robust platform for researchers to showcase their findings and achievements in this dynamic field. This book serves not only as a repository of knowledge but also as a catalyst for further research and exploration in this critical domain.

As you embark on this enlightening journey through the pages of "Advanced Computing Solutions for Healthcare," I am confident that you will gain profound insights that will inspire, inform and shape the future of healthcare. May this volume contribute significantly to the ongoing dialogue, fostering innovation and paving the way for a healthier and more technologically advanced world.

S. Sivanantham
School of Electronics Engineering
Vellore Institute of Technology
Vellore, Tamilnadu
India

FOREWORD II

In the dynamic confluence of technology and healthcare, the pages of "Advanced Computing Solutions for Healthcare" unfold a compelling narrative that transcends the boundaries of conventional medical practices. This meticulously curated anthology stands as a testament to the transformative power of advanced computing, unravelling a spectrum of innovations that are reshaping the very essence of patient care, diagnosis and medical research.

As we delve into the chapters of this volume, a diverse tableau of cutting-edge topics unfolds, each contributing a vital piece to the puzzle of the healthcare revolution. From the intricacies of patch antenna design for on-off body communication to the profound implications of the 5G revolution in healthcare, this collection epitomizes the multidimensional impact of technology on the well-being of individuals and communities.

The chapters navigate the landscape of machine learning applications, from investigating transfer learning techniques for classifying Alzheimer's disease datasets to the revolutionary impact of machine learning in women's health, spanning skin, breast, ovarian cancers and polycystic ovary syndrome (PCOS). Venturing further, the realms of augmented reality (AR) and virtual reality (VR) unveil themselves as integral tools for healthcare exploration, paving the way for emerging applications that transcend the boundaries of traditional medical practices.

As we journey through the chapters on cyber ethics for artificial intelligence-assisted healthcare systems, data-driven decision support systems, and the empowering fusion of haptic-enabled language to pulse devices for communicative impairments, it becomes evident that this volume encapsulates the diversity, complexity, and limitless potential of advanced computing in healthcare.

In the pursuit of knowledge, this compendium serves as a compass, guiding researchers, practitioners, and enthusiasts through the intricate web of technological advancements. "Advanced Computing Solutions for Healthcare" is not just a book; it is a collaborative expedition into the future of healthcare, where data science, artificial intelligence, and innovative technologies converge to chart a course toward a healthier, more connected world.

May this anthology spark inspiration, ignite curiosity and foster a collective commitment to leveraging technology for the betterment of healthcare, paving the way for a future where advanced computing solutions are synonymous with improved patient outcomes and enhanced well-being.

M. Kannan
Faculty of Electronics & Communication Engineering
Anna University
Chennai, Tamilnadu
India

PREFACE

Welcome to "Advanced Computing Solutions for Healthcare," a pioneering compilation that navigates the ever-evolving landscape at the intersection of cutting-edge technology and the healthcare sector. In this era of rapid innovation, the dynamic synergy between advanced computing and healthcare has ushered in a new era of possibilities, redefining patient care, research methodologies, and administrative frameworks.

This comprehensive volume, comprising 22 insightful chapters, is a testament to the transformative potential of technology in enhancing healthcare outcomes. From AI-driven diagnostics and predictive analytics to secure data management and telemedicine, our contributors explored a myriad of solutions that stand at the forefront of technological innovation.

As we embark on this journey through the realms of smart systems, personalized healthcare, artificial intelligence, machine learning, and data science, readers will gain a deeper understanding of the challenges and strategies shaping the future of healthcare. This exploration extends to the Internet of Things (IoT), image and signal processing techniques, wireless networks, and sustainable technologies, providing a holistic view of the intricate landscape in which advanced computing converges with healthcare.

Each chapter is a beacon of knowledge that sheds light on topics such as federated learning, neuromorphic systems, and secure, robust, and efficient computing solutions. The culmination of these insights paves the way for a healthier future, emphasizing the critical role of technology in revolutionizing the healthcare industry.

As editors, we are proud to present this indispensable guide, hoping it will inspire researchers, practitioners, and enthusiasts to explore the limitless possibilities that lie at the nexus between advanced computing and healthcare. This volume stimulates further innovation and contributes to the ongoing transformation of healthcare globally.

Chapter 1 delves into neurons, which are electrically sensitive cells vital for cellular communication. It explores neuronal classification, function, anatomy, and histology. This chapter also examines neuron models, including biological and compartmental neuron models. Experts' contributions to neuro-inspired designs and methodologies are highlighted, with a focus on applications in massively parallel systems. The chapter concludes by briefly outlining the imminent applications of neuromorphic computing.

Chapter 2 explores the pivotal role of data mining in the 21st century, propelled by technological strides and a surge in medical data. Essential for clinical decisions and innovation, medical data are harnessed using descriptive and predictive data-mining techniques, unraveling profound insights. By demonstrating significant implementation, data mining enhances diagnosis accuracy, reduces diagnosis time, and minimizes errors. This chapter underscores the transformative potential of data mining, promising advancements in healthcare systems and overall public health.

Chapter 3 explores Data-Driven Decision Support Systems (DD-DSS), vital for managing escalating data volumes. This computerized program, integrating machine learning and statistical analysis, aids informed decision-making in healthcare. By merging expert knowledge and diverse data, this chapter investigates the benefits, features, and real-world applications of DD-DSS through a blend of literature review and case studies.

Chapter 4 explores the rapid growth of deep convolutional neural networks (CNN) in recent years, particularly their application in healthcare through hardware accelerators such as Field Programmable Gate Arrays (FPGAs). Focusing on edge computing and the potential for implementing CNNs in safety-sensitive biomedical applications, this study provides a comprehensive analysis of the challenges in FPGA-based hardware acceleration. This survey offers valuable insights for researchers engaged in artificial intelligence, FPGA-based hardware accelerators, and system design for biomedical applications.

Chapter 5 discusses the revolutionary integration of smart sensors in smartwatches for health monitoring. Recent advancements include biometric sensors, environmental sensors, and activity trackers. This review evaluates the accuracy, reliability, and potential use of machine learning and addresses challenges such as privacy concerns and battery life. This is a valuable resource for researchers and healthcare professionals.

Chapter 6 explores the crucial role of design thinking in integrating data science into health care. It delves into the impact of data quality, integration, and visualization on patient outcomes, predictive modeling, unsupervised learning, and ethical considerations. This chapter envisions a future shaped by AI, precision medicine, and ethical data practices.

Chapter 7 explores the transformative impact of Internet of Things (IoT) integration in healthcare. This underscores how IoT enables real-time patient monitoring, personalized treatment plans, and preventative care through continuous data gathering. While enhancing diagnostic precision and resource utilization, challenges such as data security and interoperability require resolution for IoT in healthcare to reach its full potential. This chapter emphasizes the significant effects of the IoT on healthcare delivery and the importance of a comprehensive strategy for navigating this rapidly evolving landscape.

Chapter 8 delves into the transformative impact of 5G technology on the medical industry by revolutionizing disease diagnosis, treatment, and management. Examining the evolution of wireless networks, this article explores 5G's fundamental features—high speed, low latency, and reliability. It analyzes the synergy of 5G with disruptive technologies, such as AI and IoT in healthcare, emphasizing data security and privacy. The chapter envisions a future in which 5G transforms healthcare delivery, fosters innovation, and enhances user-friendliness, cost-effectiveness, and efficiency.

Chapter 9 explores the transformative potential of Tiny Machine Learning (Tiny ML) in healthcare, marked by low power consumption and compact size. This emphasizes real-time monitoring, early disease identification, personalized treatment plans, and improved medical imaging. While Tiny ML enhances patient outcomes and reduces healthcare costs, challenges such as data privacy, ethics, and regulatory compliance require careful consideration. The future holds promise for widespread adoption, enhanced telemedicine, improved diagnostics, and a patient-centric, efficient healthcare ecosystem, provided that ethical considerations are prioritized for Tiny ML's responsible utilization.

Chapter 10 illuminates the integration of techniques and resources, collectively known as artificial intelligence (AI), in healthcare to elevate patient care and streamline administrative tasks. Its increasing relevance stems from its potential to enhance the efficacy, accuracy, and accessibility of healthcare services. AI's capacity to analyze vast medical data aids decision-making, personalizes treatment, and forecasts disease outbreaks, ultimately improving patient outcomes and healthcare affordability. As its influence has grown, AI has become a transformative force in healthcare.

Chapter 11 explores the crucial role of cutting-edge image processing in medical research. This book comprehensively covers concepts and methodologies, emphasizing the importance of image processing in healthcare for diagnosis, treatment planning, and patient care across various medical fields.

Chapter 12 delves into recent strides in science and technology, uncovering the evolving landscape and prospects of Augmented Reality (AR) and Virtual Reality (VR) in healthcare. Highlighting their potential to transform medical education, patient treatment, and surgical procedures, this chapter provides an overview of the AR and VR principles. It emphasizes their distinct features, operations, strengths, and limitations in healthcare, focusing on medical education, patient treatment, and surgical planning. This work showcases successful applications in medical education, patient interventions, and surgical procedures through case studies, illustrations, and academic examples.

Chapter 13 discusses the widespread use of chembioinformatic tools in modern medical science. These tools play pivotal roles in genomic and proteomic data analysis, gene prediction, genome annotation, and building biological networks. This chapter underscores the clinical applications of chem-bioinformatics, revealing its significance in cancer biomarker identification, personalized therapeutics, and drug design. It explores how bioinformatics tools facilitate the study of host-pathogen interactions, diagnosis of infectious diseases, treatment of metabolic disorders, and point-of-care diagnostics. By manipulating biological datasets, these tools contribute to the control, monitoring, and modification of various clinical processes and offer a comprehensive overview of their applications in the medical sector.

Chapter 14 presents the challenging task of diagnosing white blood cell diseases, such as Leukemia and Myeloma, with an emphasis on restoring the balance of the immune system. This study introduces a Computer-Aided Diagnosis (CAD) model using a Deep Convolutional Neural Network (DCNN) to classify leukocyte types. Employing a Gaussian distribution and k-means clustering for image segmentation, the gray-level covariance matrix method extracts texture features for DCNN training. The proposed model achieved a notable classification accuracy of 97.8%, surpassing existing deep learning classifiers in terms of precision, recall, and F1 score. This chapter elucidates the efficacy of the CAD model for early-stage leukocyte cancer detection.

Chapter 15 introduces the "Haptic-Enabled Language to Pulse" device, a transformative solution for empowering those with speech impairments. Utilizing Python, TensorFlow Lite's DeepSpeech model, and Raspberry Pi, the system converts spoken language to Morse code conveyed through haptic feedback. Beyond aiding in communication, it serves as an educational tool that contributes to inclusive solutions for diverse abilities.

Chapter 16 addresses Alzheimer's disease (AD), which is a serious mental health concern that causes cognitive decline. This study employed transfer-learning techniques, including VGG16, InceptionResNet-V2, Resnet50, Resnet101, and Resnet152, to classify AD datasets. The results were compared and analyzed using metrics such as accuracy, loss, validation accuracy, and validation loss. Obtained from the Kaggle repository, this study aims to enhance the accuracy of AD prediction models through deep learning.

Chapter 17 explores the pivotal role of ML in medical diagnostics and predictions. Highlighting AI's learning capacity, it focuses on improving clinical decisions, automating healthcare tasks, and enhancing women's health by addressing specific issues, such as skin cancer, breast cancer, ovarian cancer, and PCOS.

Chapter 18 explores the extensive use of mathematics for the design of physiological models. With a rich history, mathematical modeling in physiology involves the creation of representations of real-life conditions. This chapter delves into the creation of mathematical representations for physiological systems, aiding in understanding complex biological relationships and predicting system behavior in diseased states. Recent advancements in high-throughput data production techniques have further strengthened the reliance on computational approaches and mathematical modeling in the study of biological systems.

Chapter 19 focuses on developing a healthcare web application for predicting various diseases using machine learning models, such as decision trees, SVM, KNN, and Random Forest. The proposed system aims to offer a user-friendly and accurate solution by consolidating multiple disease predictions in one accessible platform.

Chapter 20 explores on-body communication using antennas, which are crucial for applications such as healthcare monitoring and IoT connectivity. Focused on 2.4 GHz on-off body communication, this study investigates custom patch antenna designs for efficient data exchange. Using FR-4 and copper, the antenna exhibited exceptional performance in free space and on-body, demonstrating resilience against environmental factors for practical applications.

Chapter 21 presents a comparative study of an 8051-controlled syringe that uses servo and stepper motors for precise fluid injection. This study evaluates the design, performance, and functionality by comparing the continuous fluid delivery of a servo motor-driven pump to the precise steps of a stepper motor-driven pump. Experiments assess the accuracy, response time, and disturbance impact, providing insights for system selection based on application requirements, such as accuracy and speed. This study contributes to the optimization of fluid delivery systems across various industries.

Chapter 22 introduces a simple, cost-effective ECG analyzer prototype for real-time signal acquisition and display through IoT devices, such as mobiles. The prototype, equipped with a pre-trained Deep Learning model, classifies ECG signals to diagnose conditions such as Arrhythmia, Congestive Heart Failure, and Normal Sinus Rhythm. This innovative tool offers quick insights into potential medical care needs.

Sivakumar Rajagopal
Department of Sensor and Biomedical Technology
School of Electronics Engineering, Vellore Institute of Technology
Vellore, Tamilnadu-632014, India

Prakasam P.
Department of Communication Engineering
School of Electronics Engineering, Vellore Institute of Technology
Vellore, Tamilnadu-632014, India

Konguvel E.
Department of Embedded Technology
School of Electronics Engineering, Vellore Institute of Technology
Vellore, Tamilnadu-632014, India

Shamala Subramaniam
Department of Communication Technology and Networks
Faculty of Computer Science and Information Technology
University Putra Malaysia, Serdang-43400, Malaysia

Ali Safaa Sadiq Al Shakarchi
Department of Computer Science
Nottingham Trent University, Nottingham, UK

&

B. Prabadevi
School of Computer Science Engineering and Information
Systems Vellore Institute of Technology, Vellore
Tamilnadu-632014, India

List of Contributors

Abrar Abu Hamdia	Department of Medical Laboratory Science, Faculty of Medicine and Health Sciences, An-Najah National University, Nablus, Palestine
Abhishek Liju Liju	Amity Institute of Forensic Sciences, Amity University, Noida, Uttar Pradesh-201301, India
Anitej Chander Sood	School of Computer Science & Engineering, Vellore Institute of Technology University, Vellore, Tamil Nadu, India
AKM Moniruzzaman Mollah	Environmental Sciences Program, Asian University for Women, Chittagong-4000, Bangladesh
Anushka Bukkawar	School of Electronics Engineering (SENSE), Vellore Institute of Technology, Vellore, India
Aayush Singh	School of Computer Science and Engineering, Vellore Institute of Technology, Vellore, India
Bhuvaneswari M.	Department of Mechatronics Engineering, Sri Krishna College of Engineering and Technology, Coimbatore, India
Bhawesh Mishra	School of Electronics Engineering (SENSE), Vellore Institute of Technology, Vellore, India
B. Jeyapoornima	Department of Electronics and Communication Engineering, Saveetha School of Engineering, Saveetha Institute of Medical and Technical Sciences, Saveetha Nagar, Thandalam, Chennai-602105, Tamil Nadu, India
Chintan Singh	Amity Institute of Forensic Sciences, Amity University, Noida, Uttar Pradesh-201301, India
Chilakalapudi Malathi	School of Computer Science & Engg, VIT-AP University, Vijayawada, Andhra Pradesh, India
Christine Thevamirtha	Environmental Sciences Program, Asian University for Women, Chittagong-4000, Bangladesh
Chandrashish Kukrety	School of Electronics Engineering (SENSE), Vellore Institute of Technology, Vellore, India
Dhruv Jain	School of Computer Science & Engineering, Vellore Institute of Technology University, Vellore, Tamil Nadu, India
Debosree Ghosh	Department of Physiology, Government General Degree College, Kharagpur II, Paschim Medinipur, Pin 721149, West Bengal, India
D. Ravi Teja	Vellore Institute of Technology, Vellore, India
Fatema-Tuz-Zohora	Environmental Sciences Program, Asian University for Women, Chittagong-4000, Bangladesh
Fatema Khusnoor	Social Science Program, Asian University for Women, Chittagong-4000, Bangladesh
G. Jeeva	Department of Networking and Communications, School of Computing, SRM Institute of Science and Technology, Kattankulathur, India

Gunavathi C.	School of Computer Science and Engineering, Vellore Institute of Technology, Vellore - 632014, Tamil Nadu, India
Harshit Poddar	School of Electronics Engineering (SENSE), Vellore Institute of Technology, Vellore, India
Iyappan Perumal	School of Computer Science & Engineering, Vellore Institute of Technology University, Vellore, Tamil Nadu, India
J. Sumitha	Department of Computer Science, Dr. SNS Rajalakshmi College of Arts and Science, Tamil Nadu, India
Jain Ankur	Manipal University, Jaipur, India
J. Joselin Jeya Sheela	Department of Electronics and Communication Engineering, Saveetha School of Engineering, Saveetha Institute of Medical and Technical Sciences, Saveetha Nagar, Thandalam, Chennai-602105, Tamil Nadu, India
Jaswanth K.	Vellore Institute of Technology, Vellore, India
Keerthi Nalliboyina	School of Electronics Engineering (SENSE), Vellore Institute of Technology (VIT), Vellore, India
Kurapati Hemalatha	School of Electronics Engineering (SENSE), Vellore Institute of Technology (VIT), Vellore, India
K.P. Parthiban	Department of EEE, VSB College of Engineering and Technical Campus, Coimbatore, India
K. P. Sujith	School of Electronics Engineering, Vellore Institute of Technology, Chennai, India
Kavita Nampoothri	School of Electronics Engineering (SENSE), Vellore Institute of Technology, Vellore, India
M. Kavibharathi	Department of Computer Science, Dr. SNS Rajalakshmi College of Arts and Science, Coimbatore, India
M. Muthukrishnaveni	Department of Physics, Sri Ramakrishna Institute of Technology, Coimbatore, India
Mayank Kumar Dubey	School of Computer Science & Engineering, Vellore Institute of Technology University, Vellore, Tamil Nadu, India
Mahtabin Rodela Rozbu	Environmental Sciences Program, Asian University for Women, Chittagong-4000, Bangladesh
Maryam Wardeh	Environmental Sciences Program, Asian University for Women, Chittagong-4000, Bangladesh
Mosae Selvakumar Paulraj	Environmental Sciences Program, Asian University for Women, Chittagong-4000, Bangladesh
M. Vanitha	School of Computer Science Engineering and Information Systems, Vellore Institute of Technology, Vellore, India
M. Logeshwaran	Department of Electronics and Communication Engineering, R.M.K. Engineering College, RSM Nagar, Kavaraipettai, Gummidipoondi Taluk, Tiruvallur-601206, Tamil Nadu, India

Mohamed Osman Zaid K.B. School of Computer Science and Engineering, Vellore Institute of Technology, Vellore, India

Natraj N.A. Symbiosis Institute of Digital and Telecom Management, Symbiosis International (Deemed University), Pune, India

Nishant Kumar Singh School of Computer Science & Engineering, Vellore Institute of Technology University, Vellore, Tamil Nadu, India

N. Duraichi Department of Electronics and Communication Engineering, Saveetha School of Engineering, Saveetha Institute of Medical and Technical Sciences, Saveetha Nagar, Thandalam, Chennai-602105, Tamil Nadu, India

P. Mahalakshmi Department of Networking and Communications, School of Computing, SRM Institute of Science and Technology, Kattankulathur, India

Prasad J. Department of Electronics and Communication Engineering, KPR Institute of Engineering and Technology, Coimbatore, India

Poornima N.V. Faculty of Management, Symbiosis Centre for Management Studies, Bengaluru Campus - 560100, Symbiosis International (Deemed University), Pune, India

P. Vetrivelan School of Electronics Engineering, Vellore Institute of Technology, Chennai, India

P. Prakasam School of Electronics Engineering, Vellore Institute of Technology, Vellore, India

Partha Sarathi Singha Department of Chemistry, Government General Degree College, Kharagpur II, Paschim Medinipur, Pin 721149, West Bengal, India

R. Charanya School of Computer Science Engineering and Information Systems, Vellore Institute of Technology, Vellore, India

Sakthivel Ramachandran School of Electronics Engineering (SENSE), Vellore Institute of Technology (VIT), Vellore, India

S. Thenmalar Department of Networking and Communications, School of Computing, SRM Institute of Science and Technology, Kattankulathur, India

S. Muthu Vijaya Pandian Department of EEE, SNS College of Technology, Tamil Nadu, India

Suriya K. Department of Electronics and Communication Engineering, SNS College of Technology, Coimbatore, India

Sivakumar Rajagopal Department of Sensor and Biomedical Technology, School of Electronics Engineering, Vellore Institute of Technology, Vellore, India

Sheela Jayachandran School of Computer Science & Engg, VIT-AP University, Vijayawada, Andhra Pradesh, India

Samiha Nuzhat Environmental Sciences Program, Asian University for Women, Chittagong-4000, Bangladesh

Sweety Angela Kuldeep Environmental Sciences Program, Asian University for Women, Chittagong-4000, Bangladesh

S. Sundar School of Electronics Engineering (SENSE), Vellore Institute of Technology, Vellore, India

Salma Hashem   Environmental Sciences Program, Asian University for Women, Chittagong-4000, Bangladesh

Sivaraj Chandrasekaran   School of Computer Science and Engineering, Vellore Institute of Technology, Vellore, India

Sridhar Raj S.   School of Computer Science and Engineering, Vellore Institute of Technology, Vellore, India

Suvendu Ghosh   Department of Physiology, Hooghly Mohsin College, Chinsura, Hooghly, Pin 712101, West Bengal, India

Sonia Mondal   Department of Mathematics, Government General Degree College, Kharagpur II, Paschim Medinipur, Pin 721149, West Bengal, India

Sivaraj Chandrasekaran   School of Computer Science and Engineering, Vellore Institute of Technology, Vellore, India

Sridhar Raj S.   School of Computer Science and Engineering, Vellore Institute of Technology, Vellore, India

S. Sundar   Vellore Institute of Technology, Vellore, India

T. R. Sureshkumar   School of Electronics Engineering, Vellore Institute of Technology, Vellore, India

Tasnim Aktar   Social Science Program, Asian University for Women, Chittagong-4000, Bangladesh

Vinyas Shetty   Vellore Institute of Technology, Vellore, India

Velswamy Karunakaran   Sri Eshwar College of Engineering, Coimbatore, India

CHAPTER 1

A Review of Biological Neurons *Versus* Artificial Neuron Models for Neuromorphic Computing Applications

Keerthi Nalliboyina[1] and **Sakthivel Ramachandran**[1,*]

[1] *School of Electronics Engineering (SENSE), Vellore Institute of Technology (VIT), Vellore, India*

Abstract: A neuron, or nerve cell, is an electrically sensitive cell that communicates with different cells through specific associations known as neurotransmitters. Aside from wipes and placozoa, this is the principal segment of sensory tissues within the entire organism. Vegetation's well-being as an organism does not depend on nerve cells. This paper analyzes the neuron's classification, functions, anatomy, and histology. The authors also aim to analyze neuron models, such as the biological and compartmental neuron models. Several experts are involved in neuro-inspired designs, methodologies, learning approaches, and software platforms to investigate the massively parallel system and various relevant applications. A new era of medicine may emerge in neuromorphic engineering, which replicates brain-like behaviours using neural systems models in hardware and software. It suggests minimal power consumption, low latency, a smaller footprint, and large bandwidth solutions. Additionally, the applications of neuromorphic computing will be discussed shortly in this survey.

Keywords: Biological neurons, Compartmental neuron model, Dendritic, Neuromorphic computing, Spiking neurons.

INTRODUCTION

Biological neurons are of three kinds based on their dimensions. Tangible neurons respond to stimuli like touch, sound, or light, which impacts the cells of the tactile organ. They also convey messages to the vertebral string, otherwise the brain. Motor neurons receive signals from the brain and the spinal cord to control all functions, as the muscles contract to move. Interneurons connect neurons to various other neurons within a specific region of the brain or spinal cord. A group

---

* **Corresponding author Sakthivel Ramachandran:** School of Electronics Engineering (SENSE), Vellore Institute of Technology (VIT), Vellore, India; E-mail: rsakthivel@vit.ac.in

Sivakumar Rajagopal, Prakasam P., Konguvel E., Shamala Subramaniam, Ali Safaa Sadiq Al Shakarchi & B. Prabadevi (Eds.)

of interconnected neurons is known as a neural circuit. A typical neuron consists of a cell body (soma), dendrites, and a single axon, as illustrated in Fig. (**1**). This represents the anatomical and functional components of the nervous system [1].

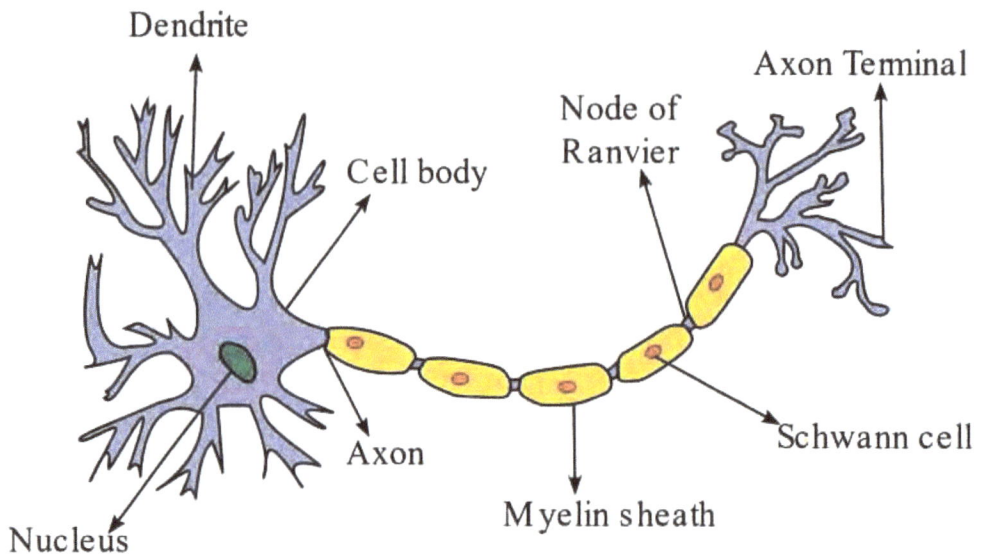

Fig. (1). Anatomy of a multipolar neuron.

Generally, the soma has minimized. The axons, as well as dendrites, have fibers that expel from them. Dendrites commonly diverge abundantly and widen two or three hundred micro meters as of the soma. The axon leaves the soma on an expanding hillock that can travel as far as one meter as humans but not as far as different species. This divergence typically maintains a consistent width. At the furthest point of the axon's branches are their terminals, through which neurons can communicate signals through neurotransmitters to additional cells. Neurons may require dendrites; otherwise, they do not have axons. The name neurite is used to describe a dendrite or an axon, especially once the cells are distinguishable [2].

Utmost neurons get signals through the dendrites and the soma, conveying signals downward through the axon. For most neurotransmitters, the signal crosses the axon of a single neuron towards the dendrite of another. However, neurotransmitters can connect an axon with another axon or a dendrite with another dendrite. The signalling process is partly electrical and primarily chemical. Neurons are electrically active due to the presence of voltage gradients across their membranes. When the voltage changes sufficiently over a short period, the neuron generates an impulse that would otherwise disrupt electrochemical signals, known as an action potential. These potentials travel rapidly along the axon and activate synaptic connections as they propagate [3]. Synaptic signs may be excitatory or inhibitory, rising or diminishing the net voltage that persists at the soma. The neuron can be produced from neurally undifferentiated organisms in youth mental health. Neurogenesis largely halts in several brain regions during adulthood.

BIOLOGICAL NEURON ANATOMY AND HISTOLOGY

Neurons have been exceptionally particular about preparing and broadcasting cellular signs. Characterized by its variety of functions acting on different pieces of the sensory system, it has an extensive assortment of shapes, sizes, and electrochemical assets. For example, the soma of a neuron could differ by 4 to 100 micrometers in measurement [4]. The biological neuron system has anatomical and functional modules.

- The soma is the physical structure of the neurons. Since it contains a nucleus, the maximum protein synthesis occurs at this point. The nucleus can range from 3 to 18 micrometers in diameter.
- Neuron dendrites contain cellular augmentations through numerous branches. These general shapes and designs are allegorically referenced, like a dendritic tree. This is where most of the contribution to the neuron happens through the dendritic vertebrae.
- The axon has been a better link as projections, which can expand ten, hundred, or even a massive number of times the length of the soma. The axon fundamentally transmits nerve signals on or after the soma and carries limited information.
- Several neurons contain only one axon; however, these axons might, as well as ordinarily, undergo extensive branching, allowing communication with several target cells.
- The portion of the axon that appears through soma is acknowledged. However, as an anatomical structure, the axon hillock attains the optimal thickness of voltage-gated sodium channels.
- It performs most effortlessly empowering parts of the neurons and spiking the commencement region of the axon. In electrophysiological terminology, this includes the most adverse edge potential. The axons and the axon hillock are two different types most commonly associated with data transmission, and these areas could similarly acquire and receive inputs from various neurons.

The acknowledged perspective on the neuron attributes specific functions to its various anatomical segments; in any case, dendrites and axons regularly act in manners despite their purported fundamental capacity [5].

Axons and dendrites in the focal sensory systems are ordinarily just around 1 micrometer wide, while those in the peripheral sensory systems are much broader. The soma is usually around 10-25 micrometers in diameter, which is not significantly larger than the cell nucleus. The longest axon of a human motor neuron can be over a meter long, extending from the spinal cord to the toes.

Sensory neurons may contain axons that extend from the toenails to the lower segment of the spinal cord, measuring over 1.5 meters in adults. A giraffe has only one axon, a few meters long, running through its neck. Quite a bit about axonal capacity originates from studying the giant squid axon, an ideal experimental setup due to its relatively large size.

Completely separated neurons are permanently postmitotic [6]. However, the stem cells present in the adult brain may regenerate functional neurons throughout the life of an organism. Astrocytes are star-formed glial cells. They can transform into neurons through the differentiation of their stem cells as a characteristic of pluripotency. Figs. (**2a** and **2b**) illustrate the schematic structures of biological neurons and artificial neurons.

Spiking Neuron Model

A biological neuron model, also known as a spiking neuron model [7], is a numerical representation of the properties of specific cells within the sensory systems that produce sharp electrical possibilities across their cell membrane about 1 m into term, known as action potentials, or spikes. Fig. (**3**) illustrates a visualization of a spiking signal.

Subsequently, spikes are communicated laterally through the axon, and neurotransmitters transmit from the neurons to numerous different neurons— spiking neurons are a significant part of the sensory system that handles observed data.

The spiking neuron model falls into various classifications: the most point-by-point numerical model is the biophysical neuron model (also termed Hodgkin-Huxley (HH) models), which describes the membrane voltages as a function of the input current and the activation of the ion channels.

Numerically, the coordinates and the Integrate and fire (I&F) neuron model have been less complex, depicting the film voltages as a component of the information current and predicting the spike time without a description of the biophysical measures that shape the duration of the action potential. Importantly, an additional unique model predicts output spikes (but not membrane voltages) as a function of the stimulation, allowing the stimulation to occur over sensory data or pharmacologically. This paper outlines various spiking neuron models and connections whenever possible to test phenomena [8].

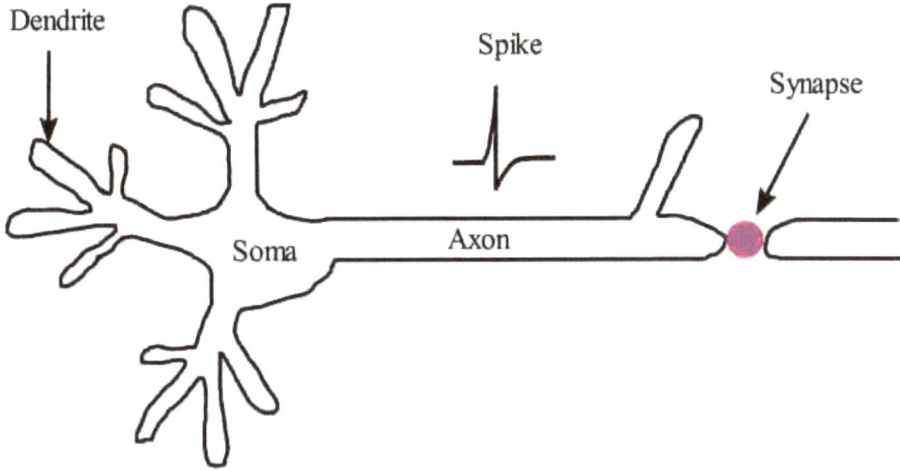

(a) Biological neuron.

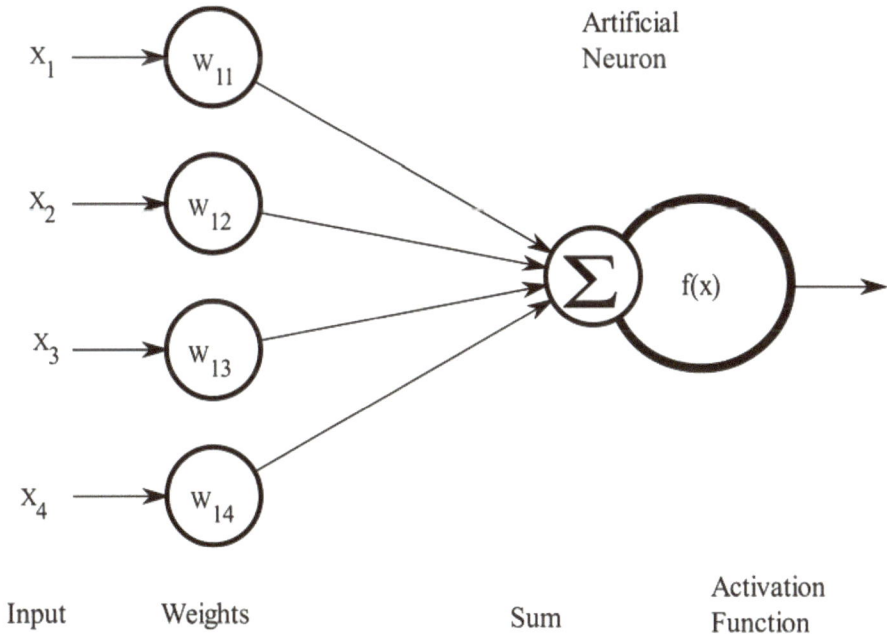

(b) Artificial neuron.

Fig. (2). Schematic structure of neurons (**a**) Biological neuron, (**b**) Artifical neuron.

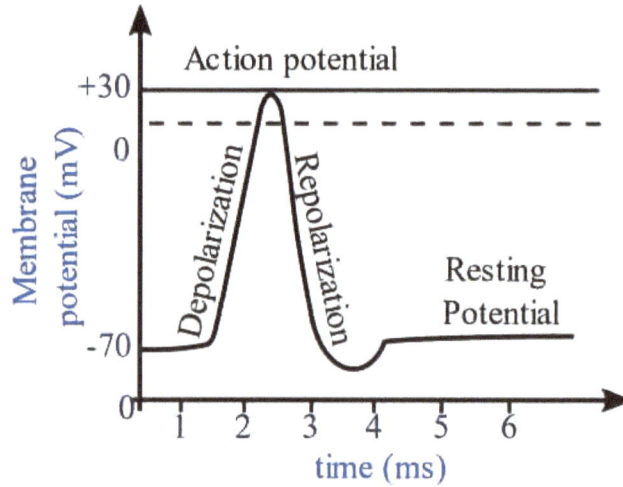

Fig. (3). Time Course of Neuronal Action Potential.

Not every one of the cells in the sensory system produces the kind of spikes that characterize the extent of the spiking neuron models. For instance, cochlear hair cells, retinal receptor cells, and retinal bipolar cells do not spike. Moreover, numerous cells within the sensory system have not yet been designated as neurons but have been classified as glia.

It incorporates deterministic and probabilistic models. Various testing strategies estimate neuronal action. The "whole cell" estimation method captures the spiking activity of a solitary neuron and delivers complete adequacy activity possibilities [9].

This architecture proposes real-time, adaptable, and highly efficient solutions for various medical applications by efficiently combining the benefits of SNNs with the demands of contemporary healthcare, as shown in Fig (**4**) [10]. The input module receives image signals to record muscle signals. Spiking neural networks convert the images and process signals from images into spike trains. From that, features related to signals are extracted. It interprets motion intentions and sends out control signals using the output module. For example, real-time control of the prosthetic limb is possible through intention decoding.

With extracellular estimation strategies, a terminal (or arrangement of several electrodes) situation occurs within the extracellular spaces. Spikes, often from several spiking sources, depend on the sizes of the terminal and their proximity to the sources and are associated with signal processing methods. Extracellular measurement has several advantages:

1. It is simpler to acquire tentatively.

2. It is powerful and goes on for a more extended time frame.

3. It can reflect the predominant impact, mainly when led in an anatomical locale with numerous comparable cells.

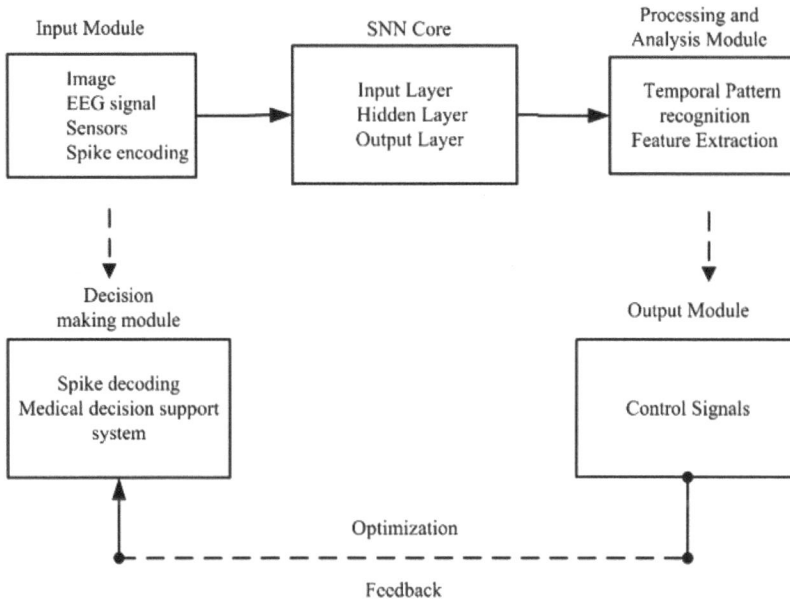

Fig. (4). Spiking Neural Network (SNN) block diagram for healthcare applications.

Overview of Neuron Models

Depending on the biological neuron, there are different types of neuron models, like biologically inspired, biologically plausible, integrated, and fire, and other neuron models available—the complexity of neuron models in Fig. (**5**) [11].

These neuron models, Hudgkin-Huxley (HH), are more complex than the other neuron models. The integrate-and-fire model is moderate, while the McCulloch-Pitts neuron model is less complex.

Neuron models are categorized into two classes based on the actual components of the models' crossing points. A generalized spiking neuron model shown in Fig. (**6**) represents a generalized leaky integrate-and-fire model. Each classification is further divided by the deliberation/detailed levels:

1. Electrical input and output activities predict membrane output voltages similar to a given electrically induced component, such as voltage or flow inputs.

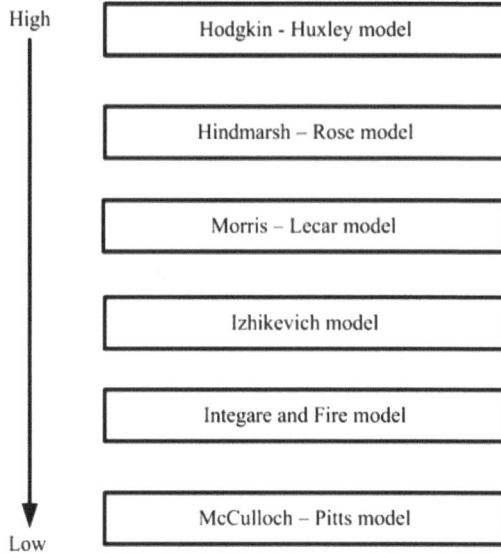

Fig. (5). Complexity of different neuron models.

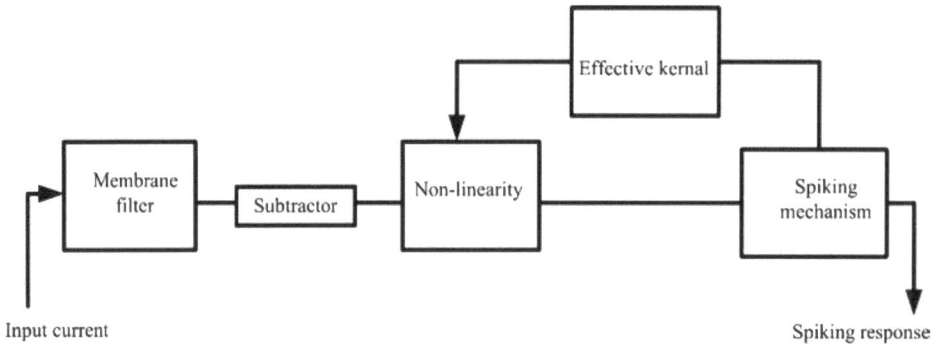

Fig. (6). Generalized Leaky-Integrate-Fire (LIF) model block diagram [12].

2. The different models within these classes vary in the specific functional connection among the input current, output voltages, and degree of subtleties.

3. A few models within these classifications foresee only a snapshot of the event of output spikes (also known as "activity potential"). Other models have been further categorized and recorded as being sub-cellular cycles. These classes include models that are either predictable or probabilistic.

4. Source neuronal simulations based on natural stimuli or pharmaceutical input. They associate the input stimulus, either pharmacological or natural, with the likelihood of a spike.

5. The input phase among those models was not electrical but pharmacological (compound) fixation units or actual units that describe an outside stimulus like light, stability, or various pressing factors. Furthermore, the output stage addresses the likelihood of a spike occasion, not an electrical voltage.

Even though it is not surprising in science and design to have a few precise models for various deliberation and brief levels, the quantity of different, at times negating, biological neuron models have been incredibly high. The present circumstance has been partly the result of the various experimental conditions and the difficulty in isolating the biological properties of individual neurons as measurement effects, in addition to the interactions of multiple cells. To speed up the convergence towards a unified theory, we list several models in each category and, where relevant, also references to supporting studies.

The membrane potential of the HH model is,

$$C_m \frac{dv}{dt} = I - I_{Na} - I_K - I_L \tag{1}$$

From equation (1), C_m is the membrane capacitance, I is the external current, and I_{Na}, I_K and I_L are the sodium, potassium, and leakage ionic currents shown in equations (2) to (4).

$$I_{Na} = g_{Na} m^3 h(V - E_{Na}) \tag{2}$$

$$I_k = g_K n^4 (V - E_K) \tag{3}$$

$$I_L = g_L (V - E_L) \tag{4}$$

Here g_{Na}, g_K, and g_L Are the sodium, potassium, and leakage conductance, and the reversal potentials E_{Na}, E_K, and E_L. The gating variables are m, n, and h.

As previously discussed, the four quadratic nonlinear differential equations, six complex functions, and seven constants that comprise the Hodgkin-Huxley model make it extremely challenging to simulate and implement.

Equation (5) illustrates how the Hindmarsh and Rose neuron model produces bursting and repetitive firing.

$$v' = u - F(v) + I - w \tag{5}$$

$$u' = G(v) - u \tag{6}$$

$$w' = \frac{(H(v) - w)}{\tau} \tag{7}$$

Equations (6) and (7): $G(v)$ is a quadratic equation, $F(v)$ is a cubic equation, and $H(v)$ is a linear equation.

Three nonlinear ordinary differential equations with the dynamical variables $x(t)$, $y(t)$ and $z(t)$ comprise the Hindmarsh–Rose model. Consequently, the three-dimensional neuron model equation is provided from equation (8) to (10).

$$x = y - ax^3 + bx^3 + I - z \tag{8}$$

$$y = c - dx^2 - y \tag{9}$$

$$z = r(s(x - x_1) - z) \tag{10}$$

where ion channels constants a, b, c, and d, the recovery variable is denoted by r.

Compared to the HH neuron model, the Hindmarsh-Rose model is comparatively more straightforward and can exhibit various spiking behavior patterns in biological neurons.

The membrane potential of the ML neuron model is described in equations (11) and (12).

$$C_m \frac{dv_m}{dt} = I_{ex} - G_{ca}m_{ss}(v_m) - (v_m - E_{ca}) - G_k.n(v_m - E_k) \tag{11}$$

$$- G_L(v_m - E_L)$$
$$\frac{dv}{dt} = I_{ex} - \lambda(v_m)(n_{ss}.(v_m) - n) \tag{12}$$

The Izhikivich neuron model is as feasible in biology as the HH neuron model and as simple as the Integrate-and-Fire (I&F) model. The four parameters a, b, c, and d in this model can replicate all possible ring patterns found in the neurons of the cortex (Appendix A). The mathematical representation of the Izhikivich neuron model is given by the following two-dimensional ordinary differential equations (13) and (14):

$$\frac{dv}{dt} = 0.04\,v^2 + 5v + 140 - u + I_{ex} \tag{13}$$

$$\frac{du}{dt} = a(bv - u) \tag{14}$$

If $v > 30$ *mv*, then $v = c$ and $u = u + d$,

Where u is a membrane recovery variable, v stands for the membrane potential, and I_{ex} represents synaptic input current. The dimensionless parameters are a, b, c, and d. Depending on the membrane potential history before the spike, the desired threshold potential can range from approximately -55 mV to -40 mV.

The simplest form of spike neuron is the I&F neuron, in which the membrane voltage reaches a value below a threshold, and the membrane capacitor generates a firing potential when it's charged voltages attain a predetermined threshold voltage. Calculating the membrane potential for the I&F model can be expressed in equation (15).

$$C\frac{dv}{dt} = I, where\ I > 0 \tag{15}$$

Capacitor current I_c is given by,

$$I_c = C_m \frac{dv_m}{dt} \tag{16}$$

Due to its simplicity, it cannot replicate the many intricate details of the electrophysiology of real neurons. It is also known as the voltage threshold model or the Non-Leaky IF (NLIF) Model.

Eventually, the biological neuron model aims to clarify the essential components of the activity of the sensory systems. Modelling assists in investigating test information and addressing queries. For example, how the spikes of neurons relate to sensory stimulation or motor actions, arm developments, and the type of neural code utilized by the sensory systems should be addressed. Models have also been significant in re-establishing lost- brain function through neuroprosthetic gadgcts.

The comparison of various networks, characterized by numerous neurons and synapses, with the density of the neuron LaCSNN (Lesion-aware Convolutional Neural Network), which has benefited from excellent biological plausibility and a vast network scale, is discussed in Table **1**. To better understand brain dynamics, [13] a study developed a thalamocortical system. However, they employed a simplified model of ionic conductance dynamics.

Minimal spiking neuron models, like those of the LIF [14 - 16] and Izhikevich models [17], do not enable a massive array of ion conductance, making them unsuitable for studying particular electric current types. For conductance-based models [18], a graphics processing unit-based system could have a practical replica of the basal ganglia network, although its scalability is restricted.

Furthermore, fundamental circuitry limitations make additional advancements using previously practical approaches increasingly challenging. Moreover, a leakage current is frequently incorporated in the I&F models, as leaking may be crucial for allowing neurons to provide a time-varying activity. This change indicates a more incredible potential than the threshold state and will gradually decrease to its resting position. The fault current determines the rate at which it occurs, so this variation is referred to as the LIF model. Several direct electronic simulations of the LIF models of a neuron are available, along with various computations and specialized hardware realizations [19, 20].

Table 1. Comparison between network scale and biological accuracy.

Models	No. of Synapses Per Core		
Biological accuracy	10^2	10^4	10^6
Integrate & Fire (IF)	Rubin & Terman model [14]	Cheris *et al.* [15]	Sapun [16]
Izhikevich model	Yang *et al.* [17]	-	Izhikevich [13]
Conductance based	-	GPU [18]	LaCSNN [19]

I&F, Leaky integrate-and-fire (LIF), and other more straightforward models are commonly employed in thoroughly leaky-integrate-and-fire Spike Response Model (SRM) SNNs for visual identification tasks [21].

The inputs used in a series of classification tasks have some temporal dependence. Therefore, they require models that can account for this dependence when producing forecasts. Sequence classification tasks can benefit significantly from SNNs due to their inherent capacity to capture temporal information, making them directly compatible with such temporal inputs.

Using the IBM DVS gesture data set, we investigate gesture recognition employing deep SNNs. According to their findings, SNNs perform on par with corresponding state-of-the-art ANN implementations. The performance analysis of different SNN models on image classification tasks from neuromorphic datasets (N-MNIST, CIFAR10-DVS [22, 23]) and frame-based image datasets (MNIST, CIFAR10, ImageNet [24, 25]) is depicted in Table **2**.

SNNs aim to maximize energy efficiency by achieving competitive accuracy while minimizing the number of time steps. Accordingly, the best overall performance is by networks that train using gradient-based backpropagation techniques and use pixel values directly as input. CIFAR10-DVS Neuromorphic-MNISTs (N-MNIST) are the dynamic vision sensor-recorded spiking versions of the MNIST and CIFAR10 datasets.

Table 2. SNNs' performance on various image-recognition datasets.

Neuron Model	Data Set	Learning Paradigm and Coding Technique	Accelerators	Accuracy
LIF [26]	MNIST	STDP and Rate coding	2 FC	95%
IF [21]	CIFAR10	ANN-to-SNN and Temporal coding	VGG16	93.63%
IF [21]	ImageNet	ANN-to-SNN and Temporal coding	VGG16	73.46%
LIF [27]	NMNIST	Event sensor Backprop	2 FC	98.74%
LIF [28]	ImageNet	Hybrid and Rate coding	VGG16	65.19%

Applications of Spiking Neurons

Spiking Neuron Models have been utilized in a variety of applications [29 - 33], which require encoding in terms of neuronal spike trains concerning neuroprosthetics and brain-computer (Personal Computer) interfaces such as retinal prostheses, artificial limb controls, and sensations. These are some applications of spiking neurons used in neuromorphic computing.

COMPARTMENTAL NEURON MODELS

Compartmental modeling of dendrites involves multiple-compartment modeling, simplifying the electrical conduct of complex dendrites. Fundamentally, the compartmental modeling of dendrites has been a valuable tool for fostering novel biological neuron models. Dendrites are vital because they possess the most membrane area in many neurons, allowing them to associate with many different cells. Initially, the dendrites exhibit steady conductance. Still, in the present day, this is perceived as having a dynamic voltage-gated ion channel that impacts the response of neurons to synaptic input—numerous numerical models exist to understand the electrical conductance of the dendrites [34]. Dendrites tend to be highly divergent and complex. Consequently, the compartmental approach to understanding the electrical conductance of dendrites enhances this valuable insight [35].

Compartmental modeling is highly effective for modeling dynamical systems that contain specific inalienable assets through conservation principles [36 - 40]. Compartmental modeling has proven to be a valuable approach, as illustrated in Fig. (**7**), which defines the state space that effectively captures the dynamical systems governed by the conservation rules [41, 42]. Regardless of whether this preserves mass, energy, a liquid stream, or a data stream, compartments exchange unsigned int (graded) state values across tree topologies, similar to a dendritic tree, and can also emit spikes to relay critical events to neighboring neurons [43].

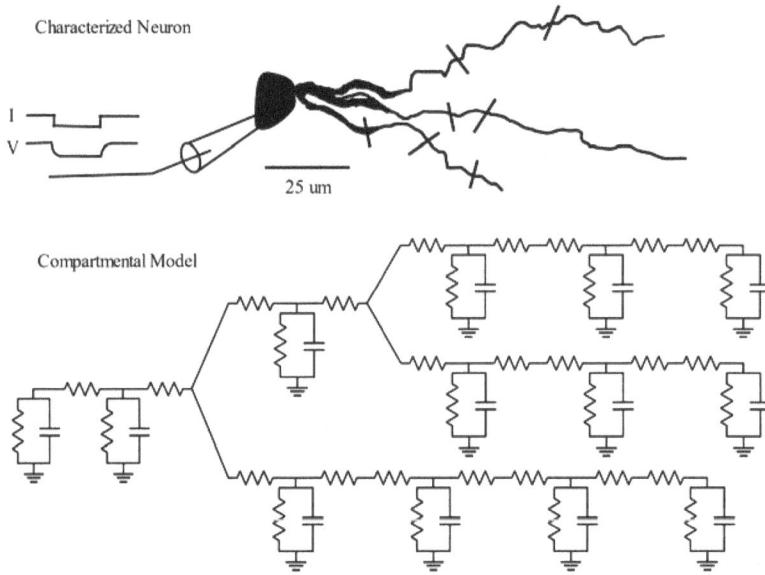

Fig. (7). Compartmental neuron model [35].

Fig. (**8**) shows the compartmental model and its implementation. Essentially, it has models where the condition factors are generally non-negative. Therefore, the conditions for mass equilibrium, energy, fixation, and the liquid stream are composed. Finally, it leads downward towards networks where the mind has been the biggest of everything, actually similar to the Avogadro number, a considerable measure of interrelated particles. The mind has intriguing interconnections. Over time, thermodynamics has been challenging to understand, yet it has a naturally visible interpretation that everyone recognizes, as it adheres to several general rules. Similarly, the mind has various interconnections, which makes it practically impossible to compose a differential condition for the development of real neurons and the empirical content of the model structure shown in Fig. (**9**).

The model's empirical substance, ranging from the quadratic model, which merely predicts the train of spikes, to the multi-compartmental model of HH, consists of membrane potential and ionic currents projected at changed locations in the neuron.

General perceptions regarding how cerebral capacities relate to the first and second laws of thermodynamics have been extensive. The brain is an exceptionally interconnected system. In some way, the neurons must function like a chemical reaction system. Therefore, they must adhere to the laws of chemical thermodynamics in some manner.

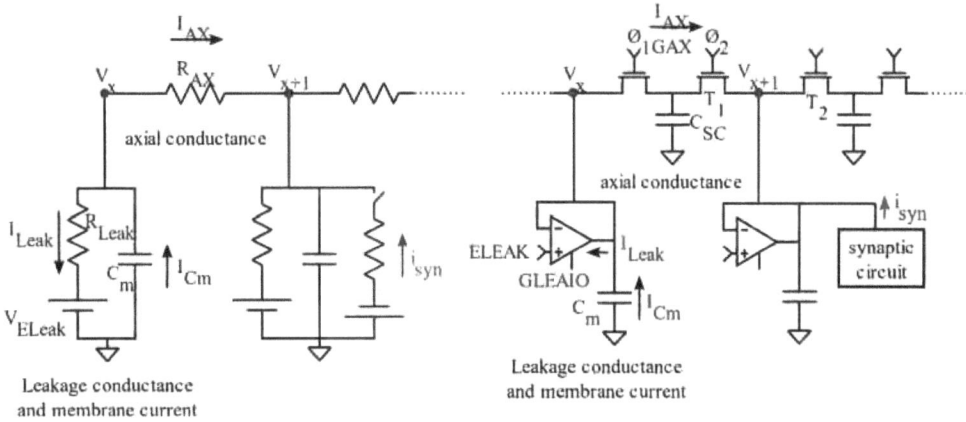

Fig. (8). Compartmental model and its implementation [44].

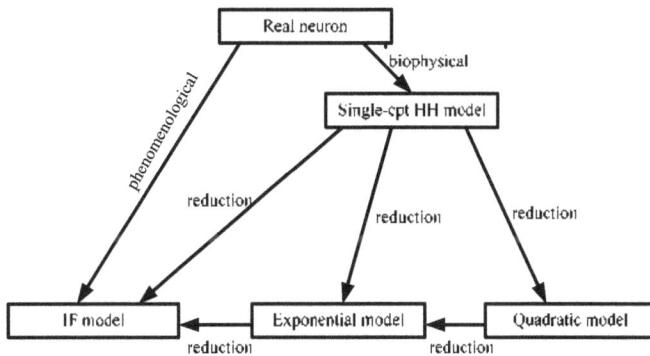

(a)

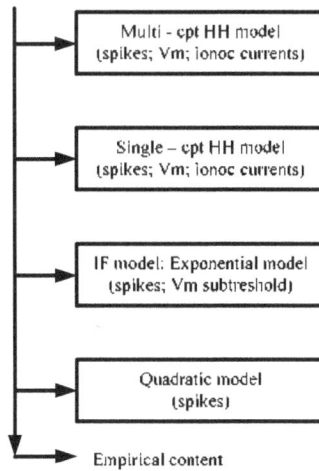

(b)

Fig. (9). Empirical content of models [45].

This methodology may lead to a more summarized model of mind. Compartment models are used to numerically calculate the membrane potential of a complicated dendritic tree. Dendritic compartments are numerically defined, with membrane capacitance and transverse resistance influenced by longitudinal resistance and external input to the compartment. Fig. (**10**) illustrates an implementation of the multi-compartment neuron model.

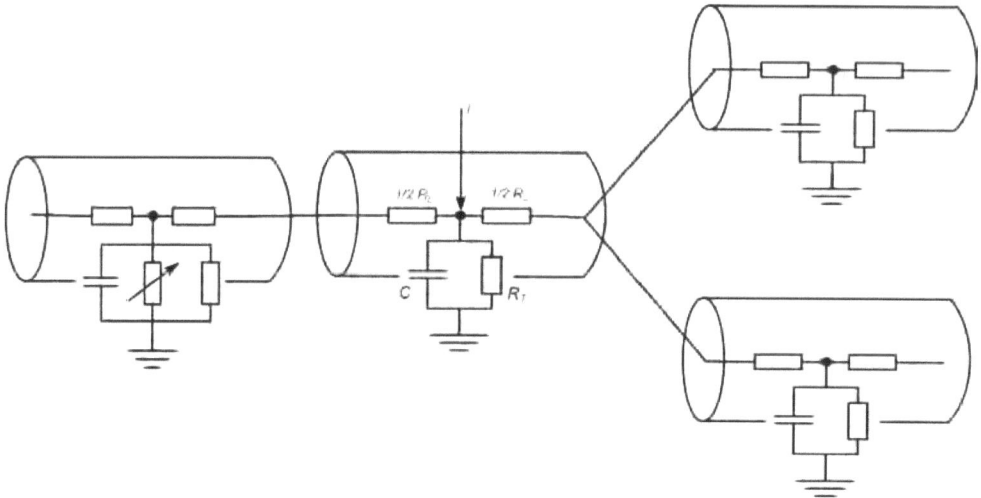

Fig. (10). Multi-compartment neuron model [46].

Some or all compartments may also contain nonlinear ion channels; variable resistors are available in most compartments. To build a neuromorphic system that can mimic the brain's abilities with artificial networks, researchers must first understand how the cortex and, more broadly, the central nervous system work and analyze information [47]. The nerve cell is a specific type of cell that serves as the basic building block of a cortex.

Applications

An exploration of hardware implementations in neuromorphic computing often requires fundamental hardware architectures, as depicted in Fig. (**11**) that lack the more exotic device components, including neuron processors, including analog, digital, or mixed approaches, as well as neurocomputers like programmable gate arrays, accelerators, *etc*.

The applications of neuromorphic computing include various neuron models, as mentioned in Fig. (**12**). Single cells and neuronal assemblies can process, encode, and convey information in different ways, according to compartmental models.

Different brain regions use solitary pyramidal nerve cells and small neural networks to interpret data involved in memory formation. *in vitro*, dopaminergic (DA) neurons in the human brainstem have meager firing rates. Finding the sources of frequency constraint: DA neurons' reactions to various pharmacological interventions [48]. In neurology, the mode-locked behaviour of the electrical system under periodic stimuli is essential. Direct numerical simulations and the expected Arnold tongue architecture agree well [49].

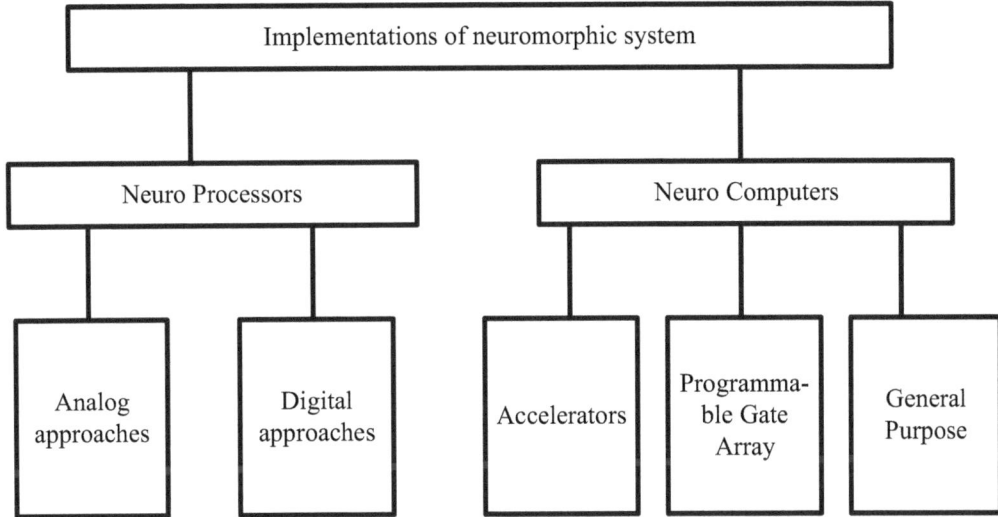

Fig. (11). Hardware implementations of neuromorphic computing.

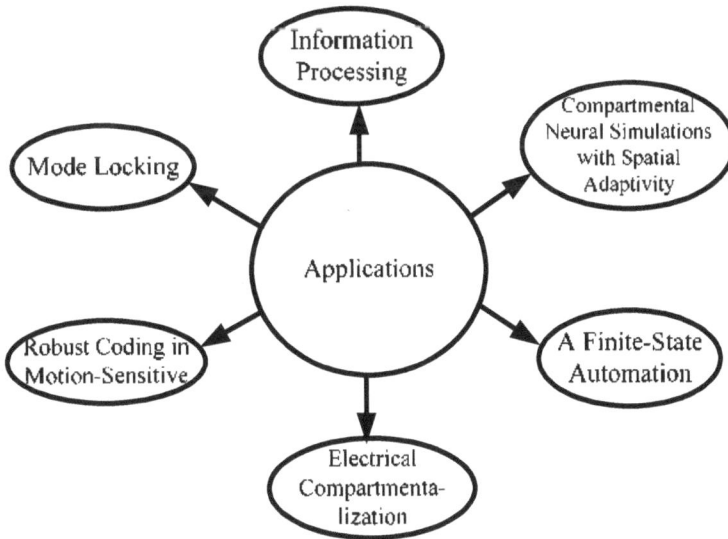

Fig. (12). Applications for various neuron models [46].

Spatial adaptivity is simple once the calculation level of branches is higher than that of cells. The active portions of a compartment are recognized, and computing activity, while performing the calculations, saves the cell membrane. In certain situations, spatial flexibility can reduce the calculation time by as much as a percentage [50, 51]. The frequency-dependent plasticity paradigm can decrease the number of hidden nodes in the networks by using this technique to reduce the complexity of the network in neuromorphic computing [52].

A finite-state automated model of transmembrane activity could minimize computational load, allowing for the evaluation and comparison of vast numbers of replicated three-dimensional networks. The automation can model the most significant properties of neuronal membranes, and it resembles the behavior of 2-equation ODE (Ordinary Differential Equations) models for wave propagation in excitable mediums [53].

The information regarding changes in the hippocampus's ability to respond to a stimulus is caused by age-induced changes in calcium-dependent membrane processes using a multi-compartmental model of a CA1 (Cornu Ammonias 1) pyramidal cell. Neurobiological information could be estimated as the parameters of the model. Computer models replicate the reduced reactivity of elderly CA1 cells, which is characterized by higher internal calcium build up, stronger post-burst, delayed afterhyperpolarization, and improved spiking frequency sensitivity. Simple coupling methods that selectively link specific types of calcium channels to particular potassium (K+) channels could simulate age-induced changes in CA1 excitability [54].

Bio-medical Applications

Time-varying signals naturally lend themselves to SNN-based processing and are the hallmarks of real-world biological impulses and patterns. In light of this, scientists have devised several methods for deciphering and categorizing these biological patterns.

These consist of studies that analyze and decode signals from various sources, including electroencephalograms (EEGs), electrocardiograms [25, 55], and so forth. Furthermore, researchers in References (ECG) Electromyography, or EMG, demonstrate that SNNs can effectively process and analyze the epileptogenic high-frequency fluctuations (HFO) tissue in recordings for Electrocorticography (ECoG) [56, 57].

Table **3** represents a comparison of different paradigms for neuromorphic processors. The non-flexibility of Neurogrid restricts its use in various network architectures with different neuron models. The Izhikevich model, which employs a small selection of unspecified ionic conductance, has been used in the SpiNNaker project. True North has significantly weaker empirical support and adaptability. It is broader because SpiNNaker can replicate any spiking neuron model. The projects of Brain Scales and HiAER generate excellent contributions, but neither can mimic the dynamics of ionic conductance using the analog method and both lack adaptability [18]. For example, neurons within a core may be dispersed throughout numerous compartments or dynamic output values, each with its filtering dynamics [43].

Table 3. Comparison of neuromorphic processors with different paradigms.

Paradigms	Neuro grid	SpiN Naker	True north	Brain ScaleS	HiAER	LaCNN
Models	HH	Izhikevich	LIF	AdEXP	*IF*	*Conductance*
Accuracy	High	Moderate	Low	Moderate	Low	High
Plasticity	-	Programmable	-	STDP	STDP	Programmable
Reconfiguration	No	possible	possible	No	No	Possible

They have postulated that dendritic spine functions as electric cells in terms of actively processing specific synaptic information. Compartmentalization is not yet compact enough to electrically disconnect a synapse, according to assessments of impedance between the spinal base and the parent dendrite [58]. Compartmental modeling reveals that spatially segregating the presynaptic cells from the dendritic synapses' conductivity enhances the gap junctions' coupling strength, allowing for this approximation. Vertical system neurons (VS cells) combine broad-range moving objects from a retinotopic matrix of small motion sensors [59]. The morphological complexities of these systems are often reduced, improving the speed and usability. Dendritic branching patterns with simpler neuronal networks may interpret signals more effectively than fully branching patterns [60].

CMOS and neuromorphic processors are appropriate for biomedical and portable healthcare applications, as presented in Tables **4** and **5** [61]. Fig. (**13**) illustrates a ranking of the applications for neuromorphic systems. These systems have various applications like image classification processing, data classification control, basic benchmark tests, biology sensor-inspired, robotics, video, sound, intelligent sensors, and implantable wearables. Current advancements in neuromorphic medicine include biomedical interfaces like motor, cognitive, and perceptual prostheses, the processing of bio signals for diagnosis, and imaging for healthcare and cancer diagnosis [62, 63]. Researchers at the University of Chicago have developed a neuromorphic device based on electrochemical transistors that are inherently flexible and ideal for accurately gathering data on health monitoring, such as body temperature and heart rate [64, 65].

Table 4. Silicon chips are appropriate for biomedical applications.

Silicon Chip	Size of the Core (mm²)	Performance evalutaion Power (GOP/s)	Power Efficiency (TOPS/W)	Applications
Eyeriss [27]	12.55	17-42	0.06–0.15	Cancer diagnosis, cardiology, gastroenterology
ConvNet processor [67]	2.5	102	0.8–10	Ultrasound processing
Neural processor [68]	5.5	1900-7000	4.5-11.5	Skin cancer
Origami [72]	3.09	196	0.8	Heart health monitoring
Intel Nervana NNP-I 1000 (Spring Hill) [70]	10	48000	4.8	Diagnosis
LNPU [71]	16	600	25	Cancer diagnosis

Table 5. An overview of biomedical applications and neuromorphic platforms [60].

Neuromorphic processor	Technology Node	Energy per SOP	Size	Biomedical Applications
DYNAP-SE	180 nm	17 pJ @ 1.8V	38.5 mm²	EMG, ECG, HFO
SpiNNaker	ARM968, 130 nm	Peak power 1W per chip	102 mm²	EMG and EEG
Loihi	14 nm FinFET	26 pJ @ 0.775 V	60 mm²	EMG
True North	28 nm	23.6 pJ @ 0.75V	0.093 mm² (core)	EEG and Local Field Potential (LFP)
sODIN	28 nm FDSOI	12.7 pJ@0.55V	0.086 mm²	EMG

Practical algorithms are more precise for medical imaging analysis with the help of neuromorphic computing. In mammograms, for instance, it has been demonstrated that neuromorphic systems can identify cancer cells more precisely than conventional algorithms [10]. Personalized treatment plans can be developed based on each patient's unique genetic and medical information through neuromorphic computing. With neuromorphic computing, wearable technology with real-time feedback and health monitoring capabilities can be developed [73, 74]. Neuromorphic devices may monitor a patient's blood pressure, glucose levels, and heart rate [55, 75].

A summary of neuromorphic computing processors used in various biomedical applications is presented in Table **3**. Loihi is a neuromorphic chip that Intel is developing specifically for medical applications. New algorithms for wearable technology, drug discovery, and medical imaging are being implemented on Loihi

[76]—True North neuromorphic chip designed by IBM is intended for medical use like disease tracking and epilepsy detection [77]. An arm prosthetic by the human brain, Brain Co, enables patients to move their prosthetic arm more naturally; Brain Co has developed a prosthetic arm that combines neuromorphic computing with machine learning [78]. Wearable technology enables large-scale ambient data collection from individuals and their surroundings, requiring flexible, small, and sensitive sensors to gather accurate body data [79].

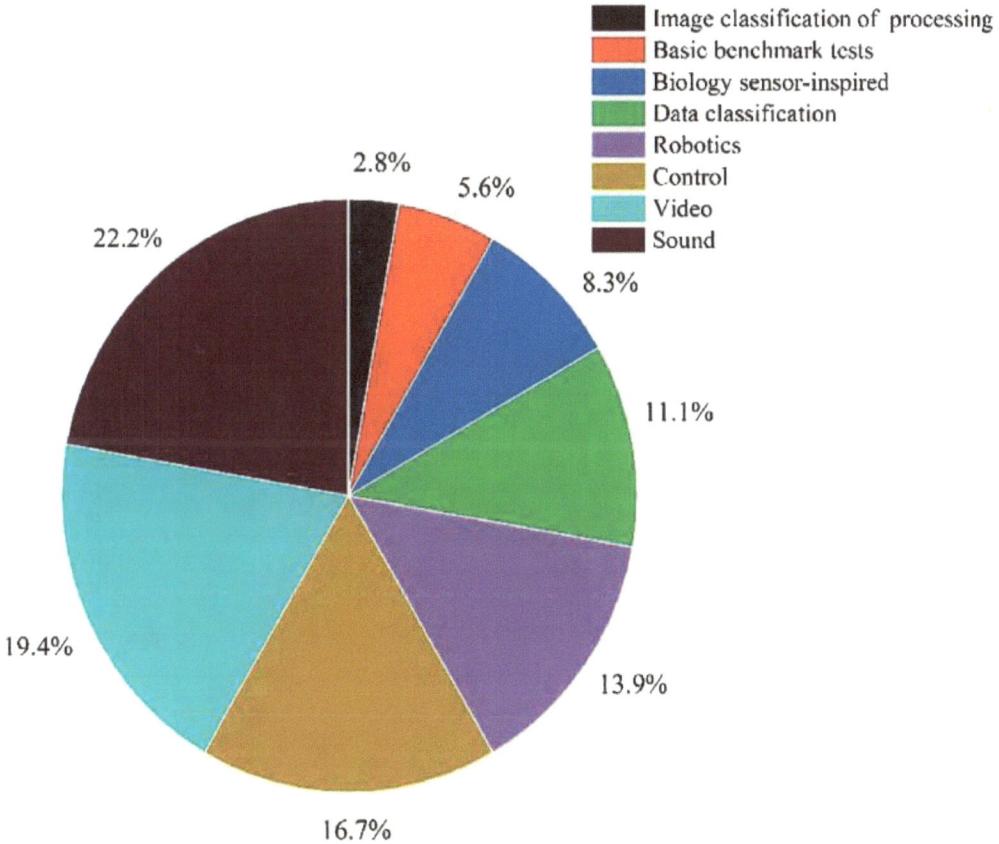

Fig. (13). Analysis of Neuromorphic System Applications [11]

Neuromorphic APIs and Libraries

It is necessary for standard software and libraries, like Pytorch, TensorFlow, Caffe, and others, to undergo a paradigm shift to accommodate neuromorphic systems with various SNN architectures. With the launch of its updated Loihi-2, Intel also revealed its LAVA software framework, which is a step in the right direction. By providing a standard set of tools, techniques, and libraries, LAVA meets the requirement to provide a frequent neuromorphic software framework

[80]. It enables scientists and developers to efficiently run neural network models on various heterogeneous architectures, including conventional and neuromorphic hardware.

Neuromorphic systems have great potential to speed up drug discovery, medical diagnosis, and personalized treatment plans in the healthcare industry. Table **6** presents medical applications for neuromorphic computing applications. For instance, neuromorphic hardware designed by scientists at the University of Manchester, England, will create sophisticated interfaces to brain-computer interfaces (BCIs) for individuals with neurological conditions [10]. By allowing direct communication between the brain and external devices, these BCIs enable individuals with disabilities to control prosthetic limbs or interact more naturally with assistive technologies.

Table 6. Future potential medical uses for neuromorphic computing applications.

Author and Reference	Objective (Health care application)	Required tools and Chip	Accuracy
Getty N *et al.* [81].	Breast cancer classification (using Image)	Memristor crossbar SNN and Loihi	85.6% accuracy
Bauer FC *et al.* [28]	Real-time ECG classification (Biosignal processing)	SRNN and DYNAP	91% true
Dethier *et al.* [83]	Monkey computer cursor control (Neural Interfaces)	SNN	SNN >94% success rate
Boi F *et al.* [84]	Rat mobile cart control (Neural Interfaces)	SNN and ROLLs	100% convergence rate
Donati E *et al.* [85]	EMG and gesture prediction (Neural Interfaces)	SNN, DYNAP, and Myo	95% accuracy
Buccelli S *et al.* [86]	Modulating and restoring network (Neural Interfaces)	SNN and FPGA Bridged	Bridged lesioned neurons

Several of the issues mentioned here involve the application of neuroscience research. Robust ensembles of experts, system developers, and math researchers will assist in identifying and decoding significant computational features in biological systems.

CONCLUSION AND FUTURE SCOPE

In this research paper, we have discussed neuron classification and their functions, anatomy, and histology. We have also analyzed neuron models such as the biological and compartmental neuron models. In this survey, we explore why neuromorphic engineering is one of the most capable topics in developing computational technologies and healthcare applications, elaborating on its fundamental concepts and approaches, as well as its existing status and potential challenges. It aims to leverage significant breakthroughs in creating new strategies and deploying device approaches when necessary and feasible. Neuromorphic systems attempt to provide a specialized basis for processing spiking neural networks in a biologically plausible manner to investigate and potentially replicate the remarkable capabilities of the biological brain system. By replicating the essential functions and structure of the human nervous system, researchers can begin to address the limitations of current medical technology by providing biologically consistent solutions that are ultra-energy-efficient, have low latency, and possess high bandwidth.

The appendix of this chapter is given in the end of this book in the appendix section.

REFERENCES

[1]   Muzio MR, Fakoya AO, Cascella M. Histology, Axon. 2022 Nov 14. In: StatPearls [Internet]. Treasure Island (FL): StatPearls Publishing 2024 Jan.
[PMID: 32119275]

[2]   Crawford LK, Caterina MJ. Functional anatomy of the sensory nervous system: updates from the neuroscience bench. Toxicol Pathol 2020; 48(1): 174-89.
[http://dx.doi.org/10.1177/0192623319869011] [PMID: 31554486]

[3]   Ryan TA, Jin Y. Editorial overview: Molecular neuroscience. Curr Opin Neurobiol 2019; 57: iii-vi.
[http://dx.doi.org/10.1016/j.conb.2019.06.002] [PMID: 31266696]

[4]   Davies M. The Neuron: size comparison. Neuroscience: A journey through the brain. 2002; 4(9): 5.

[5]   Chudler EH. Brain Facts and Figures, from Eric H. Chudler's Neuroscience for Kids: faculty. washington. edu/chudler/facts. html. Retrieved from Internet on Jun 2016; 6: 14.

[6]   Herrup K, Yang Y. Cell cycle regulation in the postmitotic neuron: oxymoron or new biology? Nat Rev Neurosci 2007; 8(5): 368-78.
[http://dx.doi.org/10.1038/nrn2124] [PMID: 17453017]

[7]   Gerstner W, Kistler WM. Spiking neuron models: Single neurons, populations, plasticity. Cambridge university press 2002; p. 15.
[http://dx.doi.org/10.1017/CBO9780511815706]

[8]   Taherkhani A, Belatreche A, Li Y, Cosma G, Maguire LP, McGinnity TM. A review of learning in biologically plausible spiking neural networks. Neural Netw 2020; 122: 253-72.
[http://dx.doi.org/10.1016/j.neunet.2019.09.036] [PMID: 31726331]

[9]   Auge D, Hille J, Mueller E, Knoll A. A survey of encoding techniques for signal processing in spiking neural networks. Neural Process Lett 2021; 53(6): 4693-710. [http://dx.doi.org/10.1007/s11063-02- -10562-2].
[http://dx.doi.org/10.1007/s11063-021-10562-2]

[10]   Tian F, Yang J, Zhao S, Sawan M. NeuroCARE: A generic neuromorphic edge computing framework for healthcare applications. Front Neurosci 2023; 171093865
[http://dx.doi.org/10.3389/fnins.2023.1093865] [PMID: 36755733]

[11]   Schuman CD, Potok TE, Patton RM, *et al.* A survey of neuromorphic computing and neural networks in hardware. 2017.

[12] Pozzorini C, Naud R, Mensi S, Gerstner W. Temporal whitening by power-law adaptation in neocortical neurons. Nat Neurosci 2013; 16(7): 942-8.
[http://dx.doi.org/10.1038/nn.3431] [PMID: 23749146]

[13] Izhikevich EM, Edelman GM. Large-scale model of mammalian thalamocortical systems. Proc Natl Acad Sci USA 2008; 105(9): 3593-8.
[http://dx.doi.org/10.1073/pnas.0712231105] [PMID: 18292226]

[14] Rubin JE, Terman D. High frequency stimulation of the subthalamic nucleus eliminates pathological thalamic rhythmicity in a computational model. J Comput Neurosci 2004; 16(3): 211-35.
[http://dx.doi.org/10.1023/B:JCNS.0000025686.47117.67] [PMID: 15114047]

[15] Chersi F, Mirolli M, Pezzulo G, Baldassarre G. A spiking neuron model of the cortico-basal ganglia circuits for goal-directed and habitual action learning. Neural Netw 2013; 41: 212-24.
[http://dx.doi.org/10.1016/j.neunet.2012.11.009] [PMID: 23266482]

[16] Eliasmith C. A large-scale model of the functioning brain (vol 338, pg 1202, 2012). Science 2012; 338(6113): 1420. [PMID: 23239717].
[PMID: 23239717]

[17] Yang S, Wang J, Li S, *et al.* Cost-efficient FPGA implementation of basal ganglia and their Parkinsonian analysis. Neural Netw 2015; 71: 62-75.
[http://dx.doi.org/10.1016/j.neunet.2015.07.017] [PMID: 26318085]

[18] Igarashi J, Shouno O, Fukai T, Tsujino H. Real-time simulation of a spiking neural network model of the basal ganglia circuitry using general purpose computing on graphics processing units. Neural Netw 2011; 24(9): 950-60.
[http://dx.doi.org/10.1016/j.neunet.2011.06.008] [PMID: 21764258]

[19] Yang S, Wang J, Deng B, *et al.* Real-time neuromorphic system for large-scale conductance-based spiking neural networks. IEEE Trans Cybern 2019; 49(7): 2490-503.
[http://dx.doi.org/10.1109/TCYB.2018.2823730] [PMID: 29993922]

[20] Olin-Ammentorp W, Cady N. Biologically-inspired neuromorphic computing. Sci Prog 2019; 102(3): 261-76.
[http://dx.doi.org/10.1177/0036850419850394] [PMID: 31829848]

[21] Han B, Roy K. Deep spiking neural network: Energy efficiency through time-based coding. European Conference on Computer Vision. Cham: Springer International Publishing 2020; pp. 388-404.
[http://dx.doi.org/10.1007/978-3-030-58607-2_23]

[22] Orchard G, Jayawant A, Cohen GK, Thakor N. Converting static image datasets to spiking neuromorphic datasets using saccades. Front Neurosci 2015; 9: 437.
[http://dx.doi.org/10.3389/fnins.2015.00437] [PMID: 26635513]

[23] Li H, Liu H, Ji X, Li G, Shi L. Cifar10-dvs: an event-stream dataset for object classification. Front Neurosci 2017; 11: 309.
[http://dx.doi.org/10.3389/fnins.2017.00309] [PMID: 28611582]

[24] Lecun Y, Bottou L, Bengio Y, Haffner P. Gradient-based learning applied to document recognition. Proc IEEE 1998; 86(11): 2278-324. [http://dx.doi.org/10.1109/5.726791].
[http://dx.doi.org/10.1109/5.726791]

[25] Deng J, Dong W. ImageNet: A large-scale hierarchical image database. IEEE Conference on Computer Vision and Pattern Recognition 248-55.

[26] Diehl PU, Cook M. Unsupervised learning of digit recognition using spike-timing-dependent plasticity. Front Comput Neurosci 2015; 9: 99.
[http://dx.doi.org/10.3389/fncom.2015.00099] [PMID: 26941637]

[27] Lee JH, Delbruck T, Pfeiffer M. Training deep spiking neural networks using back propagation. Front Neurosci 2016; 10: 508.

[http://dx.doi.org/10.3389/fnins.2016.00508] [PMID: 27877107]

[28]   Rathi N, Srinivasan G, Panda P, Roy K. Enabling deep spiking neural networkswith hybrid conversion and spike timing dependent backpropagation. 2020.

[29]   Mathieson K, Loudin J, Goetz G, *et al.* Photovoltaic retinal prosthesis with high pixel density. Nat Photonics 2012; 6(6): 391-7.
[http://dx.doi.org/10.1038/nphoton.2012.104] [PMID: 23049619]

[30]   Peterman MC, Mehenti NZ, Bilbao KV, *et al.* The Artificial Synapse Chip: a flexible retinal interface based on directed retinal cell growth and neurotransmitter stimulation. Artif Organs 2003; 27(11): 975-85.
[http://dx.doi.org/10.1046/j.1525-1594.2003.07307.x] [PMID: 14616516]

[31]   Iezzi R, Finlayson P, Yong Xu , Katragadda R. Microfluidic neurotransmitter-based neural interfaces for retinal prosthesis. Annu Int Conf IEEE Eng Med Biol Soc 2009; 2009: 4563-5.
[http://dx.doi.org/10.1109/IEMBS.2009.5332694] [PMID: 19963838]

[32]   Yoshida K, Farina D, Akay M, Jensen W. Multichannel intraneural and intramuscular techniques for multiunit recording and use in active prostheses. Proc IEEE 2010; 98(3): 432-49.
[http://dx.doi.org/10.1109/JPROC.2009.2038613].
[http://dx.doi.org/10.1109/JPROC.2009.2038613]

[33]   Bruns TM, Wagenaar JB, Bauman MJ, Gaunt RA, Weber DJ. Real-time control of hind limb functional electrical stimulation using feedback from dorsal root ganglia recordings. J Neural Eng 2013; 10(2)026020
[http://dx.doi.org/10.1088/1741-2560/10/2/026020] [PMID: 23503062]

[34]   Ermentrout B, Terman DH. Mathematical foundations of neuroscience. New York: springer 2010.
[http://dx.doi.org/10.1007/978-0-387-87708-2]

[35]   Lindsay AE, Lindsay KA, Rosenberg JR. Increased computational accuracy in multi-compartmental cable models by a novel approach for precise point process localization. J Comput Neurosci 2005; 19(1): 21-38.
[http://dx.doi.org/10.1007/s10827-005-0192-7] [PMID: 16133823]

[36]   Wijekoon JHB, Dudek P. Compact silicon neuron circuit with spiking and bursting behaviour. Neural Netw 2008; 21(2-3): 524-34.
[http://dx.doi.org/10.1016/j.neunet.2007.12.037] [PMID: 18262751]

[37]   Indiveri G, Linares-Barranco B, Hamilton TJ, *et al.* Neuromorphic silicon neuron circuits. Front Neurosci 2011; 5: 73.
[http://dx.doi.org/10.3389/fnins.2011.00073] [PMID: 21747754]

[38]   Szczęsny S, Huderek D. 60 pW 20 μm size CMOS implementation of an actual soma membrane. J Comput Electron 2020; 19(1): 242-52. [http://dx.doi.org/10.1007/s10825-019-01431-2].
[http://dx.doi.org/10.1007/s10825-019-01431-2]

[39]   Kwon MW, Baek MH, Hwang S, *et al.* Integrate-and-fire neuron circuit using positive feedback field effect transistor for low power operation. J Appl Phys 2018; 124(15)152107
[http://dx.doi.org/10.1063/1.5031929].
[http://dx.doi.org/10.1063/1.5031929]

[40]   Woo S, Cho J, Lim D, Park YS, Cho K, Kim S. Implementation and characterization of an integrateand-fire neuron circuit using a silicon nanowire feedback field-effect transistor. IEEE Trans Electron Dev 2020; 67(7): 2995-3000. [http://dx.doi.org/10.1109/TED.2020.2995785].
[http://dx.doi.org/10.1109/TED.2020.2995785]

[41]   Zohora FT, Debnath S, Rashid AH. Memristor-CMOS hybrid implementation of leaky integrate and fire neuron model. 2019; International Conference on Electrical, Computer and Communication Engineering (ECCE) 1-5.
[http://dx.doi.org/10.1109/ECACE.2019.8679259]

[42]   Babacan Y, Kaçar F, Gürkan K. A spiking and bursting neuron circuit based on memristor. Neurocomputing 2016; 203: 86-91. [http://dx.doi.org/10.1016/j.neucom.2016.03.060]. [http://dx.doi.org/10.1016/j.neucom.2016.03.060]

[43]   Davies M, Wild A, Orchard G, *et al.* Advancing neuromorphic computing with loihi: A survey of results   and   outlook.   Proc   IEEE   2021;   109(5):   911-34. [http://dx.doi.org/10.1109/JPROC.2021.3067593]. [http://dx.doi.org/10.1109/JPROC.2021.3067593]

[44]   Brette R. What is the most realistic single-compartment model of spike initiation? PLOS Comput Biol 2015; 11(4)e1004114 [http://dx.doi.org/10.1371/journal.pcbi.1004114] [PMID: 25856629]

[45]   Millner S, Hartel A, Schemmel J, Meier K. Towards biologically realistic multi-compartment neuron model emulation in analog VLSI. InESANN 2012.

[46]   Gerstner W, Kistler WM, Naud R, Paninski L. Neuronal dynamics: From single neurons to networks and models of cognition. Cambridge University Press 2014; p. 24. [http://dx.doi.org/10.1017/CBO9781107447615]

[47]   Kendall JD, Kumar S. The building blocks of a brain-inspired computer. Appl Phys Rev 2020; 7(1)011305 [http://dx.doi.org/10.1063/1.5129306]. [http://dx.doi.org/10.1063/1.5129306]

[48]   Kuznetsova AY, Huertas MA, Kuznetsov AS, Paladini CA, Canavier CC. Regulation of firing frequency in a computational model of a midbrain dopaminergic neuron. J Comput Neurosci 2010; 28(3): 389-403. [http://dx.doi.org/10.1007/s10827-010-0222-y] [PMID: 20217204]

[49]   Svensson CM, Coombes S. Mode locking in a spatially extended neuron model: active soma and compartmental   tree.   Int   J   Bifurcat   Chaos   2009;   19(8):   2597-607. [http://dx.doi.org/10.1142/S0218127409024347]. [http://dx.doi.org/10.1142/S0218127409024347]

[50]   Rempe MJ, Spruston N, Kath WL, Chopp DL. Compartmental neural simulations with spatial adaptivity. J Comput Neurosci 2008; 25(3): 465-80. [http://dx.doi.org/10.1007/s10827-008-0089-3] [PMID: 18459041]

[51]   Roh H, Cunin C, Samal S, Gumyusenge A. Towards organic electronics that learn at the body-machine   interface: A   materials   journey.   MRS   Commun   2022;   12(5):   565-77. [http://dx.doi.org/10.1557/s43579-022-00269-3]. [http://dx.doi.org/10.1557/s43579-022-00269-3]

[52]   Khan SQ, Ghani A, Khurram M. Frequency-dependent synaptic plasticity model for neurocomputing applications.   Int   J   Bio-inspired   Comput   2020;   16(1):   56-66. [http://dx.doi.org/10.1504/IJBIC.2020.109001]. [http://dx.doi.org/10.1504/IJBIC.2020.109001]

[53]   Schilstra M, Rust A, Adams R, Bolouri H. A finite state automaton model for multi-neuron simulations. Neurocomputing 2002; 44-46: 1141-8. [http://dx.doi.org/10.1016/S0925-2312(02)0043--1]. [http://dx.doi.org/10.1016/S0925-2312(02)00438-1]

[54]   Markaki M, Orphanoudakis S, Poirazi P. Modelling reduced excitability in aged CA1 neurons as a calcium-dependent   process.   Neurocomputing   2005;   65-66:   305-14. [http://dx.doi.org/10.1016/j.neucom.2004.10.023]. [http://dx.doi.org/10.1016/j.neucom.2004.10.023]

[55]   Aimone JB, Date P, Fonseca-Guerra GA, *et al.* A review of non-cognitive applications for neuromorphic computing. Neuromorphic Computing and Engineering 2022; 2(3)032003 [http://dx.doi.org/10.1088/2634-4386/ac889c].

[http://dx.doi.org/10.1088/2634-4386/ac889c]

[56]   Yan Z, Zhou J, Wong WF. Energy efficient ECG classification with spiking neural network. Biomed Signal Process Control 2021; 63102170 [http://dx.doi.org/10.1016/j.bspc.2020.102170].
[http://dx.doi.org/10.1016/j.bspc.2020.102170]

[57]   Burelo K, Sharifshazileh M, Krayenbühl N, Ramantani G, Indiveri G, Sarnthein J. A spiking neural network (SNN) for detecting high frequency oscillations (HFOs) in the intraoperative ECoG. Sci Rep 2021; 11(1): 6719.
[http://dx.doi.org/10.1038/s41598-021-85827-w] [PMID: 33762590]

[58]   Grunditz Å, Holbro N, Tian L, Zuo Y, Oertner TG. Spine neck plasticity controls postsynaptic calcium signals through electrical compartmentalization. J Neurosci 2008; 28(50): 13457-66.
[http://dx.doi.org/10.1523/JNEUROSCI.2702-08.2008] [PMID: 19074019]

[59]   Elyada YM, Haag J, Borst A. Different receptive fields in axons and dendrites underlie robust coding in motion-sensitive neurons. Nat Neurosci 2009; 12(3): 327-32.
[http://dx.doi.org/10.1038/nn.2269] [PMID: 19198603]

[60]   Hendrickson EB, Edgerton JR, Jaeger D. The capabilities and limitations of conductance-based compartmental neuron models with reduced branched or unbranched morphologies and active dendrites. J Comput Neurosci 2011; 30(2): 301-21.
[http://dx.doi.org/10.1007/s10827-010-0258-z] [PMID: 20623167]

[61]   Gromiha MM, Preethi P, Pandey M. From Code to Cure: The Impact of Artificial Intelligence in Biomedical Applications. BioMedInformatics 2024; 4(1): 542-8.
[http://dx.doi.org/10.3390/biomedinformatics4010030].
[http://dx.doi.org/10.3390/biomedinformatics4010030]

[62]   Aboumerhi K, Güemes A, Liu H, Tenore F, Etienne-Cummings R. Neuromorphic applications in medicine. J Neural Eng 2023; 20(4)041004
[http://dx.doi.org/10.1088/1741-2552/aceca3] [PMID: 37531951]

[63]   Sakthi U, Bongirwar A, Sree KB. Convolutional Neural Network for Breast Cancer Prediction using MRI Images. 2024; 2nd International Conference on Intelligent Data Communication Technologies and Internet of Things (IDCIoT) 573-7.
[http://dx.doi.org/10.1109/IDCIoT59759.2024.10467925]

[64]   Dai S, Dai Y, Zhao Z, et al. Intrinsically stretchable neuromorphic devices for on-body processing of health data with artificial intelligence. Matter 2022; 5(10): 3375-90.
[http://dx.doi.org/10.1016/j.matt.2022.07.016].
[http://dx.doi.org/10.1016/j.matt.2022.07.016]

[65]   Liu D, Shi Q, Dai S, Huang J. The Design of 3D☐Interface Architecture in an Ultralow☐Power, Electrospun Single☐Fiber Synaptic Transistor for Neuromorphic Computing. Small 2020; 16(13)1907472
[http://dx.doi.org/10.1002/smll.201907472] [PMID: 32068955]

[66]   Chen YH, Krishna T, Emer JS, Sze V. Eyeriss: An energy-efficient reconfigurable accelerator for deep convolutional neural networks. IEEE J Solid-State Circuits 2017; 52(1): 127-38.
[http://dx.doi.org/10.1109/JSSC.2016.2616357].
[http://dx.doi.org/10.1109/JSSC.2016.2616357]

[67]   Moons B, Verhelst M. An energy-efficient precision-scalable ConvNet processor in 40-nm CMOS. IEEE J Solid-State Circuits 2017; 52(4): 903-14. [http://dx.doi.org/10.1109/JSSC.2016.2636225].
[http://dx.doi.org/10.1109/JSSC.2016.2636225]

[68]   Song J, Cho Y, Park JS, et al. 7.1 An 11.5 TOPS/W 1024-MAC butterfly structure dual-core sparsityaware neural processing unit in 8nm flagship mobile SoC. In2019 IEEE international solid-state circuits conference-(ISSCC) 2019; 17: 130-2.

[69]   Cavigelli L, Benini L. Origami: A 803-GOp/s/W convolutional network accelerator. IEEE Trans Circ

Syst Video Tech 2017; 27(11): 2461-75. [http://dx.doi.org/10.1109/TCSVT.2016.2592330].
[http://dx.doi.org/10.1109/TCSVT.2016.2592330]

[70]   Hruska J. Intel details its Nervana inference and training AI cards. Extreme Tech 2019; 21: 20.

[71]   Lee J, Lee J, Han D, Lee J, Park G, Yoo HJ. 7.7 LNPU: A 25.3 TFLOPS/W sparse deep-neuralnetwork learning processor with fine grained mixed precision of FP8-FP16. In2019 IEEE International Solid-State Circuits Conference-(ISSCC) 142-4.

[72]   Covi E, Donati E, Liang X, *et al.* Adaptive extreme edge computing for wearable devices. Front Neurosci 2021; 15611300
[http://dx.doi.org/10.3389/fnins.2021.611300] [PMID: 34045939]

[73]   Vitale A, Donati E, Germann R, Magno M. Neuromorphic edge computing for biomedical applications: Gesture classification using emg signals. IEEE Sens J 2022; 22(20): 19490-9.
[http://dx.doi.org/10.1109/JSEN.2022.3194678].
[http://dx.doi.org/10.1109/JSEN.2022.3194678]

[74]   Cleary F, Srisa-an W, Gil B, *et al.* Wearable uBrain: Fabric Based-Spiking Neural Network. arXiv preprint arXiv 2022; 12984.

[75]   Naghshvarianjahromi M, Majumder S, Kumar S, Naghshvarianjahromi N, Deen MJ. Natural braininspired intelligence for screening in healthcare applications. IEEE Access 2021; 9: 67957-73.
[http://dx.doi.org/10.1109/ACCESS.2021.3077529]

[76]   Buettner K, George AD. Heartbeat classification with spiking neural networks on the Loihi neuromorphic processor. In2021 IEEE Computer Society Annual Symposium on VLSI (ISVLSI) 2021; 138-43.
[http://dx.doi.org/10.1109/ISVLSI51109.2021.00035]

[77]   Nurse E, Mashford BS, Yepes AJ, Kiral-Kornek I, Harrer S, Freestone DR. Decoding EEG and LFP signals using deep learning: heading TrueNorth. Proceedings of the ACM international conference on computing frontiers 2016; 259-66.
[http://dx.doi.org/10.1145/2903150.2903159]

[78]   Nie Y, Ren T, Shi Z. The developments and applications of brain-like computing chips. Int Conf Algo Micr Net App. 272-85.
[http://dx.doi.org/10.1117/12.2636417]

[79]   Parmar R, Yadav K, Anand G, Trivedi G. An SNN-Inspired Area- and Power-Efficient VLSI Architecture of Myocardial Infarction Classifier for Wearable Devices. IEEE Trans Circuits Syst II Express Briefs 2024; 71(6): 3191-5. [http://dx.doi.org/10.1109/TCSII.2024.3355016].
[http://dx.doi.org/10.1109/TCSII.2024.3355016]

[80]   Rathi N, Chakraborty I, Kosta A, *et al.* Exploring neuromorphic computing based on spiking neural networks: Algorithms to hardware. ACM Comput Surv 2023; 55(12): 1-49.
[http://dx.doi.org/10.1145/3571155].
[http://dx.doi.org/10.1145/3571155]

[81]   Getty N, Brettin T, Jin D, Stevens R, Xia F. Deep medical image analysis with representation learning and neuromorphic computing. Interface Focus 2021; 11(1)20190122
[http://dx.doi.org/10.1098/rsfs.2019.0122] [PMID: 33343872]

[82]   Bauer FC, Muir DR, Indiveri G. Real-time ultra-low power ECG anomaly detection using an eventdriven neuromorphic processor. IEEE Trans Biomed Circuits Syst 2019; 13(6): 1575-82.
[http://dx.doi.org/10.1109/TBCAS.2019.2953001] [PMID: 31715572]

[83]   Dethier J, Nuyujukian P, Ryu SI, Shenoy KV, Boahen K. Design and validation of a real-time spiking-neural-network decoder for brain–machine interfaces. J Neural Eng 2013; 10(3)036008
[http://dx.doi.org/10.1088/1741-2560/10/3/036008] [PMID: 23574919]

[84]   Boi F, Moraitis T, De Feo V, *et al.* A bidirectional brain-machine interface featuring a Neuromorphic hardware decoder. Front Neurosci 2016; 10: 563.

[http://dx.doi.org/10.3389/fnins.2016.00563] [PMID: 28018162]

[85]   Donati E, Payvand M, Risi N, Krause R, Indiveri G. Discrimination of EMG signals using a neuromorphic implementation of a spiking neural network. IEEE Trans Biomed Circuits Syst 2019; 13(5): 795-803.
[http://dx.doi.org/10.1109/TBCAS.2019.2925454] [PMID: 31251192]

[86]   Buccelli S, Bornat Y, Colombi I, *et al.* A neuromorphic prosthesis to restore communication in neuronal networks. iScience 2019; 19: 402-14.
[http://dx.doi.org/10.1016/j.isci.2019.07.046] [PMID: 31421595]

CHAPTER 2

A Review on Data Mining Techniques and Their Applications in Medicine

Abrar Abu Hamdia[1,*]

¹ Department of Medical Laboratory Science, Faculty of Medicine and Health Sciences, An-Najah National University, Nablus, Palestine

Abstract: Data mining is a crucial aspect of the 21st century. Technological advancements and the exponential expansion of medical data have driven it. Medical data is mandatory for clinical judgment, scientific projects, and innovation. Data-mining methods include descriptive and predictive techniques. They have been used to extract insights from massive, diverse, and erratic medical datasets. These methods have massive implementation; they enhance diagnosis accuracy and decrease diagnosis time. They also provide quantified temporal information on crucial medical behaviors, minimizing errors. They offer opportunities for healthcare transformation, improving the overall healthcare system and general health.

Keywords: Artificial intelligence, Chinese medicine, Clinical data, Comparative analysis, Data mining, Data science, Databases, Data privacy, Drug discovery, Early diagnosis, Ehr, Experimental data, Fraud detection, Machine learning, Patient record, Precision medicine, Predictive analysis, Regression analysis, Supervised learning, Survival analysis.

INTRODUCTION

The 21st century is the birth of "Big Data." Technological advancements and the subsequent exponential expansion of data have driven this concept. Big data were represented through the five Vs: volume, variety, velocity, veracity, value, and variability. Medical data are the cornerstone of the healthcare system. They direct clinical judgments, scientific projects, and innovation. These data come from several sources, including public records, patient portals, Electronic Health Records (EHRs), payer records, search engine data, generic databases, smart devices, research studies, wearable technology, and public databases [1].

---

* **Corresponding author Abrar Abu Hamdia:** Department of Medical Laboratory Science, Faculty of Medicine and Health Sciences, An-Najah National University, Nablus, Palestine; Tel: +970598238887; E-mail: abrarabuhamdia@gmail.com

Sivakumar Rajagopal, Prakasam P., Konguvel E., Shamala Subramaniam, Ali Safaa Sadiq Al Shakarchi & B. Prabadevi (Eds.)

Data-mining methods have emerged as a result of the constraints of traditional data management and analysis tools for handling big data. Data science is an interdisciplinary field that combines techniques, including math, statistics, artificial intelligence, and machine learning. Data mining is a crucial element in this field. The core of data mining is the extraction of implicit, significant, and novel insights from massive, noisy, diverse, and erratic datasets. The primary types of data-mining techniques are descriptive and predictive. Predictive methods are used to predict unknown or future values of variables; descriptive methods identify patterns, associations, and trends in data [2, 3].

Data mining has been widely used in the medical field. A recent study used the bibliometric visualization tools VOSviewer and CiteSpace. They looked at the geographical distribution, international collaborations, and citation patterns of studies that used data mining in medicine. Notably, they discovered that these publications and citations have rapidly increased since 2013. This increase demonstrates the efficiency and effectiveness of this approach in the medical domain. It has improved diagnosis accuracy, reduced diagnosis time, provided quantifiable temporal data on critical medical behaviors, and reduced medical errors. It effectively guides doctors in their daily clinical practice [1, 4].

Finally, data mining provides numerous opportunities for healthcare transformation. Through ongoing evolution, it has the potential to improve the healthcare system and overall health. This chapter offers insight into medical data and data-mining methods, including their applications in medical fields and their challenges.

MEDICAL DATA

In the twenty-first century, the concept of "Big Data" emerged due to technological advancements and the exponential growth of data. Big data refers to large and complex data collections that conventional technologies cannot manage. Douglas Laney described it using the three Vs: volume, variety, and velocity. Later, additional Vs were added to the list. It contains veracity, value, and variability. The volume emphasizes how much information there is. The variety reflects the types of data that exist. Velocity indicates how quickly the data are generated. Veracity indicates how accurate and reliable the collected data is. Values show the value of the collected data. Variability illustrates how big data can be used and organized [1].

Medical data serve as the foundation of the healthcare system. They contain massive, frequently updated, imperfect, and time-sensitive data. These data highlight the diverse nature of disease, treatment, and outcome. They also reveal the complexities of data collection, processing, and interpretation. Examples of medical data collection methods include government organizations, patient portals, electronic health records (EHRs), payer records, search engine data, generic databases, smartphones, research studies, wearable technology, image data, and public records. Medical data comprise several categories, with clinical data taking up most of the space [1].

Clinical Data

Clinical data include information about the management of a patient. They include past prescriptions, medical and family histories, symptoms, physical examination, vital signs, and treatment plans and outcomes. Medical charts, electronic patient records (EPRs), electronic health records (EHRs), and electronic medical records (EMRs) are tools for collecting medical data. Healthcare specialists use electronic medical records (EMRs). Healthcare settings such as hospitals use EPRs. They are more comprehensive but limited in scope. According to the Institute of Medicine, a National Academies of Sciences and Engineering division, EHRs are the most extensive and integrated electronic health information system. They offer a comprehensive and long-term record of a patient's health and medical history. Healthcare organizations share their EHRs.

Medical data collection tools have expanded with the emergence of wearable technology and the Internet of Things (IoT). These tools include blood glucose monitors, cell phones, and fitness tracker platforms. Multinational corporations, such as Apple and Google, invented these platforms. These platforms allow the gathering of real-time data such as heart rate, blood glucose level, calories burned, sleep cycle patterns, and cortisol levels at low cost [1, 5].

Other Types of Medical Data [1, 2]

- **Laboratory Data:** This consists of test results from several specimens, such as bone marrow, hair and nails, urine, semen, sputum, stool, tissue, saliva, cerebrospinal fluid (CSF), and nasal or throat swabs. These findings shed light on a patient's current state of health and may help with tailored medicine, specifically by using genetic testing.
- **Imaging Data:** Medical imaging is obtained using various methods, such as computed tomography (CT), ultrasound, positron emission tomography (PET), magnetic resonance imaging (MRI), and X-rays. These images aid in the diagnosis and observation of numerous medical conditions.
- **Administrative Data:** The administrative departments of the healthcare facility gather non-clinical data, including patient demographics, insurance information, billing, and scheduling.

- **Study Data, or experimental data:** They comprise the gathered data for research purposes. One or more of the following data types may be present in the research data: clinical, laboratory, imaging, and administrative.

Numerous platforms, such as databases, cloud storage, and data warehousing, have emerged for securely handling, processing, and analyzing all of these data. Large volumes of data can be gathered, stored, and analyzed using data warehousing for reporting and analytical needs. Cloud storage is a scalable and affordable tool for storing large volumes of unstructured or semi-structured data, including backup copies, documents, photos, and videos. Although databases and data warehouses offer high-performance querying capabilities, they may not offer the same accessibility, scalability, or durability for data as cloud storage. A database is a well-structured collection of data typically stored electronically in a computer system. It makes data management, retrieval, updating, and storage more efficient [1, 2, 5].

DATA MINING

Definition and Types

In front of "Big Data," how to glean insights from diverse and complex datasets is our primary concern. Data science is an interdisciplinary field that integrates mathematical and statistical methodologies, specialized programming, sophisticated analytics, artificial intelligence (AI), and machine learning. Data mining is an essential part of data science that works with other data science components to extract insights from large amounts of noisy, heterogeneous, and irregular data. Rather than replacing conventional statistical analysis techniques, data mining reinforces them. There are two categories of data-mining techniques: predictive and descriptive. Descriptive methods, which include association analysis and cluster analysis, are used to find patterns, associations, and trends in data. In contrast, predictive methods, including regression and classification, estimate the upcoming or unknown values of relevant variables [1, 2].

Descriptive Methods [1, 2, 6]

Association Analysis

Association analysis identifies patterns, relationships, correlations, or causal structures among data variables. There are two steps: (1) Make a set for every item that occurs frequently, and (2) Produce association rules. Upgrade Lift, the FP tree frequency set algorithm, and the Apriori algorithm are examples of machine-learning techniques for association analysis.

Cluster Analysis

Cluster analysis groups comparable objects into a single category. Using static categorization, which splits comparable objects into many groups or subsets based on shared characteristics, we obtain subsets, each containing comparable items. Hierarchical clustering and density-based, grid-based, and partition-based algorithms are examples of clustering algorithms.

Prediction Methods

Regression

Regression analysis determines the quantitative relationships between several variables. It consists of linear and multiple linear regression, depending on the number of independent variables; the former has just one independent variable, while the latter has two or more. Multiple regression is more practical and effective than linear regression, which is related to the complexity of diseases and phenomena. The steps in multiple regression are as follows: Using the gathered data, (1) create a regression equation, (2) use a hypothesis test to assess the regression equation. If the results are significant, (3) calculate each independent variable's partial regression coefficient, (4) eliminate variables lacking a significant partial regression coefficient, (5) reconstruct the regression equation incorporating variables possessing a substantial partial regression coefficient, and (6) repeat steps 2 through 6 until you are confident that your equation is significant and solely comprises significant variables.

Regression approaches rely on the premise of data distribution in the real world, which is difficult to define with a relatively limited mathematical formula. Constraints in regression methods have led to the development of machine learning methods, which are the primary data-mining techniques that do not require prior assumptions. Machine learning methods include the nearest neighbor algorithm, artificial neural networks, decision trees, adaptive boosting, bagging, random forests, and support vector machines.

Classification

Classification is a supervised learning method in which data must be "Tagged". Therefore, the accuracy of the classifications influences the quality of information. Examples of classification techniques are machine learning, logistic regression, probit regression, and classical discriminant analysis. In particular, machine learning is superior for complicated data and provides more accurate results.

Data Mining Methods

Data mining methods were used to construct the analysis model. These methods are divided into two groups according to whether the dependent variables are labeled: supervised or unsupervised learning. Supervised learning is applied when variables are labeled, while unsupervised learning is applied when labels are absent [1, 3].

Supervised Learning

The foundation of this approach is dataset partitioning, which prevents the model from becoming overfit. There are three steps involved:

1. A training set consisting of examples to help learn and fit the parameters.

2. A validation set consisting of examples to help tune a classifier's parameter. However, in practice, a validation set is rarely used.

3. A test set consisting of examples to evaluate the model's performance.

Two categories of supervised learning approaches can be identified based on the need for presumptions. While a competitive risk model, decision trees, the random forest (RF) method, and support vector machines (SVMs) do not require assumptions about the data, linear regression, generalized linear regression, and a proportional hazards model (the Cox regression model) must. Assumptions are challenging when dealing with medical data, particularly with clinical data. As a result, researchers have favored approaches that do not necessitate making any assumptions about the data.

Comparative Risk Model

A comparative risk model (CRM) is employed for survival analysis, which quantifies how different risk factors affect the disease burden in a community, reduce risks, and enhance public health outcomes. In contrast to CRM, conventional survival analysis techniques, such as the Cox proportional hazards model and Kaplan-Meier marginal regression, consider only one endpoint. As outcomes in the medical field typically have multiple endpoints, traditional approaches are less reliable. According to a study that used the Surveillance, Epidemiology, and End Results (SEER) database and data mining to study prognosis in patients with esophageal carcinoma, a competing risk model achieved a more accurate estimate of the cumulative incidence function (CIF) and prognostic factors than the Cox proportional hazards model and cause-specific hazard function (CS). CRM has many other uses, including prediction of prognosis and customized treatment planning. Research on penile cancer using the

SEER database has reported that the comparative risk model could be applied to estimate the risk of penile cancer.

Decision Trees

A decision tree is a fundamental technique for regression and classification. It explores the relationships between variables and groups them into discrete categories based on their properties to diagnose certain diseases and forecast disease prognosis. Decision trees serve in many areas, such as the early detection of renal calculi, investigation of the relationships between predictors and prediction of type 2 diabetes, prediction of the risk of in-hospital cardiac arrest (IHCA), and prediction of the likelihood of mortality in breast cancer patients.

The Random Forest (RF) Algorithm

It is an ensemble learning method for classification and regression tasks. The term "ensemble learning" describes the fusion of predictions from many models, which facilitates the management of various data structures and enhances overall performance, including overfitting and prediction accuracy. RF comprises several decision trees, each containing a different subset of inputs. It makes use of a method known as bagging or Bootstrap Aggregating. Every decision tree is trained using a replacement on a random sample of the dataset. For classification, RF chooses the class that each tree predicts most often, while regression uses the mean of each tree's predictions.

According to a study on the features of research using data mining for chronic diseases, random forest was the most widely used approach, performing better in risk factor assessment and hospital readmission, survival, and death prediction. Moreover, RF has been applied in genome-wide investigations, personalized medicine, protein interaction prediction, and gene expression microarray analyses. The following characteristics of RF that clarifies its widespread use:

1. Its adaptability and suitability to various data aspects, including both longitudinal and non-longitudinal research designs.

2. Its explainability approaches, such as counterfactual sets.

3. Its ability to manage a large number of variables while limiting overfitting, in contrast to the Cox Proportional Hazard Regression Model.

4. Its ability to address complex data interactions, a known difficulty in conventional regression models.

However, RF cannot manage large datasets or uncommon incidents.

The Support Vector Machines (SVMs)

Initially, SVMs are used for binary classification, in which the hyperplane that optimizes the margin between classes divides the data points into two classes. The margin is the gap between the closest data points from each class and the hyperplane. SVMs aim to locate the hyperplane that maximizes this margin to improve the robustness and generalization. Using a kernel function, SVMs can handle nonlinear data by projecting them into a higher-dimensional space. Strategies such as one-*vs.*-one or one-*vs.*-all allow the modification of SVMs for use in multiclass classification situations by breaking down the problem into a sequence of binary classifications. SVMs can also serve in regression tasks, in which the algorithm predicts a continuous value.

SVMs are resilient to managing high-dimensional and noisy datasets, nonlinear and linear data, short sample sizes, situations with more features than samples, and complex classification tasks. However, SVMs are expensive and time-consuming for large observational samples and can be computationally demanding. The choice of kernel function and hyperparameters can strongly influence performance.

SVMs serve in many different domains, such as text classification, image classification, and medical diagnosis. Numerous studies have applied SVMs in various aspects, including treatment compliance in patients with heart failure, understanding the unknown mechanisms underlying some chronic and complicated diseases, interactions between genes and the environment, and the prediction of diabetes.

Unsupervised Methods

Tagging data is a laborious and challenging task. Unsupervised learning solved this problem by enabling the classification of unlabeled data. It groups the data according to commonalities, traits, and correlations. Principal component analysis (PCA), association rules, and clustering analysis are three primary unsupervised techniques.

Clustering Analysis

This approach finds patterns and correlations in unlabeled data and accordingly groups the variables into clusters. A cluster of objects is composed of objects that are similar to or close to one another. After clusters are created, it is essential to verify their quality, typically by evaluating metrics such as within-cluster cohesion and between-cluster separation. Numerous clustering algorithms are available, and each one adopts a unique method. The K-means and hierarchical clustering algorithms are the most common.

K-Means

K-means divide the dataset into a predetermined number of clusters (K), each with similar data points. K-means procedures include: (1) First, K initial centroids are chosen frequently randomly or using a heuristic technique. Centroids represent the center of each cluster. (2) Each data point is allocated to the cluster with the closest centroid based on the Euclidean distance. (3) Following the assignment of all data points, each cluster's centroids are recalculated using the average of the data points inside the cluster. (4) Step two and three are iteratively repeated up to convergence. Convergence occurs after a predetermined number of iterations or when the centroids no longer exhibit significant variation. The final result is a collection of K clusters, each with data points close to the centroid of its corresponding cluster. It is challenging to decide on the ideal number of clusters (K); however, the elbow method and silhouette score offer a suitable value for K.

K-means is a popular choice for medical data analysis owing to its high efficiency, low time complexity, and simplicity. However, it cannot work well with high-dimensional data or when clusters have different sizes or forms.

Hierarchical Clustering

This approach arranges data points into a hierarchy of clusters, which helps to comprehend the data's hierarchical structure. Clusters are shown as dendrograms, and there is no need to identify the number of clusters in advance. However, this is complex. Hierarchical clustering can be classified into two types: agglomerative and divisive.

Agglomerative clustering is a method in which the two nearest clusters are combined into a single, larger cluster at each stage, with each data point beginning as its cluster. This procedure continues until every data point is a part of a single cluster. Compared with divisive clustering, Agglomerative clustering is used more frequently in the medical field because of its effectiveness and simplicity of interpretation.

Divisive clustering is the antithesis of the preceding method. At each stage, the cluster is split into smaller clusters, starting with all data points in one cluster. This process continues until every data point is inside the cluster.

Applications of Data Mining in Medicine [1, 2, 6]

There are a few of the numerous medical applications for data mining:

Disease Diagnosis and Prediction

Data mining, which finds patterns and risk factors in patient data, can make screening and diagnosis strategies, treatment planning, and disease prevention easier and more accurate. Integrating diverse data sets, such as genomic and image data, with advanced technologies, such as artificial intelligence, will improve accuracy, guide efficient decision-making, and subsequently improve the healthcare system and overall health [7]. Neural network and decision tree data mining techniques can be applied to data from a prospective pregnancy cohort to identify preterm risk factors that increase the risk of perinatal morbidity and mortality. Their model contributed to the early detection of high-risk pregnancies and prompt action to reduce or completely eradicate preterm births.

Personalized Medicine

Personalized treatment is made more accessible by data mining, which analyzes various medical data, including genetic information and clinical records. It assists in determining the genetic markers, treatment reactions, and individual patient outcomes. The incorporation of radiomics with imaging data has further enhanced precision medicine. Radiomics is the process of converting images into mineable data to predict disease phenotypes for precision medicine or to produce quantitative image-based phenotypes for data mining with other omics or discoveries (*i.e.*, imaging genomics).

Pharmacology and Pharmacovigilance

Databases can use data mining to identify new drug targets, predict the safety and efficacy of these candidates, and help uncover drug mechanisms of action. For instance, some studies [8, 9] identified known and novel multi-item adverse drug event (ADE) associations using association-rule mining.

Traditional Chinese Medicine (TCM)

Data mining has recently been used in TCMs to search through thousands of studies and identify combinations with therapeutic potential. Commonly employed techniques include recurrent neural networks, correlation rule mining, and frequency analysis. A study [10] used network pharmacology and data mining to identify targets, regulatory networks, and mechanisms of action for different combinations of Chinese anti-endometriosis herbs.

Moreover, data mining [11] has been used to determine the mechanisms of action of TCMs in treating stroke by examining 1679 Chinese medicinal formulas. To determine the clinical efficacy of five herbs (Ephedra, Almond, Platycodon grandiflorum, Licorice, and Scutellaria) for the treatment of Mycoplasma pneumoniae [12], combined a systematic pharmacology approach with literature data mining. Fifty-three proteins and enzymes were MPP-related targets, and 93 chemicals out of 1004 components in five herbs were identified as candidate compounds. Together, these studies advance our knowledge of disease pathogenesis and help develop phytomedicine for disease therapy.

Healthcare System

Patient-centric healthcare comprises several components, such as information about hospital visits and services, social data, medical records, web-based platforms, clinical reports, databases of claims, drug sales, resource usage, stream data, omics data, and clinical forums. Patient data management consists of effective scheduling and care delivery for patients during their hospital stay. Enhancing the scheduling system, improving clinical performance, cutting costs, and saving time are all possible with data mining [13]. An author developed an efficient clinical schedule by estimating the likelihood that a patient will not show up by using electronic medical records to create a logistic regression model.

Epidemiology

Incorporating data mining into databases that analyze public health trends, risk factors, and disease prevalence makes tracking disease outbreaks, identifying the factors contributing to their spread, and developing preventive strategies easier.

PRIVACY OF MEDICAL DATA AND FRAUD DETECTION [1]

Maintaining medical data privacy and detecting fraud, is vital, especially in light of the exponential rise in social media usage and the tendency for individuals to connect personal information to these sites. Patient data are typically protected using the K-anonymization model, but large multidimensional datasets are beyond the scope of this model. Several models have been proposed to address the limitations of conventional models.

A cloud computing framework on mobile devices was created, offering the best data security and storage. EHR data can be stored in this framework, enabling data privacy, interoperability, and large-scale data storage. One-time passwords (OTP) and two-factor authentication are examples of approved restricted access algorithms that preserve privacy.

BigQuery tools and MapReduce are two examples of Google's effective and affordable analytical approaches to big data. They also offer high data protection in contrast to conventional methods that allow for the re-identification of the data, giving hackers and fraudsters the ability to re-identify the data, recognize patients, and abuse them. A case study [14] demonstrated how hacking can connect divergent bits of data to identify patients. Although many algorithms are in place to protect people's personal data, not all data sources use them.

LANGUAGE BARRIER [1]

Electronic health records (EHRs) are mandatory in the healthcare system. An analysis of EHRs from several nations could reveal the prevalence of diseases and their management approaches. However, linguistic differences between nations give rise to what is known as a language barrier.

Numerous multilingual language models have been developed by researchers who can comprehend the context and each language's semantics, grammatical rules, and structure.

For instance,

"Let us get together tomorrow at the riverbank."

"I need to take money out of my bank account."

Here, the word "bank" has multiple meanings depending on the situation. These discrepancies can be distinguished by using cross-lingual language models. Some of these models include the following:

- mBERT is a multilingual BERT created by the Google Research team.
- XLM is a cross-lingual model, an mBERT improvisation developed by Facebook AI.
- Multift is a QRNN-based model created by fast AI. It tackles the issues faced by low-resource language models.

CONCLUSION

The medical sector has witnessed considerable use of data mining, and the future seems bright for this technology. It has the potential to significantly improve clinical decision systems, illness prediction and prevention, medication discovery and development, telemedicine and remote monitoring, healthcare fraud detection, personalized medicine, the ability to predict and control future pandemics, research, and innovation. The advancement of data mining technology will continue to improve general health and healthcare systems.

REFERENCES

[1]   Subrahmanya SVG, Shetty DK, Patil V, *et al.* The role of data science in healthcare advancements: applications, benefits, and future prospects. Ir J Med Sci 2022; 191(4): 1473-83.
[http://dx.doi.org/10.1007/s11845-021-02730-z] [PMID: 34398394]

[2]   Yang J, Li Y, Liu Q, *et al.* Brief introduction of medical database and data mining technology in big data era. J Evid Based Med 2020; 13(1): 57-69.
[http://dx.doi.org/10.1111/jebm.12373] [PMID: 32086994]

[3]   Wu WT, Li YJ, Feng AZ, *et al.* Data mining in clinical big data: the frequently used databases, steps, and methodological models. Mil Med Res 2021; 8(1): 44.
[http://dx.doi.org/10.1186/s40779-021-00338-z] [PMID: 34380547]

[4]   Hu Y, Yu Z, Cheng X, Luo Y, Wen C. A bibliometric analysis and visualization of medical data mining research. Medicine (Baltimore) 2020; 99(22): e20338.
[http://dx.doi.org/10.1097/MD.0000000000020338] [PMID: 32481411]

[5]   Paganelli AI, Mondéjar AG, da Silva AC, *et al.* Real-time data analysis in health monitoring systems: A comprehensive systematic literature review. J Biomed Inform 2022; 127: 104009.
[http://dx.doi.org/10.1016/j.jbi.2022.104009] [PMID: 35196579]

[6]   Chen K, Abtahi F, Carrero JJ, Fernandez-Llatas C, Seoane F. Process mining and data mining applications in the domain of chronic diseases: A systematic review. Artif Intell Med 2023; 144: 102645.
[http://dx.doi.org/10.1016/j.artmed.2023.102645] [PMID: 37783545]

[7]   Chen HY, Chuang CH, Yang YJ, Wu T-P. Exploring the risk factors of preterm birth using data mining. Expert Syst Appl 2011; 38(5): 5384-7.
[http://dx.doi.org/10.1016/j.eswa.2010.10.017]

[8]   Harpaz R, Chase HS, Friedman C. Mining multi-item drug adverse effect associations in spontaneous reporting systems. BMC Bioinformatics 2010; 11(S9) (Suppl. 9): S7.
[http://dx.doi.org/10.1186/1471-2105-11-S9-S7] [PMID: 21044365]

[9]   Ohra S, Sharma R, Kumar A. Repurposing of drugs against bacterial infections: A pharmacovigilance-based data mining approach. Drug Dev Res 2024; 85(4): e22211.
[http://dx.doi.org/10.1002/ddr.22211] [PMID: 38807372]

[10]   Xue D, Zhang Y, Song Z, Jie X, Jia R, Zhu A. Integrated meta-analysis, data mining, and animal experiments to investigate the efficacy and potential pharmacological mechanism of a TCM tonic prescription, Jianpi Tongmai formula, in depression. Phytomedicine 2022; 105: 154344.
[http://dx.doi.org/10.1016/j.phymed.2022.154344] [PMID: 35932605]

[11]   Ren L, Zheng X, Liu J, *et al.* Network pharmacology study of traditional Chinese medicines for stroke treatment and effective constituents screening. J Ethnopharmacol 2019; 242: 112044.
[http://dx.doi.org/10.1016/j.jep.2019.112044] [PMID: 31255722]

[12]   Sun J, Sun F, Yan B, Li J, Xin D. Data mining and systematic pharmacology to reveal the mechanisms of traditional Chinese medicine in Mycoplasma pneumoniae pneumonia treatment. Biomed Pharmacother 2020; 125: 109900.
[http://dx.doi.org/10.1016/j.biopha.2020.109900] [PMID: 32028237]

[13]   Daggy J, Lawley M, Willis D, *et al.* Using no-show modeling to improve clinic performance. Health Informatics J 2010; 16(4): 246-59.
[http://dx.doi.org/10.1177/1460458210380521] [PMID: 21216805]

[14]   Li F, Zou X, Liu P, Chen JY. New threats to health data privacy. BMC Bioinformatics 2011; 12: S7.
[http://dx.doi.org/10.1186/1471-2105-12-S12-S7] [PMID: 22168526]

CHAPTER 3

A Comprehensive Study on Data-driven Decision Support System and its Application in Healthcare

Abhishek Liju Liju[1] and **Chintan Singh**[1,*]

[1] *Amity Institute of Forensic Sciences, Amity University, Noida, Uttar Pradesh-201301, India*

Abstract: Due to the growing amount of data collected and generated, organizations are pressured to make sense of this data and use it to inform their decisions. DD-DSS (data-driven decision support system) is a computerized program that helps organizations make decisions, judgments, and plans by analyzing large amounts of data. Machine learning and statistical analysis are typically used to build these decision support systems, so combining expert knowledge with data gathered from various sources reduces the risk of making poor decisions and provides insight into the impact of different strategies before they are implemented. This article aims to enumerate the benefits of DD-DSS implementation, and the features associated with it, as well as explore the scope, frameworks used, and applications of DD-DSS in healthcare. The research was conducted through a combination of a literature review and case studies of various frameworks and applications implemented based on DD-DSS.

Keywords: Big data, Data mining, Data analytics, Data-driven model, Data science, Decision support system, Data-driven decision support, Healthcare, Healthcare informatics, Machine learning, Medical treatment, Statistical analysis.

INTRODUCTION

Data plays an important role in healthcare as well as in any other field since it can be used to create comprehensive patient views, which can be used to personalize treatments, aid in advancing treatment methods by providing insights into the effectiveness of different treatments, provide practical insights into the health system, and aid in making strategic decisions [1, 2].

With the help of advanced technologies, many laboratories and healthcare units are generating data at an unprecedented rate, and these datasets are used to make various decisions [3]. In the process of deciding, a variety of alternatives are identified and evaluated, and the most appropriate option is selected. Recent years

* **Corresponding author Chintan Singh:** Amity Institute of Forensic Sciences, Amity University, Noida, Uttar Pradesh-201301, India; E-mails: chntnsngh@gmail.com, csingh5@amity.edu

Sivakumar Rajagopal, Prakasam P., Konguvel E., Shamala Subramaniam, Ali Safaa Sadiq Al Shakarchi & B. Prabadevi (Eds.)

have seen an increase in the utilization of various types of decision support systems (DSS), which aid in making better and more informed decisions.

The DSS acts as a computerized information system, functioning autonomously or within intricate computing setups. It is designed to make reasoned choices using diverse methodologies derived from the fusion of cognitive science, artificial intelligence, and pattern recognition techniques. We are now able to combine methods from statistics and operational research [4, 5].

Providing users with reliable information is the purpose of DSS, which is done by combining knowledge and data from a variety of sources and areas. Hence, it is an information-based application that supports decisions, judgments, and analyses of large amounts of data, accumulating comprehensive information based on various data sources that can be used to solve problems [6]. With DSS, organizations can make better and more informed decisions, increase efficiency, and reduce costs, utilizing powerful augmented analytics or modelling to make analysis recommendations and identify patterns and trends in data, allowing them to predict future outcomes more accurately.

Data-Driven Decision Support System (DD-DSS) is one of the many types of DSS that use machine learning and statistical analysis to produce comprehensive information reports for pattern recognition and detection. This can be beneficial in the healthcare sector, where decisions tend to be based on the experience of physicians [7]. With DD-DSS, large amounts of data can be analysed using a computer-based approach that can aid in enhanced approaches to detecting security problems with equipment and robust techniques to analyze alternatives to symptoms, prevention, and treatment [8, 9].

The advancement of information technology in healthcare has paved the way for the convergence of healthcare and technology, which is causing a transformation in the research outlook that merges patient care, public health, and preventive health. According to a 2018 article, around 30% of the global data volume originates from the healthcare sector. By 2025, the compound annual growth rate of data for healthcare will reach 36%. That's 6% faster than manufacturing, 10% faster than financial services, and 11% faster than media and entertainment [10].

Some of the major data sources in healthcare are digital health tools (like wearable medical devices), electronic health records of patients, online patient applications of the patients (called patient portals), data collected by government and private agencies, and various research studies. As a result, big data is gathered from these sources in healthcare, which can then be analysed to make better decisions, reduce healthcare costs, and improve patient outcomes [11, 12].

Our discussion starts with what DSS is, its classification and its components and characteristics. This section explains DD-DSS, its features, benefits, subcategories, software used, and applications in the healthcare sector. The final section of this article summarizes and discusses the research's contributions and limitations. A thorough understanding of the essential elements leading to the successful implementation of DD-DSS is provided, as is its potential impact on the organization.

Decision Support System

Before diving into DD-DSSs, it is important to have a basic understanding of DSS. There is no universally agreed-upon definition for DSS, and it can vary from author to author. Following are some of the definitions provided by various authors on DSS:

1. In his description of such a model, Little stated, "It provides a model-based set of procedures for processing data and making decisions to assist managers" [13].

2. In Sprague and Carlson's statement, "DSS is an interactive computer-based system that assists decision makers in solving unstructured problems using data and models" [14].

3. As Keen and Morton explained, "Decision Support Systems combine the intellect of individuals with the capabilities of computers to enhance decision quality. This system aids management decision-makers in solving semi-structured problems" [15].

4. In Finlay's view, "a computer-based decision-support system aids decision-making" [16].

5. Described by Moore and Chang as "extensible systems capable of supporting ad-hoc analysis of data and decision modelling that is oriented toward future planning but used irregularly and unpredictably" [17].

6. Turban describes "a computer-based system that is flexible, interactive, and adaptable to address non-structured management challenges" [18].

7. According to Shim, "Computer technology solutions can be used for complex problem-solving and decision-making" [19].

8. Power describes these systems as "interactive computer-based systems that enable people to communicate, analyze documents, and make decisions as a result of computer-based information" [20].

As a result of the above definitions, DSS can generally be described as "a computer-based interactive system that supports making informed decisions." With structured and semi-structured problems, DSSs work well because they can formulate the underlying mathematical relationships easily, and to a lesser extent with unstructured problems. The original DSS concept was defined by Scott Morton in the early 1970s [21]. In the 1970s and 1980s, the concept of DSS evolved into a field of study, research, and practice [22].

Components of the Decision Support System

Various components make up a DSS, and different architectures have been proposed by different authors [23, 24]. However, the most general architecture of a DSS is composed of three key components: a database, a software system, and a user interface as shown in Fig. (**1**) [25, 26]. Together, these components improve decision-making accuracy and effectiveness.

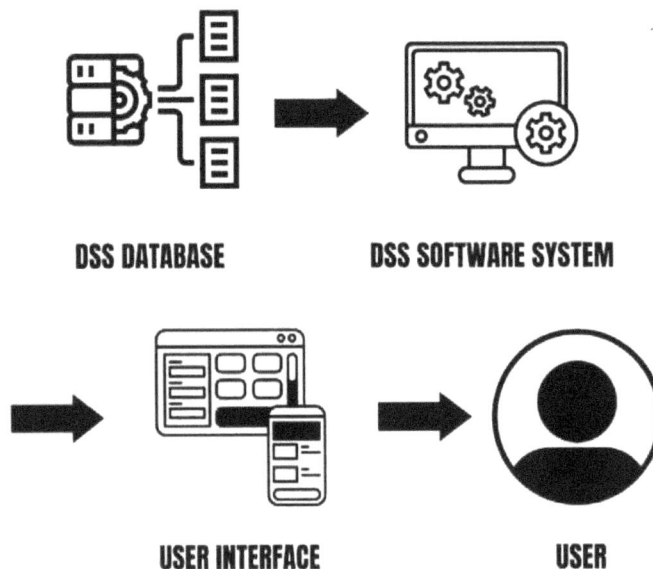

DSS DATABASE   **DSS SOFTWARE SYSTEM**

USER INTERFACE   **USER**

Fig. (1). Components of DSS.

1. DSS Database

As a database, it contains a large amount of information from various sources.

- A DSS's reasoning engine uses this library to determine which course of action to take based on information related to specific subjects.
- This includes data from internal organizational data (such as TPS) and data mined from the internet.

- Data is collected, stored, modified, and retrieved using a Transaction Processing System (TPS), which then generates reports regarding transactions within an organization [27].

2. Depending on the organizational data being stored, the database can be a small computer-based database with a small subset of data being downloaded and perhaps combined with external data, or it can be a vast data warehouse that has corporate TPS-like website transactions constantly updated.

3. DSS Software

- To analyze complex data, we use mathematical and analytical models to predict outputs based on various inputs and conditions.
- Along with finding a particular desired output, it also tries to determine what combination of inputs and conditions would produce that outcome.
- Depending on the user's needs and the software's purpose, each DSS model has a specific function, and different collections of models are available.
- Models can be categorized as:
    - I. Physical (for example, a model of a car).
    - II. Mathematical (for example, an equation).
    - III. Verbal (for example, a description of how to write a command).
- Analysing the data can be done using an OLAP system and data mining.
- A computerized technique is known as Online Analytical Processing (OLAP) for analysing data in multiple ways and allowing users to access large amounts of records easily and quickly, whereas the data mining process involves extracting and transforming raw data by analysing large sets of data to discover patterns, relationships, and insights from various sources so that informed decisions can be made [28].

4. DSS User Interface

- This component facilitates and eases communication between the user and the system.
- Graphs, charts, and graphs are some of the ways the output is displayed in a graphical interface.

Classification of Decision Support Systems

With the introduction of new technological ideas and concepts over the years, DSS classification has evolved in various ways [29]. In the same way that there is no universal definition for DSS, there is no common classification. Table **1** provides a summary of all taxonomies of DSS proposed by various authors.

Characteristics of the Decision Support System

Differentiating DSS from other computer-based information systems requires identifying certain characteristics. Over the years, many major characteristics of DSS have been put forth by various authors [31, 33, 37]. A DSS's primary function is to offer decision-relevant information and results. Hence, some of the general characteristics of DSS include:

Table 1. Classification of DSS as per various authors.

S.No.	Based on	Classification	Description	Author
1	Frequency of Decision Making	Institutional DSS	Deal with recurring decisions.	Donovan and Madnick (1977) [30]
		Ad Hoc DSS	Deals with specific problems that aren't anticipated or recur.	
2	Operation's generic nature	File drawer systems	A data store/item related to data can be accessed.	Alter (1980) [31]
		Data analysis systems	Utilizes computerized settings or tools to manipulate data.	
		Analysis information systems	Create simple models and accesses decision-oriented databases.	
		Accounting and financial models	Consider and analyze the consequences of potential actions.	
		Representational models	Analyze the consequences of actions using simulation models.	
		Optimization models	Utilize mathematically based optimization models to provide solutions.	
		Suggestion models	Analyze and propose a specific decision for a structured or well-understood situation.	
3	Support Offered	Personal DSS	Supports only one user	Hackathorn and Keen (1981) [32]
		Group DSS	Supports multiple/groups of users.	
		Organizational DSS	Supports an organization as a Whole.	

(Table 1) cont.....

S.No.	Based on	Classification	Description	Author
4	Orientation	Database oriented DSS	Stores data and information at different locations, such as in regions and make it accessible to decision-makers.	Holsapple & Whinstone (1996) [33]
		Rule oriented DSS	Provides a more intelligent approach to solutions by integrating processes and rules.	
		Spreadsheet oriented DSS	Provides decision-making support requiring aspects of knowledge creation, extraction, and improvement for analyzing data and information.	
		Solver Oriented DSS	Analyze and develop solutions in a variety of subject areas.	
		Text-oriented DSS	Data that can be accessed, used, and evaluated based on text.	
		Component Oriented DSS	Including two or more of the above five basic structures.	
5	Scope	Enterprise-wide DSS	Serves many users in the organization and is connected to large data warehouses.	Power (2000) [34]
		Desktop DSS	User-friendly, small system (desktop) that runs on the user's computer.	
6	Mode of assistance	Model-driven DSS	Provides access to, and manipulation of, models that deal with statistics, financials, optimization, or simulations.	Bhargava & Power (2001) [35]
		Data-driven DSS	Focus on accessing and manipulating time series data.	
		Knowledge-driven DSS	Gathers and stores 'expertise' that can be used to make decisions when needed.	
		Communication-driven DSS	Facilitates communication and collaboration, as well as shared decision-making support.	
		Document-driven DSS	Provides unstructured information on a variety of electronic formats and retrieves manage, and manipulates it.	
7	User relationship	Active DSS	Offers explicit suggestions or solutions for decision-making.	Heathenish-willer (2001) [36]
		Passive DSS	Aids the decision-making process but cannot provide explicit suggestions or solutions.	
		Cooperative DSS	Modifies the system's decision suggestions and repeats the process until a satisfactory solution is reached.	

1. Providing support for data access, modelling, and analysis.

2. Adaptable to user needs and changing environments.

3. Flexible and can be adapted according to the needs of the work process.

4. Providing support to decision-makers at all organizational levels.

5. Aims to improve a specific decision's accuracy, timeliness, quality, and overall effectiveness.

6. Customized ad hoc knowledge provision.

A DETAILED OVERVIEW OF DD-DSS

Data-driven DSS is a type of DSS that provides insight into decisions that are made by analysing time-series data and returning new information depending on those analyses. The DD-DSS focuses on analysing large amounts of structured data. In other words, a DD-DSS allows users to access and manipulate a vast database of structured data that contains numeric and short character strings, particularly a time series of internal and external data [38]. Through the years, DD-DSS has been referred to as a data-oriented DSS and a retrieval-only DSS [31, 39].

Data warehouse appliances, report writing tools, OLAP tools, and business intelligence and performance management software are used to create this DD-DSS [40]. It is possible to provide basic functionality using simple file systems that can be accessed *via* query and retrieval tools [41]. The most basic functionality is provided by databases accessed through ad-hoc query tools. Data from various sources is easily accessed and analyzed in an easy-to-access and easy-to-understand DD-DSS database. In order to extract data in a DD-DSS, all appropriate internal as well as external databases are utilized [42]. A data warehouse system that supports data analysis offers additional functionality. The DD-DSS integrates OLAP and spatial DSS functionalities to offer the highest level of capability and decision support [43].

History of DD-DSS

The DD-DSS journey begins with Jay Forrester and George Valley, two MIT Lincoln Lab professors, working on the first large-scale DD-DSS in the mid-1950s [43, 44]. Jay Forrester finalized and activated the SAGE air defense command and control system in 1963, marking the debut of the initial computerized data-driven system. For more than 20 years, it has provided real-time decision support. The operators managed the status of incoming hostile aircraft *via* cathode ray tube displays [45, 41]. AAIMS, or Analytical Information Management System, was one of the first APL-based DD-DSSs. It was created by Richard Klaus and Charles Weiss at American Airlines between 1970 and 1974

[46]. With the advancement of technology, DD-DSS has evolved into a more sophisticated tool.

Typically, the DD-DSS is conceived as a real-time interactive system capable of managing both spontaneous and scheduled information requests. In a study by Power, an overview of DD-DSS was provided, including its initial development, major features, and software evaluation criteria for the development of DD-DSS. According to the modern classification of DSS, DD-DSS is one of five major types [41]. DD-DSS has evolved and become increasingly sophisticated since being first implemented in 1963 by the US SAGE air-defines command and control system [46]. Now, the software can be used for several purposes, including analytics, visibility, and monitoring of performance in real-time and static modes. It is essential to have access to accurate, well-organized multidimensional data as quickly and easily as possible to ensure the success of a DD-DSS. Further, the author explains the necessary features of DD-DSS from a user perspective, which helps the reader understand the wide range of benefits and applications of this information system. The author also implies that the users will have access to better-informed decisions that will be consistent, reliable, and of high quality, hence aiding in better decision-making. It is essential to have a proper DSS with a database, a user interface with effective and consistent data gathering, and appropriate software to achieve these results. It is worth noting that organizations seeking to implement a DSS should clearly define the problem they are trying to solve as well as consider the models and data needed to support the DSS, as well as the infrastructure needed to build and maintain it. Along with these, the author lays down certain criteria required for evaluating DD-DSS software, followed by some specific issues related to it. In this paper, we will provide a basic overview of DSS, which will help us gain a deeper understanding of the DD-DSS; however, we will focus on developing a forensic perspective of the DD-DSS to gain a deeper understanding of the system.

Features and Benefits of DD-DSS

There are features associated with a general category of DSS that will not be present in a specific implementation of a DSS [47]. According to several publications, the following are some of the important features and benefits of DD-DSS [39, 48]:

1. Decision-making can be made more consistently and accurately by retrieving relevant information more quickly and providing better ways to view or solve problems.

2. A user can configure email notifications and predefined actions in some systems.

3. Integrated with Excel, users can export and download data for further analysis, whereas some systems allow the uploading of data.

4. As part of a DD-DSS, DSS designers can create and store predefined, periodic reports, allowing users to easily access these reports at any time.

5. Data can be summarized or described using descriptive statistics, trend lines can be generated, and relationships can be identified by mining the data.

6. Information can be extracted, designed, and presented interactively, usually in the form of charts, graphs, tables, and other visuals in a formal report. The user can save and reuse the format for a report once it is created, and the documents can be distributed in print, on the Web, or as PDF files.

Other features of DD-DSS include viewing predefined data displays, metadata creation and retrieval, data management, data summarization, creating data displays, increased organizational control, *etc.*

Subcategories of the Data-Driven Decision Support System

As defined by Power [20], there are four major subcategories under which DD-DSS are classified: data warehouses, OLAP systems with multidimensional databases, (EIS), and spatial data management systems [20].

Data Warehouses

A data warehouse, as defined by Bill Inmon as "the father of data warehousing," is a collection of non-volatile, subject-oriented, integrated, time-varying, integrated data used to aid decision-making [49]. The four characteristics of a data warehouse, as stated by Inmon [50]:

a. **Subject-oriented:** focuses on topics related to organizational activities, such as customers, employees, and suppliers.
b. **Integrated:** Data is stored invariably by employing naming conventions, domain constraints, physical attributes, and measurements.
c. **Time-variant:** linking data to specific time points.
d. **Non-volatile:** Once stored in the warehouse and used for decision support, the data does not change.

To complete a DD-DSS's architecture and design, DSS analysts or data modelers are required to create a data warehouse schema as well as identify analytics and end-user presentation software [51].

OLAP

Also known as On-Line Analytical Processing, is software that allows users to manipulate multidimensional data from multiple sources. According to Nigel Pendse, multidimensional information can be accessed quickly, consistently, and interactively with an OLAP system [52]. Further, he outlines five characteristics that he refers to as the FASMI test (Fast Analysis of Shared, Multidimensional Information) [53].

a. **Fast:** Users typically receive responses within five seconds of submitting their request.
b. **Analysis:** Regardless of the application's business logic or statistical analysis, the system can handle it.
c. **Shared:** Data can be shared among users with the software's security features.
d. **Multidimensional:** Analyzing metrics in different dimensions, including time, geography, gender, and product.
e. **Information:** Managers can access all the necessary data and derive information through the software.

In some cases, BI can be confused with OLAP and other synonyms for DSS. In 1989, Gartner Group's Howard Dresner introduced the concept of business intelligence as an approach to improving business decision-making through fact-based information systems [54, 55]. Business intelligence systems are examples of DD-DSS. Reporting and querying a database are the most common uses of "business intelligence" software.

Spatial DSS

A computer-based system that uses traditional data, spatially referenced data, and decision logic to help a human decision-maker [56]. Peter Keenan describes "spatial DSSs as those that are based on Geographic Information Systems (GISs). GIS provides several tools and techniques that can be incorporated into a DSS that uses spatial or geographic data" [57]. A spatial data management system (DSS) helps users access, display, and analyze geographic data.

EIS

The Executive Information System is a computerized system that provides accurate and up-to-date information to support management decision-making [58]. Information from the database is presented using interactive displays and a simple interface. In addition to preparing reports and briefing books, they provide drill-down capabilities and strong reporting capabilities to top-level executives [59].

Generally, senior managers use EIS to monitor performance and identify opportunities and challenges across the organization by analysing, comparing, and highlighting trends in variables. An EIS may help an organization reduce the number of management levels by monitoring a wide range of activities to make it easier for senior executives.

APPLICATION OF DD-DSS IN HEALTHCARE

Scientists and technologists are examining and analysing data with a data-driven approach, using increasing amounts of raw data and cheaper and faster computation power to uncover scientific and technological information [60]. DD-DSS can play a crucial role in the healthcare sector and can forever change the way we use and interpret the data obtained. The progress of DD-DSSs in this field is an active area of research. It has been reported by Zhang and Hu (2006) that big data can be used in the healthcare industry for mining, filtering, and extracting data other than providing medical analysis and ideal treatment plans for patients [61]. The application of DD-DSS in healthcare is mainly to improve the management and administration activities of the healthcare field and, in the end, provide better service to patients. These methods may include:

Descriptive Analytics

In general, descriptive analytics is an analysis of historical data used to identify patterns and relationships. It provides businesses with the ability to map trends based on what has happened in the past [62]. Descriptive analytics refers to the analysis of historical patient data in health care. The data includes information on patient visits, prescriptions, surgeries, and other treatments, as well as patient outcomes [63]. Additionally, it includes payment and billing information, geographical data (which can be used in population health metrics), and more. As an example, it can be used to determine how many patients were hospitalized in the last month based on existing information. An automated system for monitoring and analysing patterns in seniors' activities of daily living was suggested by Luo *et al* to improve caregivers' ability to assist the elderly. A novel computer vision-based descriptive analytics approach that uses state-of-the-art computer vision techniques and non-intrusive, privacy-compliant multimodal sensors to continuously detect seniors' activities and provide descriptive analytics over a long period of time [64].

Diagnostic Analytics

Diagnostic analytics is a type of advanced analytics that examines data to answer the question, "Why did this happen?". Drill-down, data discovery, data mining, and correlation techniques are used in diagnostic analytics [65]. Through diagnostic analytics, healthcare providers can gain a deeper understanding of patients, administrators, and healthcare executives. It can provide information to clinicians, administrators, and healthcare executives on many complex medical challenges. To analyze the vast amounts of data involved at speeds much faster than a human could, diagnostic analytics requires artificial intelligence (AI) running on powerful computers. As well as collecting data, AI and machine learning use algorithms for finding connections and speeding up decision-making [66]. For instance, by reading research articles, test results, clinical trials, and other data, AI can compile all of this information, allowing drug development to be sped up. When it comes to identifying false positives in mammograms, artificial intelligence outperforms radiologists, and its accuracy is comparable to that of radiologists.

Predictive Analytics

In predictive analytics, historical data and current data are used to create a model of future behavior. In other words, it is the process of identifying patterns and trends in historical healthcare data that may predict future events by analysing historical healthcare data [67]. To determine the future, it requires machine learning, in which algorithms are built to extract knowledge from existing data and combine it. Two main types of algorithms can be distinguished here: supervised learning and unsupervised learning. Later, a third category was added, called semi-supervised learning [68, 69].

1. **Unsupervised Learning:** For extracting characteristics from data, which means that it can be used for a variety of analyzes since it automatically identifies structure. Clustering is one of the major unsupervised learning methods because it groups similar entities. The goal of clustering is to find similarities in data points and group similar data points together.

2. **Supervised Learning:** We typically use supervised learning in classification when mapping inputs to output labels or regression when mapping inputs to continuous outputs. Clinically relevant results can be obtained through supervised learning, which comprises several techniques, including linear regression, logistic regression, naive Bayes, decision trees, nearest neighbors, random forests, SVMs, and neural networks.

3. **Semi-Supervised Learning:** The algorithm combines unsupervised and supervised learning and is trained using labeled and unlabeled data, making it suitable for situations where results are lacking.

Using this method, healthcare organizations, hospitals, and doctors can access, analyze, and process patient information to provide accurate diagnoses and personalized treatments based on the data. Hayn *et al.*, describe a toolset named Predictive Analytics Toolset for Health and Care (PATH), which aids in the process of objective definition, data cleaning, pre-processing, feature engineering, evaluation, result visualization, interpretation, validation, and deployment [70]. A study by Boukenze *et al.*, discussed in detail how predictive analytics can predict diseases in advance and anticipate possible cures. Through the use of the Decision Tree (C4.5) algorithm, they demonstrated that chronic kidney diseases can be predicted [71]. A study by Ashfaq *et al.* adopted a hybrid machine learning approach for predicting the side effects of drugs, categorizing side effects into various intervals, and adopting appropriate measures to reduce these effects [72]. Predictive analytics can also aid the healthcare sector in the following ways [73]:

1. Helping hospitals plan for a surge in patients by looking at bed capacity, payroll data, and nurse-to-patient ratios.

2. In addition to relieving staff of some of the burden of information overload caused by EHRs, predictive analytics can also be very effective in collecting patient care data. However, it can be difficult to use since it includes clinical, billing, and other patient information. Misuse of EHR systems can result in safety problems.

3. The revenue cycle can be improved by using predictive analytics, as hospitals use models to improve insurance reimbursements.

Prescriptive Analytics

The goal of prescriptive analytics is to analyze data and content to recommend the best course of action. It is closely associated with optimization problems and requires sophisticated algorithms to find an optimal solution. This type of problem can be solved in four distinct categories: blind search, local search, search based on population, and multi-objective optimization [74].

1. **Blind Search:** Assuming all alternatives have been exhausted, the blind search ensures that all solutions within the results space have been tested, thereby guaranteeing the best solution is found.

2. **Local Search:** From existing solutions, new solutions can be generated, including several methods that focus on a local neighborhood.

3. **Search based on Population:** By exploring more distinct areas of the search space, population-based search can achieve greater diversity in terms of defining new solutions, which can be created not only by changing each search point slightly but also by combining attributes from multiple search points.

4. **Multi-objective Optimization:** Multi-objective optimization consists of balancing multiple objectives in a single algorithm, especially if achieving one of them involves losing another, with the importance of achieving the best balance possible.

Using Artificial Intelligence and machine learning, prescriptive analytics parses the vast amounts of data available and recommends actions based on algorithms that can be automated, such as to generate delivery routes or help patients find appropriate treatment options, as with other data analytics processes. A study by Kaur *et al.*, proposes an architecture that makes use of artificial intelligence in the healthcare sector with the help of predictive analytics. They created a mobile application using this method that asks the patients about their symptoms and provides corresponding details on their health.

LIMITATIONS

Since this information system is evolving, various improvements and modifications are required. Since data-driven decision support systems are evolving and research in this field is ongoing, they are expected to be enhanced in the future to improve their effectiveness. Nevertheless, it may never be possible for data-driven decision-support systems to reach their full potential. As it is in its infancy, some various limitations and challenges need to be addressed for it to be improved [75]. These limitations can include:

1. **Data quality and availability:** DD-DSS are based on data sources that can be incomplete, inaccurate, outdated, or inconsistent. When data quality is poor, erroneous or misleading results can occur, affecting the validity and reliability of decisions [76].

2. **Ethical and legal issues:** Privacy, security, ownership, and consent issues may arise with DD-DSS [77].

3. **Sensitive Information:** Healthcare decisions deal with highly sensitive information, requiring timely action and sometimes having life-or-death consequences [78].

4. **User acceptance and trust:** In addition to a lack of awareness, training, or skills, fear of change or loss of control, organizational culture, politics, or ethical or moral concerns, data-driven decision support systems may face resistance or rejection from users. Data-driven decision support systems must be accepted and trusted by users before they can be successfully adopted and implemented [77].

5. **High Cost**: An investment in DD-DS systems may require significant resources, time, and cash since they may require data acquisition, storage, processing, and analysis, as well as system development, maintenance, and security [79].

6. **Complexity of Healthcare Data**: Healthcare data comes from various sources such as personal health information, disease registries, population health statistics, wellness apps, wearable devices, insurance, geographic, and financial data. Harnessing this vast amount of data can be challenging [75].

CONCLUSION

Using data, models, and analytical tools, the DD-DSS supports decision-makers in solving problems and making decisions. Its use can provide various benefits, including improving decision-making efficiency, accuracy, and quality, improving intelligence and performance monitoring, and facilitating innovation and knowledge discovery.

DDSS can support physicians in clinical decision-making and improve healthcare by using machine learning techniques. The power of data must be fully harnessed for better decision-making despite the advancements in data gathering, aggregation, and analysis. Data-driven DSSs should be designed with ethical considerations in mind. Additionally, data-driven DSSs need to be integrated with other systems to ensure accuracy and reliability. Finally, the usability of data-driven DSSs must be improved to ensure that practitioners can use the system effectively. The use of data-driven systems is gaining traction but they have some limitations, including data access and quality and user interface issues.

In conclusion, while DDSS has its limitations, the technology is still evolving, with new developments being made to improve analysis accuracy and speed. Despite its faults, research in this field is still moving forward rapidly. This technology's potential is enormous, and it will only grow more significant in the years to come. Its ability to influence beneficial healthcare decision-making to improve patient outcomes needs to be explored further.

REFERENCES

[1]   Subrahmanya SV, Shetty DK, Patil V, *et al.* The role of data science in healthcare advancements: applications, benefits, and future prospects. Irish Journal of Medical Science (1971). 2022; 191(4): 1473-83.
[http://dx.doi.org/10.1007/s11845-021-02730-z]

[2]   Sakovich N. The importance of data collection in healthcare and its benefits [Internet]. Sam Solutions; 2023 [cited 2023 Oct 111]. Available from: https://www.sam-solutions.com/blog/the-importance-of-data-collection-in-healthcare/
[http://dx.doi.org/10.1023/B:JOMS.0000044969.66510.d5]

[3]   Venkatraman SS. DSS: is it just an alias for MIS. Comput Pers 1989; 12(2): 4-11.
[http://dx.doi.org/10.1145/1036387.1036388]

[4]   Singh, C., Khajuria, H. and Nayak, B.P. A Study of Implementing a Blockchain-Based Forensic Model Integration (BBFMI) for IoT Devices in Digital Forensics. In International Conference on Computer Science, Engineering and Education Applications 2023; 318-327.
[http://dx.doi.org/10.21533/pen.v7i2.550]

[5]   Contributor T. What is a decision support system (DSS)? [Internet]. TechTarget; 2021 [cited 2023 Oct 11]. Available from: https://www.techtarget.com/searchcio/definition/decision-support-system

[6]   Kriegova E, Kudelka M, Radvansky M, Gallo J. A theoretical model of health management using data-driven decision-making: the future of precision medicine and health. J Transl Med 2021; 19(1): 68.
[http://dx.doi.org/10.1186/s12967-021-02714-8] [PMID: 33588864]

[7]   Gaynor M, Seltzer M, Moulton S, Freedman J. A dynamic, data-driven, decision support system for emergency medical services. In Computational Science–ICCS 2005: 5th International Conference, Atlanta, GA, USA, May 22-25, 2005. Proceedings 2005; 5(Part II): 703-11. [Springer Berlin Heidelberg.].

[8]   Azzi S, Gagnon S, Ramirez A, Richards G. Healthcare applications of artificial intelligence and analytics: a review and proposed framework. Appl Sci (Basel) 2020; 10(18): 6553.
[http://dx.doi.org/10.3390/app10186553]

[9]   Wiederrecht G, Callaway A, Darwish S. RBC Capital Markets: Navigating the changing face of healthcare episode [Internet]. RBC Capital Markets | Navigating the Changing Face of Healthcare Episode; [cited 2023 Oct 13]. Available from: https://www.rbccm.com/en/gib/healthcare/episode/the_healthcare_data_explosion

[10]   Big Data in health care: What it is, benefits, and jobs [Internet]. Coursera; [cited 2023 Oct 14]. Available from: https://www.coursera.org/articles/big-data-in-healthcare

[11]   Ilmudeen A. Big data-based frameworks for healthcare systems. In Demystifying Big Data, Machine Learning, and Deep Learning for Healthcare Analytics. Academic Press 2021; pp. 33-56.
[http://dx.doi.org/10.1016/B978-0-12-821633-0.00003-9]

[12]   Little JDC. Models and managers: The concept of a decision calculus. Manage Sci 1970; 16(8): B-466-85.
[http://dx.doi.org/10.1287/mnsc.16.8.B466]

[13]   Sprague RH Jr, Carlson ED. Building effective decision support systems. Prentice Hall Professional Technical Reference 1982 Feb 1.

[14]   Keen PG, Scott Morton MS. Decision support systems: an organizational perspective. 1978.

[15]   Singh, C., Khajuria, H., & Nayak, B. P. A Comprehensive iot security framework using edge computing: approaches and challenges. Network Optimization in Intelligent Internet of Things Applications, 274-297.

[16]   Moore JH, Chang MG. Design of decision support systems. ACM SIGOA Newsletter 1980; 1(4-5): 8-

14.
[http://dx.doi.org/10.1145/1017672.1017658]

[17] Chi RT, Turban E. Distributed intelligent executive information systems. Decis Support Syst 1995; 14(2): 117-30.
[http://dx.doi.org/10.1016/0167-9236(94)00006-E]

[18] Shim JP, Warkentin M, Courtney JF, Power DJ, Sharda R, Carlsson C. Past, present, and future of decision support technology. Decis Support Syst 2002; 33(2): 111-26.
[http://dx.doi.org/10.1016/S0167-9236(01)00139-7]

[19] Power DJ. Decision support systems: concepts and resources for managers. Quorum Books 2002.

[20] Sprague RH Jr. A framework for the development of decision support systems. Manage Inf Syst Q 1980; 4(4): 1-26.
[http://dx.doi.org/10.2307/248957]

[21] Burstein F. A framework for the development of decision support systems. MIS quarterly. 1980; 1: 1-26.

[22] Mir SA, Qasim M, Arfat Y, *et al.* Decision support systems in a global agricultural perspective-a comprehensive review. Int J Agric Sci 2015; 7(1): 403-15.

[23] Gachet A. A framework for developing distributed cooperative decision support systems: inception phase. Department of Informatics-University of Fribourg 2000.

[24] What is a decision support system? (DSS) what kinds of DSS exist? [Internet]. Chegg; 2018 [cited 2023 Oct 17]. Available from: https://www.chegg.com/homework-help/questions-and-answers/decision-support-system-dss-kinds-dss-exist-q28902229

[25] Systems for decision support [Internet]. [cited 2023 Oct 17]. Available from: https://paginas.fe.up.pt/~acbrito/laudon/ch13/chpt13-2main.htm

[26] White D. What is a transaction processing system? [Internet]. 2020 [cited 2023 Oct 17]. Available from: https://www.techfunnel.com/fintech/transaction-processing-system/

[27] Taylor D. What is OLAP? cube, analytical operations in Data Warehouse [Internet]. 2023 [cited 2023 Oct 17]. Available from: https://www.guru99.com/online-analytical-processing.html

[28] Aqel MJ, Nakshabandi OA, Adeniyi A. Decision support systems classification in industry. Periodicals of Engineering and Natural Sciences (PEN) 2019; 7(2): 774-85.
[http://dx.doi.org/10.21533/pen.v7i2.550]

[29] Donovan JJ, Madnick SE. Institutional and ad hoc DSS and their effective use. Data Base Adv Inf Syst 1977; 8(3): 79-88.
[http://dx.doi.org/10.1145/1017583.1017599]

[30] Alter S. Decision support systems: current practice and continuing challenges. No Title 1980.

[31] Hackathorn RD, Keen PGW. Organizational strategies for personal computing in decision support systems. Manage Inf Syst Q 1981; 5(3): 21-7.
[http://dx.doi.org/10.2307/249288]

[32] Kearns GS, Holsapple CW, Whinston AB, Benamati JH. Instructor's manual with test bank to accompany decision support systems: a knowledge-based approach.

[33] Power DJ. Web-based and model-driven decision support systems: concepts and issues. AMCIS 2000 Proceedings. 2000; 1: 387.

[34] Bhargava H, Power D. Decision support systems and web technologies: a status report.

[35] Haettenschwiler P. Neues anwenderfreundliches konzept der entscheidungsunterstützung. Zurich, vdf Hochschulverlag AG. 2001; 189-208.

[36] Power DJ. What are the characteristics of a Decision Support System. DSS News. 2003; 4(7).

[37]   Bernus P, Blazewicz J, Schmidt G, Shaw M. International handbooks on information systems.

[38]   Bonczek RH, Holsapple CW, Whinston AB. Foundations of decision support systems. Academic Press; 2014; 10.

[39]   Watson RT. Chapter 14 Organizational Intelligence. In: Watson RT, editor. Data Management: Databases and Analytics. 7th ed. Athens, GA: eGreen Press; 2022.

[40]   Power DJ. Understanding data-driven decision support systems. Inf Syst Manage 2008; 25(2): 149-54. [http://dx.doi.org/10.1080/10580530801941124]

[41]   Mallach EG. Understanding Decision Support Systems and Expert Systems. Irwin 1998.

[42]   Liu L, Özsu MT, Eds. Encyclopedia of database systems. New York, NY, USA: Springer 2009. [http://dx.doi.org/10.1007/978-0-387-39940-9]

[43]   Power D. What was the first computerized decision support system (DSS). DSS News. 2006; 31: 7(27).

[44]   Louw RE. Decision Support Systems. University of Missouri St. Louis; 2002 [cited 2023 Oct 20]. Available from: https://www.umsl.edu/~sauterv/analysis/488_f02_papers/dss.html

[45]   Power DJ. A brief history of decision support systems. DSSResources. com. 2007 Mar 10;3A brief history of decision support systems. DSSResources. com. 2007; 10: 3.

[46]   Power D. What are the features of a data-driven DSS. DSS News 2007; 8(4): 6.

[47]   Codd EF. 1993. roviding OLAP (on-line analytical processing) to user-analysts: An IT mandate. Available from: http://www. arborsoft. com/papers/coddTOC. html

[48]   Inmon WH. What is a data warehouse. Prism Tech Topic 1995; 1(1): 1-5.

[49]   Data Warehousing - Overview [Internet]. [cited 2023 Oct 22]. Available from: https://www.tutorialspoint.com/dwh/dwh_overview.htm

[50]   Power DJ. What is data warehousing? [Internet]. DSSResources.com; 2010 [cited 2023 Oct 22]. Available from: https://dssresources.com/faq/index.php?action=artikel&id=202

[51]   Pendse N, Creeth R. The OLAP report: succeeding with on-line analytical processing. Business Intelligence 1995.

[52]   Bhetwal MK. Data warehouse and business intelligence: comparative analysis of olap tools.

[53]   Nylund A. Tracing the BI family tree. Knowledge Management 1999; 60: 70-1.

[54]   Mitra S. Business intelligence [Internet]. DWBI.org; 2020 [cited 2023 Oct 23]. Available from: https://dwbi.org/index.php/pages/8/business-intelligence

[55]   Singh, C. and Das, A.K., 2024. Transforming digitization in supply chain management using artificial intelligence and machine learning. In Supply Chain Management (pp. 119-139). CRC Press.

[56]   Keenan PB. Spatial decision support systems: a conning of age. Control Cybern 2006; 35(1): 9-27.

[57]   Houdeshel G, Rainer RK, Watson HJ, Eds. Executive information systems: emergence, development, impact. Wiley 1992.

[58]   Power DJ. Are Executive Information Systems (EIS) needed? [Internet]. DSSResources.com; 2008 [cited 2023 Oct 22]. Available from: https://dssresources.com/faq/index.php?action=artikel&id=175

[59]   Kitchin R. Big Data, new epistemologies and paradigm shifts. Big Data Soc 2014; 1(1): 2053951714528481. [http://dx.doi.org/10.1177/2053951714528481]

[60]   Zhang SJ, Hu SQ. Status quo of medical information education in China's higher education institution: an investigation. Chin J Med Libr Inform Sci 2006; 15(06): 51-4.

[61] Wolniak R. The concept of descriptive analytics. scientific papers of silesian university of technology. Organization & management/zeszyty naukowe politechniki slaskiej. Seria Organizacji i Zarzadzanie. 2023 Mar 1(172).
[http://dx.doi.org/10.29119/1641-3466.2023.172.42]

[62] Muneeswaran V, Nagaraj P, Dhannushree U, Ishwarya Lakshmi S, Aishwarya R, Sunethra B. A framework for data analytics-based healthcare systems. InInnovative Data Communication Technologies and Application. Proceedings of ICIDCA 2020; 2021: 83-96. [Springer Singapore].

[63] Luo Z, Hsieh JT, Balachandar N, *et al.* Computer vision-based descriptive analytics of seniors' daily activities for long-term health monitoring. Machine Learning for Healthcare (MLHC). 2018; 2(1).

[64] Singh C, Khajuria H, Nayak BP. Rapid Advancement and Trends of Big Data Analytics and Cyber-Physical System Embedded in Healthcare and Industry 40 Intelligent Security Solutions for Cyber-Physical Systems. Chapman and Hall/CRC 2024; pp. 216-33.

[65] Alghamdi A, Alsubait T, Baz A, Alhakami H. Healthcare Analytics: A Comprehensive Review. Engineering, Technology &. Appl Sci Res 2021; 11(1): 6650-5.

[66] Predictive analytics in healthcare - benefits & regulation [Internet]. [cited 2023 Oct 25]. Available from: https://www.foreseemed.com/predictive-analytics-in-healthcare#:~:text=What%20is%20predictive%20analytics%20in,be%20predictive%20of%20future%20events

[67] Jiang F, Jiang Y, Zhi H, *et al.* Artificial intelligence in healthcare: past, present and future. Stroke Vasc Neurol 2017; 2(4): 230-43.
[http://dx.doi.org/10.1136/svn-2017-000101] [PMID: 29507784]

[68] Brownlee J. Supervised and unsupervised machine learning algorithms [Internet]. 2023 [cited 2023 Oct 25]. Available from: https://machinelearningmastery.com/supervised-and-unsupervised-machine-learning-algorithms/

[69] Hayn D, Veeranki S, Kropf M, *et al.* Predictive analytics for data driven decision support in health and care. it-Information Technology. 2018; 60(4): 183-94.

[70] Boukenze B, Mousannif H, Haqiq A. Predictive analytics in healthcare system using data mining techniques. Comput Sci Inf Technol 2016; 1: 01-9.
[http://dx.doi.org/10.5121/csit.2016.60501]

[71] Ashfaq A, Nowaczyk S. Machine learning in healthcare-a system's perspective. arXiv preprint arXiv:190907370 2019.

[72] Predictive analytics in healthcare: Three real-world examples [Internet]. Philips; 2022 [cited 2023 Oct 26]. Available from: https://www.philips.com/a-w/about/news/archive/features/20200604-predictive-analytics-in-healthcare-three-real-world-examples.html

[73] Alharthi H. Healthcare predictive analytics: An overview with a focus on Saudi Arabia. J Infect Public Health 2018; 11(6): 749-56.
[http://dx.doi.org/10.1016/j.jiph.2018.02.005] [PMID: 29526444]

[74] Data-driven decision-making for Health Administrators [Internet]. 2022 [cited 2023 Oct 26]. Available from: https://publichealth.tulane.edu/blog/data-driven-decision-making/

[75] TIBCO Content Team. How to become a data-driven healthcare organization in 7 steps [Internet]. 2022 [cited 2023 Oct 26]. Available from: https://www.tibco.com/blog/2021/11/01/how-to-become-a-data-driven-healthcare-organization-in-7-steps/

[76] Rehman J. Advantages and disadvantages of Decision Support System [Internet]. 2021 [cited 2023 Oct 26]. Available from: https://www.itrelease.com/2021/03/advantages-and-disadvantages-of-decision-support-system/#google_vignette

[77] Ginsburg PB, de Loera-Brust A, Brandt C, Durak A. The opportunities and challenges of data analytics in health care [Internet]. 2018 [cited 2023 Oct 26]. Available from: https://www.brookings.edu/articles/the-opportunities-and-challenges-of-data-analytics-in-health-care/

[78]   MSG Management Study Guide [Internet]. [cited 2023 Oct 26]. Available from: https://www.managementstudyguide.com/limitations-and-disadvantages-of-decision-support-systems.htm

[79]   Dinh-Le C, Chuang R, Chokshi S, Mann D. Wearable health technology and electronic health record integration: scoping review and future directions. JMIR Mhealth Uhealth 2019; 7(9): e12861. [http://dx.doi.org/10.2196/12861] [PMID: 31512582]

CHAPTER 4

Review on FPGA-based Hardware Accelerators of CNN for Healthcare Applications

Kurapati Hemalatha[1] and **Sakthivel Ramachandran**[1,*]

[1] School of Electronics Engineering (SENSE), Vellore Institute of Technology (VIT), Vellore, India

Abstract: In recent years, Deep Convolutional Neural Network (CNN) has been the fastest-growing area of Artificial Neural Network (ANN). In addition to image classification and segmentation, CNN can detect objects in video and recognize speech. This is because CNNs take a lot of computation. The CNN function also lends itself to programmable hardware such as Field Programmable Gate Arrays (FPGAs). Recently, hardware accelerators have become incredibly popular for a broad spectrum of healthcare applications. The emergence of edge computing has made it possible to combine a large number of sensors and process information using lightweight computing. Deep learning algorithms have advanced significantly over time, providing intriguing prospects for their use even in safety-sensitive biomedical and healthcare applications. This study presents a thorough analysis and discussion of several difficulties in the implementation of FPGA-based hardware acceleration for healthcare applications. There are some clear advantages that a variety of generalized new architectures and devices have over traditional processing units. This survey is expected to be useful for researchers in the area of artificial intelligence, FPGA-based hardware accelerators of CNN for Biomedical applications, and system design.

Keywords: Convolutional neural networks, FPGA, Hardware accelerator, Power efficiency, Reliability, Security.

INTRODUCTION

Digital circuit design has evolved rapidly over the last 25 years. Earlier digital circuits were designed with vacuum tubes and transistors. ICs were then involved where logic gates were placed on a single chip. IC designers have been relying on different types of semiconductor scaling to achieve high performance. In the complexity of circuit design, the number of cores and the complex system on chips, such as mobile phone processors, will combine application processors, GPUs, and DSPs, therefore, performance improvements are needed. Engineering

---

* **Corresponding author Sakthivel Ramachandran:** School of Electronics Engineering (SENSE), Vellore Institute of Technology (VIT), Vellore, India; E-mail: rsakthivel@vit.ac.in

applications such as AI are demanding heavy computational performance that cannot be met by conventional architectures, therefore, for a fixed task or limited task, energy scaling works better for a wide range of tasks, it leads to domain-specific hardware accelerators. A hardware computing engine is specialized for a particular domain of applications, that is termed domain-specific accelerator or domain-specific architecture. Future opportunities in computer architectures are evolving towards the concept of domain-specific architectures because of the inefficiency of GPU architectures.

A hardware accelerator shown in Fig. (**1**) is a special kind of hardware unit that performs a set of tasks with higher efficiency than a general purpose CPU. AI accelerator is a powerful machine learning hardware chip that is specifically designed to run artificial intelligence and machine learning applications smoothly. Examples include GPU, VPU, FPGA, ASIC, and TPU.

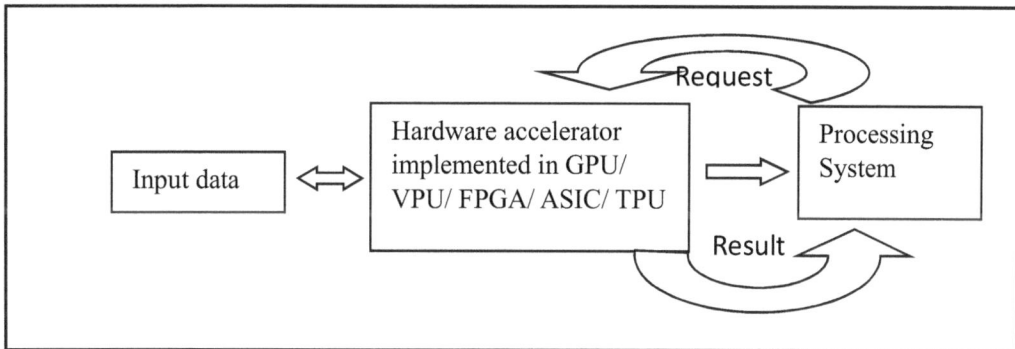

Fig. (1). Block diagram of hardware accelerator. .

Modern-age healthcare systems heavily rely upon electronics and internet technology to enable accurate, rapid diagnosis and advanced treatments, whereas electronic hardware is critical in healthcare systems for processing medical data, *e.g.*, compression, decompression, and filtering. The size of medical data (images) generated from MRI or CT scans is very large and therefore requires a large storage capacity to store and process them locally [1]. This demands a compression of medical images for low-capacity storage and low bandwidth transmission. The systolic array may be used as a coprocessor in combination with a host computer where the data samples received from the host computer pass through the processing elements (PE) and finally, reset is returned to the host computer [2]. So, you can imagine there is a lot of traffic from the accelerator to the memory. Hence, suitable CNN hardware accelerators with compression techniques are used to reduce traffic and improve performance.

Applications in the biomedical and healthcare fields are a key area where various AI techniques can be very helpful [3 - 5]. These uses include monitoring [5], early-stage prediction [6], prognosis [7], diagnosis [8], and even long-term treatment [9] regimen planning.

Fig. (2) illustrates how a smart deep learning (DL) system might handle patient data, including bio-samples, medical imaging, movement, temperature, and so on, to monitor the patient for abnormalities and/or to anticipate diseases. Prognosis and treatment choices can be recommended by DL systems, which further influences prediction and monitoring in a closed-loop setting. From the deep learning system, it is clear that based on the needs of the end user, the predictive models are customized and treatment plans are facilitated. The cloud, edge nodes, and edge devices are the three IoT tiers on which the various components of the CNN-based healthcare system can operate.

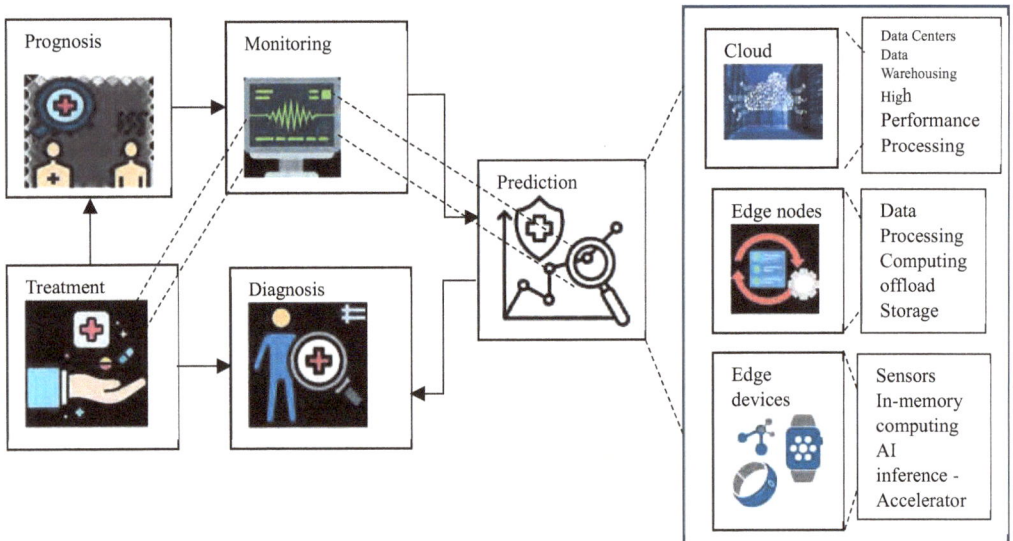

Fig. (2). The closed-loop scenario of healthcare and biomedical system.

The new AI techniques have given rise to computer systems that are sophisticated in their perception and understanding of the visual environment, and even smarter than humans in several activities. The ability to be intelligent is mostly based on both the methods and the data produced by a computer vision system and their hardware implementations, which allows us to offer a class from vision, that a computer can comprehend the physical environment. Thanks to AI and deep learning, the field of medical science has made impressive strides in the last few years that have resulted in a precise classification of brain tumors. CNN is implemented in image processing methods to detect and classify brain tumors as well as to segment, recognize, and categorize MRI images.

Greater enthusiasm for the healthcare industry and illness detection to improve the deployment of E-Health services has characterized the past several years. The author offers fresh methods and remedies [11, 12] to improve the performance of image reconstruction and recognition. To detect and diagnose conditions including heart diseases, lung diseases, diabetes, cancer, Alzheimer's disease, hepatitis, and liver disease, among others, computer science experts have created many deep learning algorithms. Convolutional neural networks (CNNs), a potent method for learning practical representations, primarily of images and other structured input, have drawn attention to deep learning. Deep artificial neural networks called CNNs are mostly utilized for object identification, picture segmentation, and image classification. CNN has demonstrated notable improvements in image recognition [12]. It is currently garnering attention across numerous sectors and has made significant progress in numerous fields. Recently, other medical specialties like neurosurgery have also attracted the attention of innovative technologies.

Authors [14, 15] suggested employing a deep supervised training technique based on the CNN model to identify and categorize liver disorders. A methodology for categorization was put forth in a study [13] and entails enhancing image processing and segmenting liver lesions. The authors created a two-step categorization process [14]. Gathering an adequate amount of isolated training samples is the first stage. Training two CNNs with the same design but distinct optimization techniques is the second phase. Approximately 95% of the architectures [14, 15] have been classified correctly. The COVID-19 pandemic has exposed the world to a virus whose behavior is unknown. As a result, numerous investigations have been launched to identify those infected with this virus [15]. With a 96% accuracy rate, the Convolutional Neural Network technique has proven to be successful. Regarding Ghulam [16], he recommended doing a deep learning-based study to create a precise classification algorithm that would divide breast cancer into eight categories.

The rest of the paper is organized as follows. CNN architectures for object detection are discussed in section 2. CNN accelerators towards biomedical and healthcare applications are explained in section 3. Evaluation metrics to improve the performance and efficiency of an accelerator are explained in section 4. Applications and future directions of FPGA-based CNN accelerators are discussed in Section 5. Finally, we conclude in section 6.

CNN ARCHITECTURES FOR OBJECT DETECTION

Several CNN designs have been proposed in the last ten years [18, 19]. The architecture of a model is an important aspect in increasing the performance of many applications. From 1989 to the present, various adjustments to CNN architecture have been made. Structure reformulation, regularisation, parameter optimizations, and so on are examples of such alterations. On the other hand, it should be emphasized that the major improvement in CNN performance was largely attributable to the restructuring of processing units and the development of new blocks [19].

The utilization of network depth was used to perform the most novel breakthroughs in CNN designs. We go over the most common CNN architectures in this part, starting with the AlexNet model in 2012 and concluding with the High-Resolution (HR) model in 2020. The key to assisting researchers in selecting the appropriate architecture for their goal task is to examine the attributes of different architectures (such as input size, depth, and resilience).

AlexNet

With the emergence of LeNet [20], the history of deep CNNs commenced. The CNNs were limited to handwritten digit identification tasks at the time, which could not be scaled to all image classes. AlexNet is a well-known deep CNN architecture that has achieved ground-breaking achievements in the fields of image recognition and classification.

AlexNet was first proposed by Krizhevesky *et al.* [21], who then improved the CNN learning ability by increasing the depth of the network and adopting multiple parameter optimization algorithms. The AlexNet architecture is depicted in Fig. (**3**) as a simple design.

B. LeNet

Due to hardware limitations, the deep CNN's learning power was limited at the time. Two GPUs (NVIDIA GTX 580) were utilized in parallel to train AlexNet to overcome these hardware restrictions. Furthermore, the number of feature extraction steps was increased from five in LeNet to seven in AlexNet to improve the CNN's adaptability to different image categories. From the architecture of LeNet shown in Fig. (**4**), it is clear that depth improves generalization for a variety of image resolutions, overfitting was the main negative factor associated with depth. To solve this difficulty, Krizhevesky *et al.* applied Hinton's theory [23, 24]. Krizhevesky *et al.* proposed an approach of randomly passing over various transformational units during the training step to ensure that the case studies by the system were extra resilient. ReLU [24] could also be used as a non-saturating activation function to improve the convergence rate [25] by decreasing the vanishing gradient problem. To improve generalization by reducing overfitting, local response normalization and overlapping subsampling were also used. Other changes were made by utilizing large-size filters (5×5 and 11×11) in the earlier layers to increase the performance of previous networks. AlexNet is significant in recent CNN developments, as well as the start of a new age of CNN research.

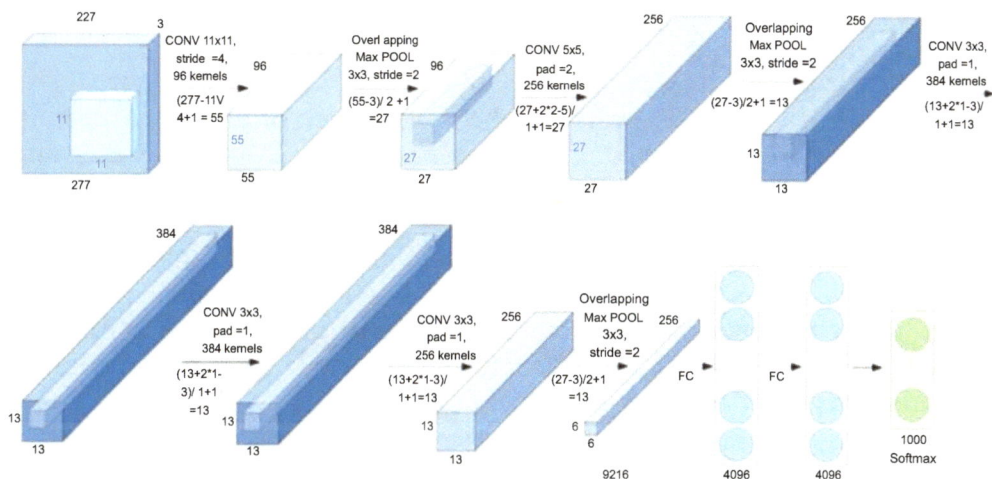

Fig. (3). The architecture of AlexNet [22, 20].

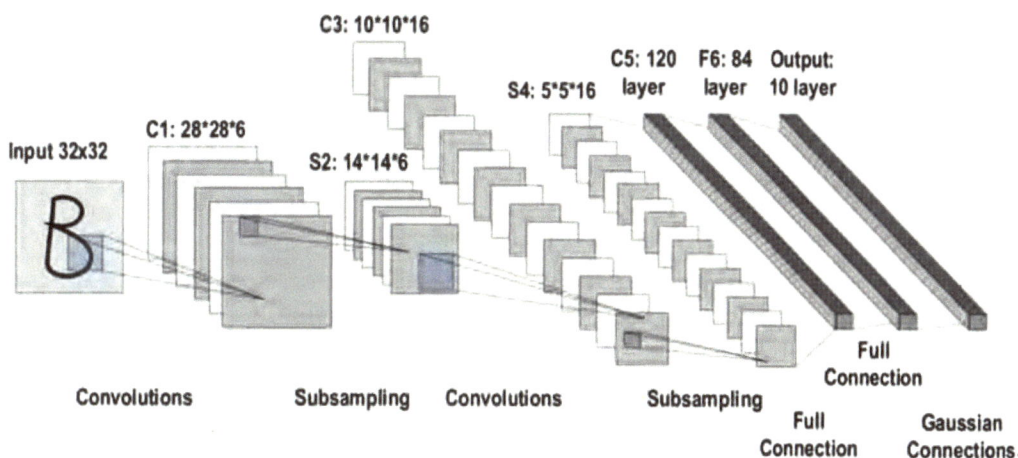

Fig. (4). The architecture of LeNet [20, 21].

ZefNet

Before 2013, the CNN learning process was largely built through trial and error, making it difficult to determine the actual goal of the upgrade. On complex images, this issue limited the performance of deep CNN. DeconvNet was introduced by Zeiler and Fergus in 2013 [26] as a reaction. This technology was later dubbed ZefNet, and it was created to quantitatively visualize the network. The network activity visualization was created to monitor CNN performance by understanding neuron activation. Erhan *et al.*, on the other hand, used this exact concept to optimize deep belief network (DBN) performance by displaying the buried layers' features [27].

In addition to this, Le *et al.* evaluated the performance of a deep unsupervised auto-encoder (AE) by displaying the image's produced, utilizing the output neurons [28]. DenconvNet mimics a forward-pass CNN by reversing the execution order of the convolutional and pooling layers. The architecture of this model is shown in Fig. (5). This type of reverse mapping sends the output of the convolutional layer backward, resulting in visually discernible image forms that correspond to the neural interpretation of the internal feature representation learned at each layer [29]. ZefNet's main premise was to keep track of the learning schematic as it progressed through the training stage.

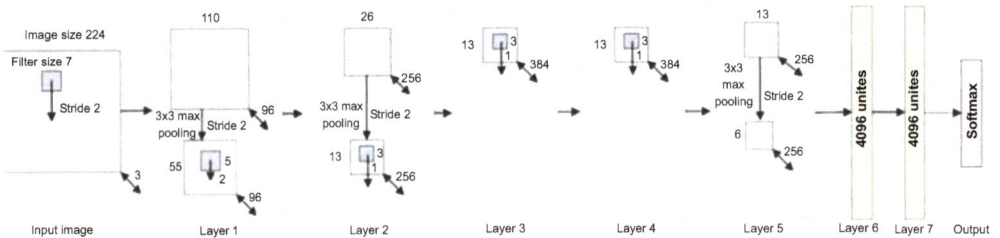

Fig. (5). The architecture of a ZefNet [19].

VGG

Simonyan and Zisserman suggested a straightforward and efficient design principle for CNN after it was determined to be useful in the field of image recognition. Visual Geometry Group was the name of this groundbreaking design (VGG). It was a multilayer model [30] with nineteen more layers than ZefNet [19] and AlexNet, allowing it to replicate the network representational capacity relations in greater depth. ZefNet, on the other hand, was the frontier network in the 2013-ILSVRC competition, proposing that small-size filters may improve CNN performance. In ZefNet, VGG inserted a layer of the heap of 3×3 filters instead of the 5×5 and 11×11 filters based on these results.

This experimentally demonstrated that assigning these small-size filters in parallel can have the same effect as assigning large-size filters. In other words, the receptive field efficiency of these small-size filters was comparable to that of the large-size filters (7×7 and 5×5). Using small-size filters added the extra benefit of reducing computational intricacy by reducing the number of parameters. These findings ushered a new research trend at CNN: working with small-size filters. VGG also adjusts network complexity by introducing 1×1 convolutions in the convolutional layers. It learns a linear grouping of the feature maps that follow.

A max-pooling layer [31] is placed after the convolutional layer in terms of network tuning, and padding is used to keep the spatial resolution. In general, VGG produced significant results for picture classification and localization challenges. Although it did not win the first place in the 2014-ILSVRC competition, it gained a reputation for its increased depth, homogeneous topology, and ease of use. VGG's computational cost, on the other hand, was exorbitant due to its use of about 140 million parameters, which was its fundamental flaw. The network's structure is depicted in Fig. (**6**).

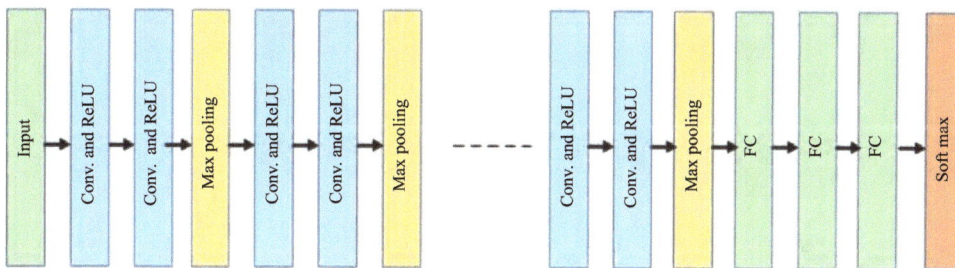

Fig. (6). The architecture of a VGG [20, 31].

GoogleNet

GoogleNet (also known as Inception-V1) was the winner of the 2014-ILSVRC competition [32]. The GoogleNet architecture's main goal is to achieve high-level accuracy at a low computing cost. It suggested a novel inception block (module) concept in the CNN context, because it employs merge, transform, and split functions for feature extraction to combine multiple-scale convolutional transformations. The conception block architecture is depicted in Fig. (**7**).

CNN ACCELERATORS TOWARD HEALTHCARE APPLICATIONS

In this section, we cover the hardware accelerators based on CMOS, FPGA, and memristors for healthcare and biomedical applications. We deliberate how they use different approaches to achieve high performance and efficiency. The power needs for computations using hardware accelerators rise dramatically to attain improved performance. In the era of cloud computing, data centers may meet enormous power needs for processing; however, this is not practical for edge devices, such as portable patient monitoring devices or wearable structures connected within networks. Because these edge-connected devices are used in important applications, they must have very strict limits on their power consumption and environmental data latency. Table **1** lists the noteworthy hardware accelerators for these kinds of applications.

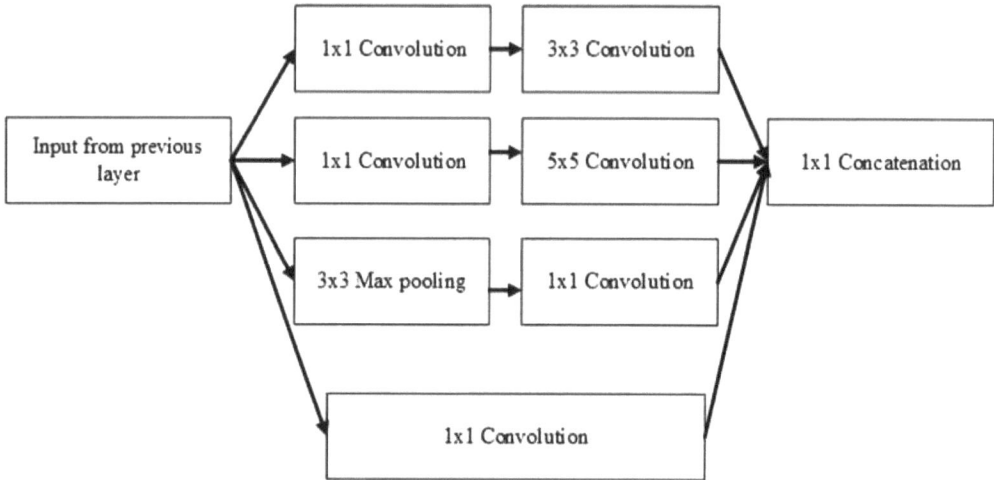

Fig. (7). The architecture of GoogleNet [33, 20].

Table 1. Hardware implementations of cnn accelerators used for healthcare and biomedical applications.

Hardware Technology	Training Model	Application	Reference
FPGA	MLP	ECG for heart monitoring	[4]
FPGA	MLP	Real-time cancer detection	[33]
FPGA and CMOS	CNN	ECG encoder for health monitoring	[34]
FPGA and CMOS	LSTM	EEG processing for neurofeedback devices	[35]
FPGA and CMOS	CNN/LSTM	Heart rate estimation	[36]
FPGA and CMOS	CNN	Physical activity monitoring	[37]
Memristor-CMOS	MLP	Breast cancer diagnosis	[38]
Memristor	CNN	ECG signal processing	[39]
CMOS	CNN	EMG signal processing	[40]
CMOS	CNN	Image-based breast cancer diagnosis	[41]
CMOS	CNN	Heartbeat detection	[42]
CMOS	CNN	ECG feature extraction	[43]

CMOS Based Accelerators

AI chips that utilize Complementary Metal Oxide Semiconductor (CMOS) technology have been extensively employed to expedite AI inference through the utilization of Multiply-Accumulate (MAC) operations, which are well-suited for convolutional Neural Networks. This makes it possible to reduce memory access, which is advantageous for both computational acceleration in data centers and AI applications on edge devices [3]. The power requirements of AI-specific processors powered by CMOS technology and meant for portable edge device applications range from a few hundred milliwatts to tens of Watts [44].

The flexibility of these designs, which enables the same chip to be utilized for numerous applications with little adjustment from the external environment, is one of the crucial features. One such option in medical imaging is the Google Tensor Processing Unit (TPU), which uses software frameworks like TensorFlow Lite and has a strong edge hardware inference engine [46, 47]. Low power consumption is promised by brain-inspired neuromorphic processors built on Spiking Neural Networks (SNN). Similarly, CMOS-based neuromorphic computers provide improved performance and power savings of several orders of magnitude [48, 49]. These characteristics enable their application in the always-on duties that biological systems often need to do.

Although there are many more CMOS accelerator chips, for example, we have selected a few well-known ones here that are made especially for CNN. Table **2** displays a few sample biomedical and healthcare applications that were chosen based on these accelerators' proven ability to run (or train [55]) a variety of - known CNN architectures, including VGG, LeeNet, AlexNet, ResNet, GoogleNet, MobileNet, and Inception.

The table displays a few representative biomedical and healthcare applications that were chosen due to the accelerators' proven ability to execute a variety of well-known CNN architectures.

FPGA Based Accelerators

Hardware acceleration uses computer hardware to do computations with less delay and higher throughput than traditional software implementations on general-purpose CPUs. To check the performance of hardware accelerators, the metrics to be considered are: accuracy, programmability, energy/power, and throughput/latency.

As the days pass, the technology is enhancing and the parameters required for the particular application are also increasing, for example, in the case of machine learning applications such as computer vision, speech recognition, and medical, high-end machine learning algorithms work with great number of parameters.

Table 2. CMOS-based accelerators for healthcare and biomedical applications.

CMOS based accelerator	Power (mW)	Performance (GOP/s)	Energy efficiency (TOPS/w)	Applications	Reference
ConvNet Processor	25-287	102	0.3-2.7	PoC ultrasound processing	[49]
Eyeriss	278	17-42	0,06-0.15	Thyroid cancer diagnosis	[50]
Cambricon-x	954	544	0.5	PoC diagnosis of cardiovascular diseases	-
Origami	654	196	0.8	Heart monitoring	[51]
Envision	7.5-300	76-408	0.8-10	Epilepsy Diagnosis	[52]
LNPU	43-367	600	25	Cancer Diagnosis	[53]
Neural Processor	39-1500	1900-7000	4.5-11.5	Skin cancer	[54]
UNPU	3.5-297	346-7372	3.08-50.6	Respiratory sound classification	[55]
DNPU	35-279	300-1200	2.1-8.1	ECG analysis	[56]

In neural network processors, general-purpose processors have been taken as the first preference because of their large computation capacity and easy-to-use development frameworks.

FPGA-based convolutional neural network accelerators, on the other hand, are becoming a study focus. FPGA is the next feasible answer to surpass General-purpose processors in terms of speed and energy efficiency, thanks to specially built hardware. To achieve high speed and energy efficiency, various FPGA-based accelerator architectures have been proposed, along with software and hardware optimization strategies [2]. The use of FPGA-based hardware accelerators for a specific CNN model is currently a major research topic in the field of Artificial Intelligence and machine learning. There are many attractive features for FPGA-based hardware accelerator architectures.

CNN architectures exhibit several sources of parallelism. However, because of the resource restriction of FPGA devices, it is not possible to fully develop all the parallelism patterns, especially with the large number of operations involved in deep learning technologies. This problem can be addressed by using a limited number of processing elements(PEs) on the FPGA [57].

The performance of the architectures is improved with the systolic arrays. The basic principle of a static systolic array is shown in Fig. (**8**). The systolic array may be used as a coprocessor in combination with a host computer where the data samples received from the host computer pass through the PEs and the final result is returned to the host computer. Systolic architecture designs can map high-level computations into hardware structures.

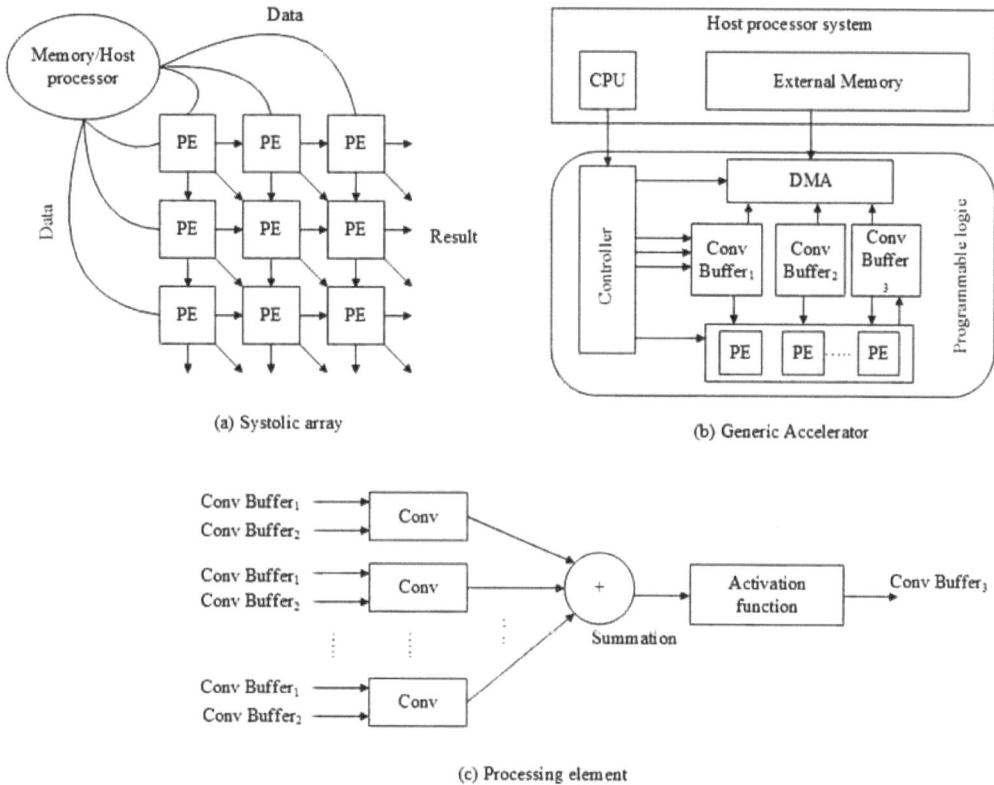

(a) Systolic array

(b) Generic Accelerator

(c) Processing element

Fig. (8). Data paths of FPGA-based CNN accelerators.

However, FPGA-based convolutional neural network hardware accelerators are specifically designed for high speed and energy efficiency to overcome the problems with CPUs and GPUs. A variety of FPGA-based hardware accelerators are designed with software and hardware optimization techniques to accomplish high speed and energy efficiency.

Only a small number of FPGA-based CNN accelerators have been created expressly for biomedical applications, such as real-time mass spectrometry data inspection for cancer detection [33] and ECG anomaly detection [4]. In these cases, the authors demonstrate that application-specific parameter quantization and customized network design can lead to a significant inference speed-up when compared to both CPU and GPU. Furthermore, the authors of a study [58] have created an FPGA-based BCI that uses an MLP to reconstruct ECog signals.

In a study [35], the authors scaled an EEG processing and neurofeedback prototype on an advanced Ultra-scale Virtex-VU9P, which has achieved 215 and 8 times power efficiency relative to CPU and GPU, respectively, after first implementing it on a low-cost, low-power FPGA. An LSTM inference framework was built for the analysis of EEG data. To perform the computations efficiently, an optimal network is compiled and run on the FPGA hardware accelerator [59].

Memristor Based Accelerators

Memristor-based CNN accelerators adapt the transconductances to variations in applied current and voltage by utilizing the memristor's property [60]. The transconductance variability can be employed as integrals that simulate how neural synapses function when learning about their surroundings, a function that is replicated by spike neural networks (SNNs). It can also be translated to weights in CNNs. Its capacity to move these core operations to the device level lowers power consumption by a factor of ten and boosts performance to the order of ten when compared to specialized devices.

Numerous flaws, such as non-linear and stochastic transconductance, which are employed as variable weights in CNN accelerators, might arise during the construction of memristive crossbars [62, 63]. When used for real-world accelerators, crossbar array performance is severely reduced by circuit yield, device variations, and spatial and temporal fluctuations.The accuracy may be impacted by these nonidealities, which is unacceptable for biomedical and clinical applications.

PROPOSED CNN ARCHITECTURE

Using MRI scans, convolutional neural networks (CNNs) have been widely employed as a deep learning technique to identify brain tumors. Owing to the small dataset, CNNs and deep learning algorithms need to be enhanced for greater efficiency. Thus, data augmentation is one of the most well-known methods for enhancing model performance [63]. Fig. (**9**) describes the CNN architecture of brain tumor detection. The model's result is "NO" for the normal brain and "YES" for the aberrant brain. The sigmoid activation function is taken in our proposed model to predict the tumor. Sigmoid activation function g(z) is best suitable for binary classification problems. Based on the activation value of the output layer mentioned in Equation 1, the tumor is predicted. $\overline{w}_j^{[t]}$ and $b_j^{[t]}$ are the weight and bias of the layer '*l*' and unit 'j'. If the activation value of the output layer $a_j^{[l]} \geq 0.5$ then the predicted output $\hat{y}$ is "YES" otherwise the predicted output $\hat{y}$ is "NO".

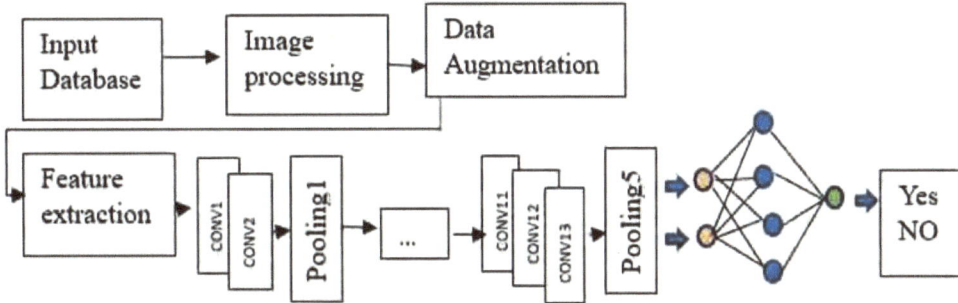

Fig. (9). CNN architecture of data augmentation-based brain tumor detection. .

$$a_j^{[l]} = g\left(\vec{w}_j^{[l]}.\vec{a}^{[l-1]} + b_j^{[l]}\right) \tag{1}$$

The performance of the model is improved by measuring the cost function. The cost function tells us how well the model is doing. So that we can try to get it to do better. The cost function takes the predicted output $\hat{y}$ and compares it with the target output 'y'. This difference is called the error. The squared error cost function is represented in equation 2 where m represents the number of training examples. Find 'w' and 'b' such that $\hat{y}^{(i)}$ is close to $y^{(i)}$ for all $x^{(i)}$ and $y^{(i)}$. Where $x^{(i)}$ and $y^{(i)}$ are the input and output features of ith training example.

$$J(w, b) = \frac{1}{2m}\sum_{i=1}^{m}(\hat{y}^{(i)} - y^{(i)})^2 \tag{2}$$

The gradient descent algorithm used in this model to reduce the cost function Jw,b by keep on changing the 'w' and 'b'. Simultaneous update of 'w' and 'b' gives the best performance results. Take the current value of 'w' and 'b' and adjust them to a small amount using equations 3 and 4 to reduce the cost function where α is the learning rate. The learning rate is a small positive number between 0 and 1.

$$w = w - \alpha\,\frac{d}{dw}J(w, b) \tag{3}$$

$$b = b - \alpha\frac{d}{db}J(w, b) \tag{4}$$

EVALUATION METRICS

Object detection is the process of identifying and locating object instances, either from a vast number of predetermined natural categories or from a given particular object. Traditional object detection models with ML do the object detection by using 3 stages informative region selection (sliding window), feature extraction with CNN, and classification. Hence, the time taken for object detection is more. In recent years, the research community has begun to develop large-scale, multi-class datasets to design a general-purpose, robust object detection model [1]. Average precision (AP) and mean Average Precision (mAP) are the two popular metrics used to scale the performance of the models. To compare different CNN models, we need to draw precision and recall curves as shown in Fig. (**10**).

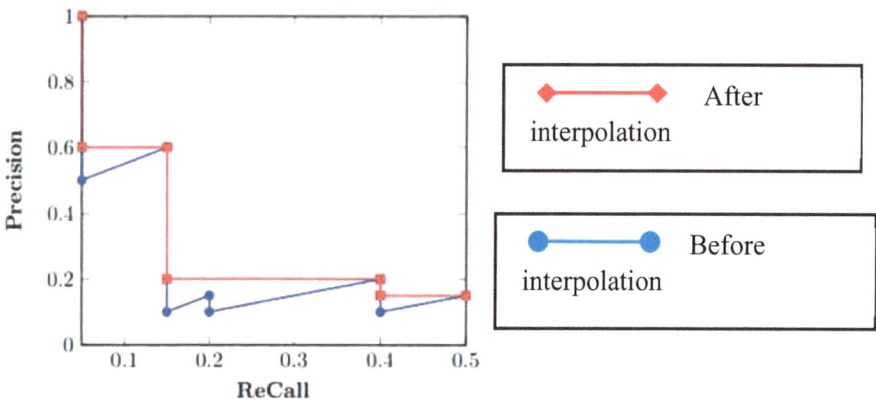

Fig. (10). A graph between precision and recall.

Precision

In a positive class, precision is used to calculate the positive patterns that are correctly predicted by all predicted patterns (Eq. 1).

$$\text{precision} = \frac{\text{True Positive}}{\text{Total true predictions}} = \frac{\text{TP}}{\text{TP+FP}} \tag{5}$$

Recall

The recall is determined by dividing the total number of Positive samples by the number of Positive samples accurately categorized as Positive (Eq. 2).

$$\text{Recall} = \frac{\text{TP}}{\text{Actual true value}} = \frac{\text{TP}}{\text{TP+FN}} \tag{6}$$

Error rate

It is the ratio of false predicted samples to the total number of samples evaluated (Eq. 3).

$$\text{Error rate} = \frac{FP+FN}{TP+TN+FP+FN} \tag{7}$$

Accuracy

It is defined as the true predicted classes to the total number of samples evaluated (Eq. 4).

$$\text{Accuracy} = \frac{TP+TN}{TP+TN+FP+FN} \tag{8}$$

Intersection over union (IOU)

In the object detection to calculate the average precision, the bounding boxes are selected based on the IOU threshold value. It is described as follows (Eq. 5):

$$\text{IOU} = \frac{\text{Area of overlap}}{\text{Area of union}} = \frac{TP}{TP+FP+FN} \tag{9}$$

F_1 Score:

It is calculated to make a trade-off between precision(P) and recall(R) (Eq. 10).

$$\text{F}_1 \text{ Score} = 2\frac{PR}{P+R} \tag{10}$$

The performance of the FPGA-based hardware accelerator is analyzed based on these evaluation metrics with the help of available datasets and pre-trained object detection algorithms. The best object detection datasets are shown in Table **3**.

Image classification object detection and localization can be done with the pre-trained object detection algorithms shown in Table **4**. Among these, YOLO [64] is the best one because it can detect objects very fast in a single step. YOLO can achieve a real-time speed of 45 Frames Per Second (FPS).

R-CNN [65] was the first two-stage method of pre-trained object detection algorithm among the earliest CNN-based generic object detection techniques. It is a region-based proposal and extracts more regions for one image. Hence it takes more time, more memory, and computer resources. Due to its multistage complex pipeline and redundant CNN feature extraction from various region proposals, it is neither elegant nor efficient. Fast R-CNN [66], before the fully connected layer, includes an ROI pooling layer. For each proposed zone, we obtain a fixed-length feature vector so that only for the input image, only one convolution process is necessary. Fast R-CNN considerably improves the detection efficiency over R-CNN. Faster R-CNN [67] presents a Region Proposal Network(RPN) to generate a region of interest and overcome the aforementioned computational complexity.

Table 3. Benchmark Datasets.

Dataset	Num. of classes	Applications	Link to dataset
ImageNet	1000	Image classification, object localization, object detection, *etc.*	http://www.image-net.org/
CIFAR10/100	10/100	Image classification	https://www.cs.toronto.edu/~kriz/cifar.html
MNIST	10	Classification of handwritten digits	http://yann.lecun.com/exdb/mnist/
Pascal VOC	20	Image classification, segmentation, object detection	http://host.robots.ox.ac.uk/pascal/VOC/voc2012/
Microsoft COCO	80	Object detection, semantic segmentation	https://cocodataset.org/#home
YFCC100M	8M	Video and image understanding	http://projects.dfki.unikl.de/yfcc100m/
YouTube-8M	4716	Video classification	https://research.google.com/youtube8m/
UCF-101	101	Human action detection	https://www.crcv.ucf.edu/data/UCF101.php
Kinetics	400	Human action detection	https://deepmind.com/research/open-source/kinetics
Google Open Images	350	Image classification, segmentation, object detection	https://storage.googleapis.com/openimages/web/index.html
CalTech101	101	Classification	http://www.vision.caltech.edu/Image_Datasets/Caltech101/
Labeled Faces in the Wild	-	Face recognition	http://vis-www.cs.umass.edu/lfw/
MIT-67 scene dataset	67	Indoor scene recognition	http://web.mit.edu/torralba/www/indoor.htm

Table 4. Pre-trained object detection algorithms.

Architecture	Year	Method	No of stages
RCNN	2014	Region-based	3 stages
Fast R-CNN	2015	selective search	3 stages
Faster R-CNN	2016	Region proposal network	3 stages
YOLO	2016	Fixed grid regression	1 stage
SSD	2016	Computation in a single network	1 stage
Mask R-CNN	2017	Pixel level segmentation	2 stages
YOLOv3	2018	Feature pyramid networks	1 stage
EfficientDet	2019	Bi-directional Feature Pyramid Network	1 stage
YOLOv6/ YOLOv7	2022	Optimized and enhanced multimedia models	1 Stage
YOLOv8	2023	Advanced technique	1 Stage
DFU_FNet & DFU_TFNet	2024	Domain transfer learning	1 Stage

As we further investigate Faster R-CNN, Mask R-CNN [68] was later proposed and it combines object detection and pixel-level instance segmentation based on Faster R-CNN. Mask R-CNN exhibited the best detection accuracy on MS-COCO in 2017. To classify the normal and abnormal features of the diabetic foot ulcer (DFU), two CNN models DFU_FNet and DFU_TFNet are proposed, which give the best performance results [69].

The most common frameworks and libraries which are suitable for the CNN hardware accelerator applications are listed in Table **5**.

APPLICATIONS AND FUTURE DIRECTIONS

- FPGAs have become more popular in recent decades. Image processing, artificial intelligence (AI), data center hardware accelerators, enterprise networking, and enhanced driver assistance systems in automobiles are all examples of emerging technologies (ADAS).
- The ability to design the FPGA's CLBs into hundreds or even thousands of comparable processing blocks benefits everyone. They can be utilized in random logic, SPLDs, device controllers, communication encoding, and filtering, to name a few applications.
- FPGAs are ideal for signal processing applications because of their high operating frequency, parallel processing capacity, and low cost.
- FPGA Design is frequently used in the construction of convolutional neural networks, or CNNs, which will lead to the development and support of artificially intelligent systems in the future.

- High-performance FPGA Design, as compared to GPUs, may provide even more aid to an application and is hence preferred for the creation of deep neural networks in machine learning technologies.
- Because of their vast computing parallelism, ability to change the computational units to the bit-width needed, and low latency, FPGA Designs are used in the disciplines of encrypting/decrypting and post-quantum cryptography. Radio astronomy is the science of capturing electromagnetic waves from space to study the processes that take place there.
- Speaker recognition is a technique that is utilized in security, information extraction systems, and other domains, and its scope of use is expected to grow in the future. In circumstances when a person's voice is compared to previously stored patterns, the FPGA Design is extremely effective.
- Due to their favorable qualities, FPGA Design is also used in a range of aviation and defense applications.
- PET tests, CT scans, X-rays, 3-D imaging, and other approaches are increasingly being employed to handle biological pictures created by FPGA Design.
- The advantages of FPGA Design and parallel processing are well suited to these requirements since medical vision systems are increasingly requiring higher resolution and processing capacity and many of them must be created in real-time.
- The use of FPGA Design to boost performance in large-scale data systems can be beneficial. FPGAs provide specialized high-bandwidth, low-latency connectivity to network and storage systems, allowing for speedier data processing.

Table 5. List of most common frameworks and libraries.

Framework	License	Core language	Homepages
Torch	BSD	C & Lua	http://torch.ch/
Theano	BSD	Python	http://deeplearning.net/software/theano/
MatConvNet	Oxford	MATLAB	http://www.vlfeat.org/matconvnet/
DL4j	Apache 2.0	Apache 2.0	https://deeplearning4j.org/
TensorFlow	Apache 2.0	C++ & Python	https://www.tensorflow.org/
Keras	MIT	Python	https://keras.io/
Caffe	BSD	C++	http://caffe.berkeleyvision.org/
MXNet	Apache 2.0	C++	https://github.com/dmlc/mxnet
CNTK	MIT	MIT	https://github.com/Microsoft/CNTK
Gluon	AWS Microsoft	C++	https://github.com/gluon-api/gluon-api/
OpenDeep	MIT	Python	http://www.opendeep.org/

CONCLUSION

In this chapter, we looked at CNN architectures for object detection and CNN hardware accelerator architectures for healthcare and biomedical applications to improve the performance of hardware accelerators. We looked at the evaluation metrics to assess the performance of the proposed CNN model. We divided the works into categories based on a variety of criteria to emphasize their similarities and distinctions. We selected the most important study directions and summed up the main points of various publications. A brief discussion of future research issues concludes this report.

REFERENCES

[1]   Feng X, Jiang Y, Yang X, Du M, Li X. Computer vision algorithms and hardware implementations: A survey. Integration (Amst) 2019; 69: 309-20.
[http://dx.doi.org/10.1016/j.vlsi.2019.07.005]

[2]   Guo K, Zeng S, Yu J, Wang Y, Yang H. [DL] A survey of FPGA-based neural network inference accelerators. ACM Trans Reconfig Technol Syst 2019; 12(1): 1-26.
[http://dx.doi.org/10.1145/3289185]

[3]   Azghadi MR, Lammie C, Eshraghian JK, *et al.* Hardware implementation of deep network accelerators towards healthcare and biomedical applications. IEEE Trans Biomed Circuits Syst 2020; 14(6): 1138-59.
[http://dx.doi.org/10.1109/TBCAS.2020.3036081] [PMID: 33156792]

[4]   Wess M, Manoj PDS, Jantsch A. Neural network based ECG anomaly detection on FPGA and trade-off analysis. 2017 IEEE Int Symp Circuits Syst 2017; 1-4.
[http://dx.doi.org/10.1109/ISCAS.2017.8050805]

[5]   Jindal V. Integrating mobile and cloud for PPG signal selection to monitor heart rate during intensive physical exercise. Proc Int Conf Mob Softw Eng Syst 2016; 36-7.
[http://dx.doi.org/10.1145/2897073.2897132]

[6]   Shi B, Grimm LJ, Mazurowski MA, *et al.* Prediction of occult invasive disease in ductal carcinoma *in situ* using deep learning features. J Am Coll Radiol 2018; 15(3): 527-34.
[http://dx.doi.org/10.1016/j.jacr.2017.11.036] [PMID: 29398498]

[7]   Zhu W, Xie L, Han J, Guo X. The application of deep learning in cancer prognosis prediction. Cancers (Basel) 2020; 12(3): 603.
[http://dx.doi.org/10.3390/cancers12030603] [PMID: 32150991]

[8]   Liu X, Faes L, Kale AU, *et al.* A comparison of deep learning performance against health-care professionals in detecting diseases from medical imaging: a systematic review and meta-analysis. Lancet Digit Health 2019; 1(6): e271-97.
[http://dx.doi.org/10.1016/S2589-7500(19)30123-2] [PMID: 33323251]

[9]   Liu F, Yadav P, Baschnagel AM, McMillan AB. MR-based treatment planning in radiation therapy using a deep learning approach. J Appl Clin Med Phys 2019; 20(3): 105-14.
[http://dx.doi.org/10.1002/acm2.12554] [PMID: 30861275]

[10]   Wang C, Wang X, Xia Z, Ma B, Shi YQ. Image description with polar harmonic Fourier moments. IEEE Trans Circ Syst Video Tech 2020; 30(12): 4440-52.
[http://dx.doi.org/10.1109/TCSVT.2019.2960507]

[11]   Wang C, Wang X, Xia Z, Zhang C. Ternary radial harmonic Fourier moments based robust stereo image zero-watermarking algorithm. Inf Sci 2019; 470: 109-20.
[http://dx.doi.org/10.1016/j.ins.2018.08.028]

[12] Jyotiyana M, Kesswani N. A study on deep learning in neurodegenerative diseases and other brain disorders. Proc FICR-TEAS 2020 2021; 791-9.
[http://dx.doi.org/10.1007/978-981-15-6014-9_95]

[13] Hassan TM, Elmogy M, Sallam ES. Diagnosis of focal liver diseases based on deep learning technique for ultrasound images. Arab J Sci Eng 2017; 42(8): 3127-40.
[http://dx.doi.org/10.1007/s13369-016-2387-9]

[14] Arjmand A, Angelis CT, Christou V, *et al.* Training of deep convolutional neural networks to identify critical liver alterations in histopathology image samples. Appl Sci (Basel) 2019; 10(1): 42.
[http://dx.doi.org/10.3390/app10010042]

[15] Tabrizchi H, Mosavi A, Szabo-Gali A, Felde I, Nadai L. Rapid COVID-19 diagnosis using deep learning of the computerized tomography Scans. 2020 IEEE 3rd Int. Conf. Work. Óbuda Electr. Power Eng. 2020; 173-8.

[16] Murtaza G, Shuib L, Mujtaba G, Raza G. Breast cancer multi-classification through deep neural network and hierarchical classification approach. Multimedia Tools Appl 2020; 79(21-22): 15481-511.
[http://dx.doi.org/10.1007/s11042-019-7525-4]

[17] Khan A, Sohail A, Zahoora U, Qureshi AS. A survey of the recent architectures of deep convolutional neural networks. Artif Intell Rev 2020; 53(8): 5455-516.
[http://dx.doi.org/10.1007/s10462-020-09825-6]

[18] Shrestha A, Mahmood A. Review of deep learning algorithms and architectures. IEEE Access 2019; 7: 53040-65.
[http://dx.doi.org/10.1109/ACCESS.2019.2912200]

[19] Alzubaidi L, Zhang J, Humaidi AJ, *et al.* Review of deep learning: concepts, CNN architectures, challenges, applications, future directions. J Big Data 2021; 8(1): 53.
[http://dx.doi.org/10.1186/s40537-021-00444-8] [PMID: 33816053]

[20] LeCun Y, Jackel LD, Bottou L, Cortes C, Denker JS, Drucker H, *et al.* Learning algorithms for classification: A comparison on handwritten digit recognition. Neural Networks Stat Mech Perspect 1995; 261: 2.

[21] Krizhevsky A, Sutskever I, Hinton GE. ImageNet classification with deep convolutional neural networks. Commun ACM 2017; 60(6): 84-90.
[http://dx.doi.org/10.1145/3065386]

[22] Srivastava N, Hinton G, Krizhevsky A, Sutskever I, Salakhutdinov R. Dropout: a simple way to prevent neural networks from overfitting. J Mach Learn Res 2014; 15: 1929-58.

[23] Dahl GE, Sainath TN, Hinton GE. Improving deep neural networks for LVCSR using rectified linear units and dropout. 2013 IEEE Int. Conf. Acoust. speech signal Process., 2013, p. 8609–13.

[24] Xu B, Wang N, Chen T, Li M. Empirical evaluation of rectified activations in convolutional network. ArXiv Prepr ArXiv150500853 2015.

[25] Hochreiter S. The vanishing gradient problem during learning recurrent neural nets and problem solutions. Int J Uncertain Fuzziness Knowl Based Syst 1998; 6(2): 107-16.
[http://dx.doi.org/10.1142/S0218488598000094]

[26] Zeiler MD, Fergus R. Visualizing and understanding convolutional networks. Comput. Vision--ECCV 2014 13th Eur. Conf. Zurich, Switzerland, Sept. 6-12, 2014. Proceedings 2014; 13(Part I): 818-33.

[27] Erhan D, Bengio Y, Courville A, Vincent P. Visualizing higher-layer features of a deep network. Univ Montr 2009; 1341: 1.

[28] Le QV. Building high-level features using large-scale unsupervised learning. 2013 IEEE Int. Conf. Acoust. speech signal Process., 2013, p. 8595–8.

[29] Grün F, Rupprecht C, Navab N, Tombari F. A taxonomy and library for visualizing learned features in

convolutional neural networks. ArXiv Prepr ArXiv160607757 2016.

[30]　Simonyan K, Zisserman A. Very deep convolutional networks for large-scale image recognition. 3rd Int. Conf. Learn. Represent. ICLR 2015 - Conf. Track Proc. 2015.

[31]　Ranzato M, Huang FJ, Boureau Y-L, LeCun Y. Unsupervised learning of invariant feature hierarchies with applications to object recognition. 2007 IEEE Conf. Comput. Vis. pattern Recognit., 2007, p. 1–8.

[32]　Szegedy C, Liu W, Jia Y, Sermanet P, Reed S, Anguelov D, *et al.* Going deeper with convolutions. Proc. IEEE Conf. Comput. Vis. pattern Recognit., 2015, p. 1–9.
[http://dx.doi.org/10.1109/CVPR.2015.7298594]

[33]　Xing F, Xie Y, Shi X, Chen P, Zhang Z, Yang L. Towards pixel-to-pixel deep nucleus detection in microscopy images. BMC Bioinformatics 2019; 20: 1-16.

[34]　Garg B, Yadav S, Sharma GK. An area and performance-aware ecg encoder design for wireless healthcare services. 2016 20th Int. Symp. VLSI Des. Test, 2016, p. 1–6.

[35]　Chen Z, Howe A, Blair HT. Cong. 2018; pp. Proc Int Symp Low Power Electron Des 2018; 1-6.

[36]　Rocha LG, Biswas D, Verhoef BE, *et al.* Binary CorNET: Accelerator for HR estimation from wrist-PPG. IEEE Trans Biomed Circuits Syst 2020; 14(4): 715-26.
[http://dx.doi.org/10.1109/TBCAS.2020.3001675] [PMID: 32746344]

[37]　Jafari A, Ganesan A, Thalisetty CSK, Sivasubramanian V, Oates T, Mohsenin T. Sensornet: A scalable and low-power deep convolutional neural network for multimodal data classification. IEEE Trans Circuits Syst I Regul Pap 2019; 66(1): 274-87.
[http://dx.doi.org/10.1109/TCSI.2018.2848647]

[38]　Cai F, Correll JM, Lee SH, *et al.* A fully integrated reprogrammable memristor–CMOS system for efficient multiply–accumulate operations. Nat Electron 2019; 2(7): 290-9.
[http://dx.doi.org/10.1038/s41928-019-0270-x]

[39]　Hirtzlin T, Bocquet M, Penkovsky B, *et al.* Digital biologically plausible implementation of binarized neural networks with differential hafnium oxide resistive memory arrays. Front Neurosci 2020; 13: 1383.
[http://dx.doi.org/10.3389/fnins.2019.01383] [PMID: 31998059]

[40]　Moothedath S, Sahabandu D, Allen J, Clark A, Bushnell L, Lee W, *et al.* Dynamic information flow tracking for detection of advanced persistent threats: A stochastic game approach. ArXiv Prepr ArXiv200612327 2020.

[41]　O'Leary G, Groppe DM, Valiante TA, Verma N, Genov R. NURIP: Neural interface processor for brain-state classification and programmable-waveform neurostimulation. IEEE J Solid-State Circuits 2018; 53(11): 3150-62.
[http://dx.doi.org/10.1109/JSSC.2018.2869579]

[42]　Chen YH, Juan Y. Very-large-scale integration implementation of a convolutional neural network accelerator for abnormal heartbeat detection. Electron Lett 2020; 56(7): 330-1.
[http://dx.doi.org/10.1049/el.2019.3752]

[43]　Janveja M, Parmar R, Trivedi G, Jan P, Nemec Z. An energy efficient and resource optimal VLSI architecture for ECG feature extraction for wearable healthcare applications. 2022 32nd Int. Conf Radioelektronika 2022; 1-6.

[44]　Yin S, Ouyang P, Tang S, *et al.* A high energy efficient reconfigurable hybrid neural network processor for deep learning applications. IEEE J Solid-State Circuits 2018; 53(4): 968-82.
[http://dx.doi.org/10.1109/JSSC.2017.2778281]

[45]　Jouppi NP, Young C, Patil N, Patterson D, Agrawal G, Bajwa R, *et al.* In-datacenter performance analysis of a tensor processing unit. Proc. 44th Annu. Int. Symp. Comput. Archit. 2017; 1-12.
[http://dx.doi.org/10.1145/3079856.3080246]

[46] Civit-Masot J, Luna-Perejón F, Corral JMR, Domínguez-Morales M, Morgado-Estévez A, Civit A. A study on the use of Edge TPUs for eye fundus image segmentation. Eng Appl Artif Intell 2021; 104: 104384.
[http://dx.doi.org/10.1016/j.engappai.2021.104384]

[47] Corradi F, Indiveri G. A neuromorphic event-based neural recording system for smart brain-machine-interfaces. IEEE Trans Biomed Circuits Syst 2015; 9(5): 699-709.
[http://dx.doi.org/10.1109/TBCAS.2015.2479256] [PMID: 26513801]

[48] Ceolini E, Frenkel C, Shrestha SB, *et al.* Hand-gesture recognition based on EMG and event-based camera sensor fusion: A benchmark in neuromorphic computing. Front Neurosci 2020; 14: 637.
[http://dx.doi.org/10.3389/fnins.2020.00637] [PMID: 32903824]

[49] Moons B, Verhelst M. An energy-efficient precision-scalable ConvNet processor in 40-nm CMOS. IEEE J Solid-State Circuits 2017; 52(4): 903-14.
[http://dx.doi.org/10.1109/JSSC.2016.2636225]

[50] Chen YH, Krishna T, Emer JS, Sze V. Eyeriss: An energy-efficient reconfigurable accelerator for deep convolutional neural networks. IEEE J Solid-State Circuits 2017; 52(1): 127-38.
[http://dx.doi.org/10.1109/JSSC.2016.2616357]

[51] Cavigelli L, Benini L. Origami: A 803-GOp/s/W convolutional network accelerator. IEEE Trans Circ Syst Video Tech 2017; 27(11): 2461-75.
[http://dx.doi.org/10.1109/TCSVT.2016.2592330]

[52] Moons B, Uytterhoeven R, Dehaene W, Verhelst M. 14.5 envision: A 0.26-to-10tops/w subword-parallel dynamic-voltage-accuracy-frequency-scalable convolutional neural network processor in 28nm fdsoi. 2017 IEEE Int Solid-State Circuits Conf 2017; 246-7.
[http://dx.doi.org/10.1109/ISSCC.2017.7870353]

[53] Lee J, Lee J, Han D, Lee J, Park G, Yoo H-J. 7.7 LNPU: A 25.3 TFLOPS/W sparse deep-neural-network learning processor with fine-grained mixed precision of FP8-FP16. 2019 IEEE Int Solid-State Circuits Conf 2019; 142-4.
[http://dx.doi.org/10.1109/ISSCC.2019.8662302]

[54] Song J, Cho Y, Park J-S, Jang J-W, Lee S, Song J-H, *et al.* 7.1 An 11.5 TOPS/W 1024-MAC butterfly structure dual-core sparsity-aware neural processing unit in 8nm flagship mobile SoC. 2019 IEEE Int Solid-State Circuits Conf 2019; 130-2.
[http://dx.doi.org/10.1109/ISSCC.2019.8662476]

[55] Lee J, Kim C, Kang S, Shin D, Kim S, Yoo HJ. UNPU: An energy-efficient deep neural network accelerator with fully variable weight bit precision. IEEE J Solid-State Circuits 2019; 54(1): 173-85.
[http://dx.doi.org/10.1109/JSSC.2018.2865489]

[56] Shin D, Lee J, Lee J, Lee J, Yoo HJ. DNPU: An energy-efficient deep-learning processor with heterogeneous multi-core architecture. IEEE Micro 2018; 38(5): 85-93.
[http://dx.doi.org/10.1109/MM.2018.053631145]

[57] Abdelouahab K, Pelcat M, Serot J, Berry F. Accelerating CNN inference on FPGAs: A survey. ArXiv Prepr ArXiv180601683 2018.

[58] Shrivastwa RR, Pudi V, Chattopadhyay A. An FPGA-based brain computer interfacing using compressive sensing and machine learning 2018 IEEE Comput. Soc. Annu. Symp. VLSI 2018; pp. 726-31.

[59] Peng P, Jiang K, You M, *et al.* Design of an Efficient CNN-Based Cough Detection System on Lightweight FPGA. IEEE Trans Biomed Circuits Syst 2023; 17(1): 116-28.
[http://dx.doi.org/10.1109/TBCAS.2023.3236976] [PMID: 37018680]

[60] Ambrogio S, Narayanan P, Tsai H, *et al.* Equivalent-accuracy accelerated neural-network training using analogue memory. Nature 2018; 558(7708): 60-7.
[http://dx.doi.org/10.1038/s41586-018-0180-5] [PMID: 29875487]

[61]   Rahimi Azghadi M, Chen YC, Eshraghian JK, *et al.* Complementary metal-oxide semiconductor and memristive hardware for neuromorphic computing. Adv Intell Syst 2020; 2(5): 1900189.
[http://dx.doi.org/10.1002/aisy.201900189]

[62]   Xia Q, Yang JJ. Memristive crossbar arrays for brain-inspired computing. Nat Mater 2019; 18(4): 309-23.
[http://dx.doi.org/10.1038/s41563-019-0291-x] [PMID: 30894760]

[63]   Alsaif H, Guesmi R, Alshammari BM, *et al.* A novel data augmentation-based brain tumor detection using convolutional neural network. Appl Sci (Basel) 2022; 12(8): 3773.
[http://dx.doi.org/10.3390/app12083773]

[64]   Redmon J, Divvala S, Girshick R, Farhadi A. You only look once: Unified, real-time object detection. Proc IEEE Conf Comput Vis pattern Recognit 2016; 779-88.

[65]   Girshick R, Donahue J, Darrell T, Malik J. Rich feature hierarchies for accurate object detection and semantic segmentation. 2014.
[http://dx.doi.org/10.1109/CVPR.2014.81]

[66]   Girshick R. Fast r-cnn. Proc IEEE Int Conf Comput Vis 2015; 1440-8.

[67]   Ren S, He K, Girshick R, Sun J. Faster r-cnn: Towards real-time object detection with region proposal networks. Adv Neural Inf Process Syst 2015; 28.

[68]   He K, Gkioxari G, Dollár P, Girshick R. Mask r-cnn. Proc IEEE Int Conf Comput Vis 2017; 2961-9.

[69]   Fadhel MA, Alzubaidi L, Gu Y, Santamaría J, Duan Y. Real-time diabetic foot ulcer classification based on deep learning & parallel hardware computational tools. Multimed Tools Appl 2024.
[http://dx.doi.org/10.1007/s11042-024-18304-x]

## CHAPTER 5

Advancements in Smart Sensor Technology for Enhanced Health Monitoring in Smart Watches

G. Jeeva[1,*]**, P. Mahalakshmi**[1] **and S. Thenmalar**[1]

[1] *Department of Networking and Communications, School of Computing, SRM Institute of Science and Technology, Kattankulathur, India*

Abstract: The integration of smart sensors in wearable devices, particularly smart watches, has revolutionized the landscape of personal health monitoring. This review paper provides a comprehensive analysis of recent advancements in smart sensor technology and their application in smartwatches for health monitoring. The paper begins with an overview of the evolution of smartwatches and their transition from timekeeping devices to sophisticated health monitoring tools. It then delves into the key components of smart sensor technology, encompassing biometric sensors, environmental sensors, and activity trackers. The review extensively covers the diverse range of health parameters that can be monitored by smartwatches, including physical activity levels, oxygen saturation, blood pressure, and heart rate. Furthermore, the paper evaluates the accuracy and reliability of these sensors, considering factors such as sensor placement, calibration, and data processing techniques. The paper also explores the potential integration of machine learning and artificial intelligence in data analysis and interpretation, highlighting their potential to enhance the effectiveness and efficiency of smartwatch health monitoring. In addition, the review addresses challenges and limitations associated with smartwatch health monitoring, including privacy concerns, data security, and battery life. This paper provides an up-to-date overview of smart sensor technology as applied to health monitoring in smartwatches. It serves as a valuable resource for researchers, healthcare professionals, and technology enthusiasts interested in understanding the potential and limitations of this rapidly evolving field.

Keywords: Healthcare, Machine learning, Smart watch, Smart sensor, Wearable.

INTRODUCTION

The fast proliferation of healthcare monitoring systems in hospitals and other health centers over the past ten years has led to a great deal of interest in wireless healthcare monitoring devices employing various technologies in many nations

---

* **Corresponding author G. Jeeva:** Department of Networking and Communications, School of Computing, SRM Institute of Science and Technology, Kattankulathur, India; E-mail: jg9473@srmist.edu.in

Sivakumar Rajagopal, Prakasam P., Konguvel E., Shamala Subramaniam, Ali Safaa Sadiq Al Shakarchi & B. Prabadevi (Eds.)

across the world. For example, taking a blood sample for laboratory analysis might be an inconvenience for patients, but wearable smart health applications aim to continuously monitor vital physiological indicators so that patients can go about their day as normal. For instance, the essential indicator of ventilation efficacy, reflecting respiratory acid-base status, is referred to as the partial pressure of arterial carbon dioxide, which is measured invasively from the arteries. Therefore, we can briefly track it in a clinical context when we draw a blood sample from the arterial system.

Transcutaneous carbon dioxide monitoring is a non-invasive surrogate approach for determining the partial pressure of arterial carbon dioxide, however, it is currently only used in specialized settings such as intensive care units and requires a bulky bedside device [1]. Polluted air poses a serious threat to human health, and lowering pollution levels could reduce the prevalence of diseases including asthma, cancer, and stroke. The MQ family of gas sensors is a useful tool for detecting pollution in the air and enforcing other safety measures. This article aims to construct an affordable air quality monitoring system, suitable for both indoor and outdoor use, using MQ sensors and an Arduino Mega. It also uses techniques such as discriminant analysis (DA) and the probabilistic neural network (PNN) to track the data back to its source. The system consists of four MQ fuel sensors and one Arduino Mega. We assess the MQ-2, MQ-3, MQ-7, and MQ-135 sensors based on their reactions in both indoor and outdoor environments [2]. On-device DL finds its application in various fields such as computer vision, image processing, NLP, and audio categorization. There has been a rise in interest in mobile and wearable sensing applications. Given that these devices incorporate a wide range of sensors and generate copious quantities of data, on-device DL can be of great assistance to them [3].

A pulse oximeter, a portable health monitor, can track an individual's heart rate and the percentage of oxygen in their blood. However, heart rate variability provides a great deal more information than just the heart rate itself. We need to put more effort into developing high-tech pulse oximeters that can not only condense HRV data but also analyze the effects of exercise, to better monitor one's health. This study addressed urgent issues at hand by combining a sophisticated (Internet of Things and artificial intelligence-friendly) programming language, Python, with a low-cost photoplethysmogram (PPG) sensor [4]. The installation-specific nature of accelerometers magnifies the number of sensors required to detect whole-body motion, as they can only measure acceleration signals at their installation sites. Since they are inherently noisy, processing them takes more time and is more difficult. For the first time, this research offers a strain sensor system integrated into a body-worn suspender, which would record the periodicity of body movement and allow for less noisy readings and non-localized observations [5].

The Internet of Things, or IoT, is now a rapidly developing field. Diverse technological fields have played a role in its advancement. A great deal of research has gone into expanding this field. A smartwatch's many sensors now allow it to track the wearer's health status. As a result, developing a health monitoring system is feasible. The development of a health monitoring system is thus feasible. Using this sphere, you can realize benefits like reduced costs and wireless data transmission. Oxygen and intensive care unit beds were also scarce in the country [6]. For many people with nicotine dependence, quitting smoking is an extremely difficult task. Using self-reporting or sensor monitoring approaches, cell phones have become the primary data-gathering tool for studies investigating the effects of quitting smoking on health. Over the past five years, the proliferation of smartwatches has prompted studies to investigate whether the accelerometer of the device can infer a user's habitual motions. The primary goal of earlier smoke detection techniques was to classify users' actual smoking habits [7].

The various IoT capabilities and devices that must continuously monitor a patient's health indicators necessitate constant advancements in healthcare monitoring. Due to the importance of healthcare, a variety of concepts and methods have led to the creation of numerous devices. Several nations have introduced air ambulances as examples of these devices to meet the growing demand for healthcare during emergencies and expedite patients' recoveries [8]. Incredible technological progress in wearable electronics has enabled a wide variety of health monitoring multi-functions, but this has increased power requirements, making larger batteries and more frequent charging a necessity. However, the downtime caused by battery replacement or charging is unacceptable for health monitoring. Despite the potential for thermoelectric power generation from body heat, wearable devices have not been able to generate enough consistent power for the continuous operation of commercial health monitoring sensors [9]. Doctors can safely store patient clinical data in the cloud, simplifying their retrieval when needed. Problems such as communication lag, insecure connections, and power loss in IoT sensors can impact the service quality. A remote patient monitoring system employs cost-effective and energy-efficient IoT sensors to address patient safety concerns [10].

RELATED WORK

Nowadays, patient health status monitoring for a specific risk is a demanding task. Obstacles arise when doctors have to constantly monitor their patients and treat them at the optimal time. This architecture utilizes numerous wearable sensors to track a person's well-being in real time. In today's healthcare system, these sensors perform a crucial function. There are a variety of wearables that could do this, including but not limited to helmets, smart bands, smart textiles, necklaces, portable smart gadgets, *etc*. Each device may have a unique form factor, function, and required size [11]. Utilizing an incremental merge segmentation (IMS) technique that uses PPG signals to extract induced respiration amplitude variations, demonstrates that robust respiration rate estimates in real-time can be achieved utilizing a low-cost microcontroller. Additionally, we employ a uniform interval interpolation approach to address the non-uniformity of the RIAV signal, and a finite impulse response (FIR) band-pass filter to eliminate non-respiratory frequencies. The Fast Fourier transform signal is used to find the main frequency that is connected to the regular interval (RR) in the analysis of the breathing amplitude variation sequence [12]. We build wearable physiological sensors with multiple functions and a constant power source through system-level integration and careful circuit design for energy management and low-power sensing. Wearable electronics could benefit greatly from using energy harvested from body movements and solar or home illumination instead of batteries or plug-in power sources [13].

The enzyme-enzyme is stuck in place using a siloxane-perfluoro sulfonated ionomer composite membrane. This makes the biosensor flux independent over the full range of physiological sweat secretion rates (0.025-2 1 cm-2 min-1). However, at physiological sweat lactate concentration, the siloxane-membran--based biosensor's current response saturates; beyond this point, the biosensor's performance greatly varies depending on the flow rate [14]. Sheela and Varghese (2020) proposed a smart health monitoring system integrated with machine learning techniques [15]. It is an excellent tool for remote patient monitoring and prompting doctors to take action if necessary. We found five metrics: the electrocardiogram (ECG), pulse rate, pressure, temperature, and position detection using wearable sensors. The system uses two circuits to accomplish this. A receiver circuit can connect both the patient and the medical staff to the system, with the former remaining with the patient at all times. There are two types of wearable technology: the passive kind and the active kind. Since passive wearables do not necessitate user input, they are more inextricably bound to the mobile devices that run the dedicated app that operates them. Indra Kumari *et al.* (2020) deployed wearable electronics that collect data from the user and relay it to the user's mobile device or other linked devices [16]. Therefore, they can communicate with other devices, either nearby or far away. The proposed sensor patch continuously estimates blood pressure (BP) by utilizing the pulse arrival time. This is because the patch integrates PPG and ECG sensors into one device.

Wu *et al.* (2020) proposed a rigid-flex wearable health monitoring sensor patch. The sensor patch includes a main board for receiving and processing signals, a power board for powering the device and charging the batteries, and three sensors for keeping tabs on the body's vitals [17]. The rigid-flex construction design allows for easy attachment of all components to the body for remote health monitoring. You can remove the peripheral boards from the main board to test a specific physiological signal (like an ECG) without using the entire complement of sensors. Mamun and Yuce (2019) developed sensors and systems for wearable environmental monitoring for IoT-enabled applications [18]. The steady advancement of microelectronics, communication technologies, and miniature environmental sensing devices has spurred the development of wearables for environmental monitoring applications. These gadgets can collect objective, high-resolution, geotagged data. This article aims to provide an overview of current research on wearable Internet of Things (IoT) environmental applications. For the elderly, falls are a serious problem since they can cause serious injury or even death.

Bharathiraja *et al.* (2023) deployed thermal sensors for real-time fall detection [19]. Preventing falls is admirable, but it is difficult to attain because of the difficulty of completely eradicating falls. One option is to use non-wearable sensors such as the AMG8833 infrared thermal sensor. The sensor is on a ceiling or wall, ideally high where it can see a large area. Calibration is required to establish a thermal signature of the environment before employing the sensor for fall detection. The AMG8833 sensor continuously captures real-time thermal data. The sensor typically returns a matrix of temperatures that map to individual image pixels. According to the principles of prevention and prediction, the modern healthcare system places a premium on the medical IoT (Internet of Things). Effective parameters from the behavioral, environmental, and physiological domains, as the most affecting sectors of interest in healthcare, must be monitored on a large scale and in a broad context. Wearables play a significant role in data measurement and collection for individualized healthcare monitors. Haghi *et al.* (2020) proposed a flexible and pervasive IoT-based healthcare platform for physiological and environmental parameter monitoring [20]. The physician, as the patient's real-time observer, can also use the IoT gateway to turn on and off the wearables' sensors as needed. As a result, medical professionals can adjust the measurement setup settings based on the goals of the inquiry, the patients' conditions, the doctors' expectations, and the patient's needs.

Wearable gadgets, such as smartwatches and fitness bands, continue to proliferate, and as they evolve, so do the capabilities and health-related features they offer. Otta (2022) discusses health software and supporting platforms for wearable devices [21]. The Smart Continuous Glucose Monitoring and Control System is implemented through a wearable device. This system can currently function on devices running Windows, macOS, Linux, Raspberry Pi, or Android. The sensor-based health monitoring system provides a vital mechanism for real-time diagnosis and management to predict and prevent the emergence of diseases. We will explore the sensors, main application areas, difficulties, and solutions encountered by researchers and practitioners during the installation of health monitoring systems. Anikwe *et al.* (2022) reviewed mobile and wearable sensors for data-driven health monitoring systems [22]. We identified homogeneous sensors, dual sensors, and heterogeneous sensors as the three main types of health monitoring sensors found in mobile and wearable devices. Due to their capacity to combine several sensors from different domains, heterogeneous sensor-based systems are widely used and the most effective in health monitoring.

Many kinds of applications require the use of accelerometers, from monitoring vehicles to measuring conditions. Smartphones and tablets use accelerometers to monitor movement. Due to their limited space, modern accelerometers have had to become increasingly accurate, cheap, small, and power-efficient. The smart device industry widely uses accelerometers based on microelectromechanical systems (MEMS) due to their reliable ability to meet these stringent standards. Koene *et al.* (2020) discuss IoT-connected devices for vibration analysis and measurement [23]. Contemporary MEMS accelerometers are becoming cost-effective enough to use in demanding measuring applications, thanks to their precision and bandwidth.

The IoT-based remote healthcare monitoring system gathers and analyzes a vast range of data and documents, largely utilizing AI and ML. Clinical decision support systems and other healthcare delivery types also use machine learning approaches to create analytic representations. Clinical decision support systems provide tailored suggestions for a patient's care, including medicine, lifestyle modifications, and other interventions, based on an analysis of several criteria. The technology aids in the analysis of activities, body temperature, heart rate, blood glucose, *etc.*, and supports healthcare applications [24]. It is challenging to estimate one's risk of heart disease because it requires both specific information and real-world experience. Recently, healthcare systems have incorporated IoT to gather sensor data for heart disease diagnosis and prognosis. The wearable smartwatch and heart monitor device track the patient's blood pressure and ECG. The modified deep convolutional neural network is used to determine if incoming sensor data is normal or abnormal [25].

COMPARATIVE ANALYSIS OF BIOMETRIC SENSORS AND DESIGNS

Health information technology is one of today's most robust and rapidly expanding IT fields. Most people use this technology for illness prediction and rapid drug delivery, as obtaining a pathology report from a doctor can be time-consuming and costly. Many scientists have responded to this need by creating or enhancing disease prediction systems. The 'MedAi' wearable application employs machine learning algorithms to predict the onset of numerous health problems, such as hypertension, respiratory disorders, myocardial infarction, stroke, renal failure, gallstones, diabetes, dyslipidemia, and cardiovascular disease. The "Sense O" Clock is a prototype smartwatch that gathers biological statistics using eleven sensors. A machine learning model analyzes the data, and a mobile application displays the anticipated value [26]. The authors took ethical measures, such as obtaining informed consent from patients and doctors, to compile a dataset from a nearby hospital that includes medical statistics. There are several different machine learning techniques to determine which one produces the best results. These algorithms include Support Vector Regression and Machine, K-NN, Extreme Gradient Boosting, Long Short-Term Memory, Random Forest, and the schematic diagram of common disease prediction shown in Fig. (1).

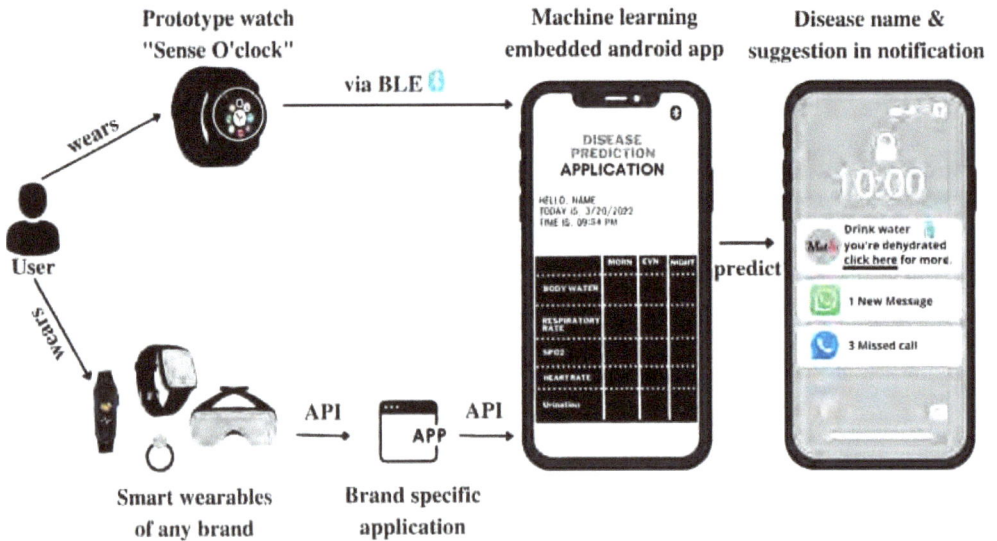

Fig. (1). Schematic diagram of machine learning based mobile application framework for common diseases prediction systems.

LoRa wireless technology is quickly emerging as one of the most power-efficient answers for Wireless Body Area Networks. As part of an adaptive patient monitoring procedure, we suggest using a LoRa-based low-power healthcare

WBAN technology named HeaLoRa. Doctors can monitor vital signs like temperature, blood pressure, heart rate, and oxygen saturation of a patient remotely. In addition, a power consumption optimization technique that prioritizes the reduction of redundant data governs data capture and transmission [27, 36]. A fuzzy logic controller decides how long the system sleeps in its current setup and how quickly data is transmitted based on the Early Warning Score (EWS). Additionally, there is an analytical model of the power required for reliable LoRa transmission. The system architecture of the health care monitoring system of LoRaWAN is shown in Fig. (**2**).

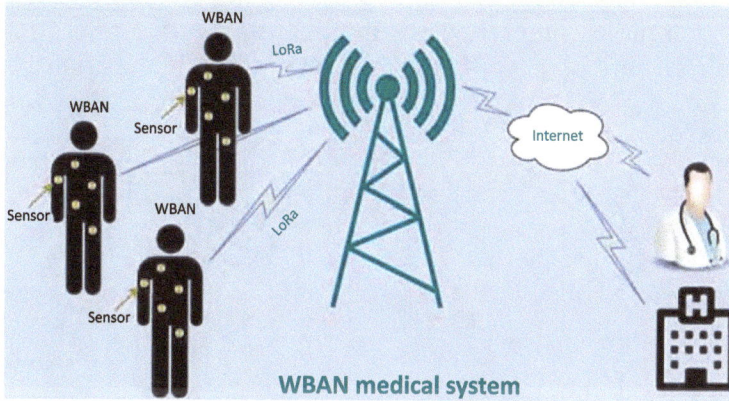

Fig. (2). System architecture of health monitoring system of LoRaWAN.

Machine learning models, on the other hand, are notoriously computationally intensive, necessitating the transfer of obtained data to external cloud servers for inference. This is not ideal from a system's point of view. Tiny Machine Learning (TinyML) is a new subfield of AI that aims to find ways to replace cloud servers with inference devices located closer to the sensing platform [28, 37]. Fig. (**3**) explains the wearable healthcare system. The system captures data from various wearable sensors and transfers it to one of three computing tiers. Cloud services are essential for cloud-based computation to begin. This is typically required because of extensive computing and data storage requirements. Although there are many benefits to sending data to remote servers, there are also possible negatives, such as the necessity for network connectivity, greater threats to privacy and security, and, in certain cases, power consumption restricting usability.

The mobile health (mHealth) industry has relied heavily on wearable technology for the diagnosis, treatment, and rehabilitation of a wide range of diseases and disorders. Parkinson's disease (PD), for example, is a neurodegenerative disease. We can categorize symptoms in PD patients into two groups: those that impair motor function and those that do not. A person living with PD faces significant reductions in quality of life. Despite the lack of a known treatment for Parkinson's disease, an early diagnosis and the provision of suitable supportive services can significantly enhance a patient's capacity to carry out their everyday activities [29]. Only a small number of healthcare facilities currently utilize wearable sensors; in these cases, patients receive a small package in the mail, containing a body-worn sensor, a few weeks before their routine clinical check-up. The sensor devices will upload their collected data to the cloud, process it, summarize it, and send it to the doctor for assessment after a week of use and simple instructions. At this point, the patient will also receive feedback. The doctor's assessment and this information would enhance the physical exam during the clinic visit, enabling the development of a more comprehensive treatment plan. A more efficient meeting and individualized care plan would result from modifying the clinical visit and the systematic diagram of the remote symptom monitoring process shown in Fig. (**4**).

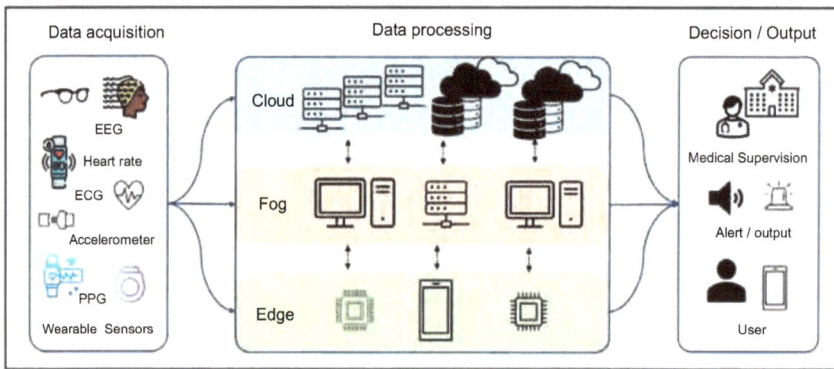

Fig. (3). System Diagram of the wearable healthcare system.

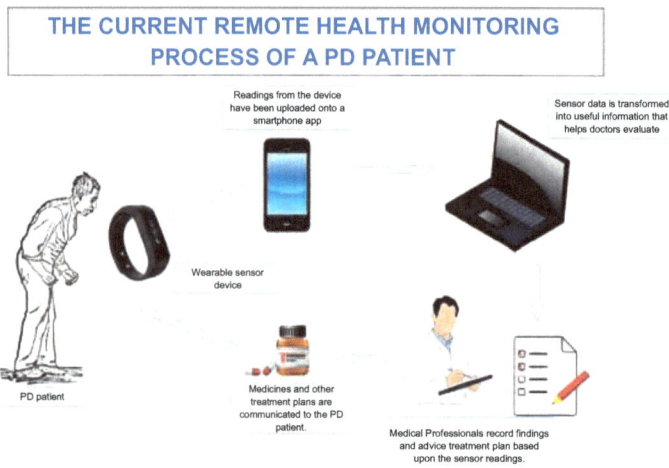

Fig. (4). System Diagram of the remote symptom monitoring process.

Daily monitoring of vital signs related to cardiovascular diseases (CVD), such as electrocardiography (ECG), heart rate monitoring, pulse oximetry (SpO2), and continuous blood pressure measurement, is an integral part of home care for patients with CVD. The authors developed a highly integrated, minimally intrusive sensor to address the lack of a wearable that can simultaneously monitor all these variables. An analog front-end (AFE) integrated chip (IC) in this sensor can pick up one-lead ECG and two-wavelength photoplethysmography (PPG) data [30]. Fig. (**5**) shows an overview of the wearable wireless multimodel vital signs. When worn on the wrist, the sensor's adjustable sensitivity enables it to detect PPG signals in both reflected and transmitted modes. Also, it can pick up both the electrocardiogram (ECG) and the photoplethysmogram (PPG) signals at once (red and infrared). Both a nRF52832 (a Bluetooth mode with an ARM Cortex M4 MCU and a Bluetooth low energy (BLE) transmitter) and an AFE IC (an analog front end) are present in the system.

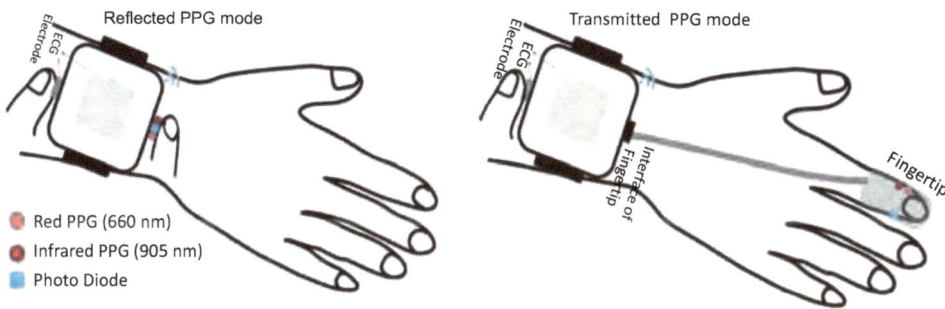

Fig. (5). System Diagram of the remote symptom monitoring process.

The prospect of scaling their heights or wandering their attractive landscapes has drawn many adventurers to the mountains for years. A team of mountaineers carries equipment for navigating steep slopes and rocky terrain. The main control centers monitor the action. Periodic reporting occurs either on an as-needed basis or as a daily summary. If the team encounters an avalanche, the snow will likely bury them alive [31]. A seasoned leader directs the mountaineering team's efforts and ensures that everyone stays safe on the mountain. The master node, complete with satellite modem and LoRa radio module, is in his possession at all times. The other members carry slave nodes equipped only with a LoRa radio module. Each node has its own battery and health-monitoring sensors. This process is assisted by an ATMega2560 microcontroller board and other components. Fig. (**6**) displays the block diagram of the health monitoring system.

To specifically evaluate patients remotely and prevent the premature spread of the COVID-19 pandemic, we need a remote monitoring system. Using the ontology approach with sensory 1D biomedical signals like ECG, PPG, temperature, and accelerometer, an Internet of Things-based remote access and alarm-enabled bio-wearable sensor system may detect COVID-19 at an early stage [32]. The integration of IoT servers allows for remote access and data synchronization *via* communication with the control unit. The one-dimensional biomedical signals are acquired *via* a WIFI module and then processed with artificial intelligence-based image processing methods. After that, the model is tested with the data shown in Fig. (7).

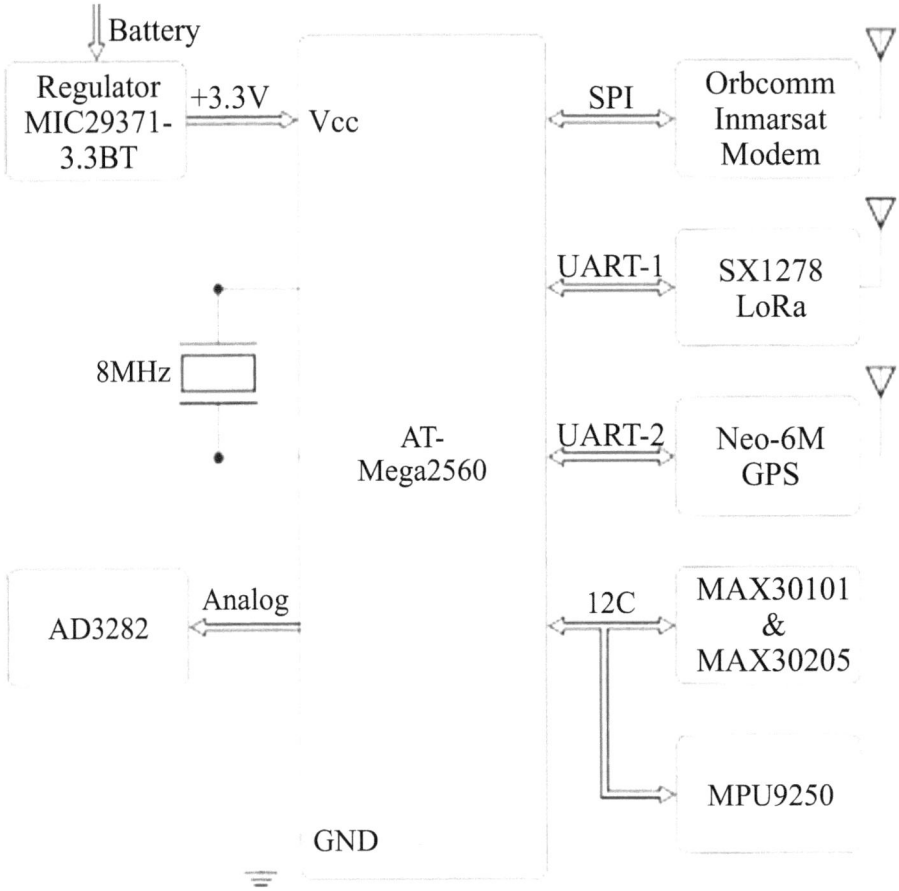

Fig. (6). Block diagram of health monitoring system.

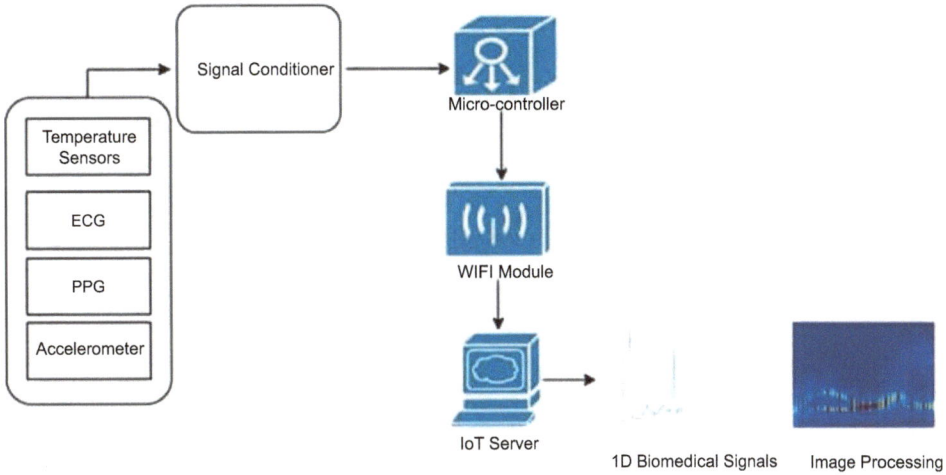

Fig. (7). Architecture diagram of wearable sensor system.

The authors have developed a system based on the Internet of Things (IoT) that uses low-power wireless sensing networks, big data, cloud computing, and smart gadgets to detect indoor falls in elderly individuals. For this, we use a wearable LowPAN device equipped with a three-axis accelerometer to track the location and velocity of senior citizens in real time [33]. A machine learning model on a cutting-edge IoT gateway processes and analyzes the sensor's signals, resulting in optimal fall detection performance. The four main components of the fall detection system are a portable device, a wireless communication network, a smart IoT gateway, and cloud computing, as shown in Fig. (8).

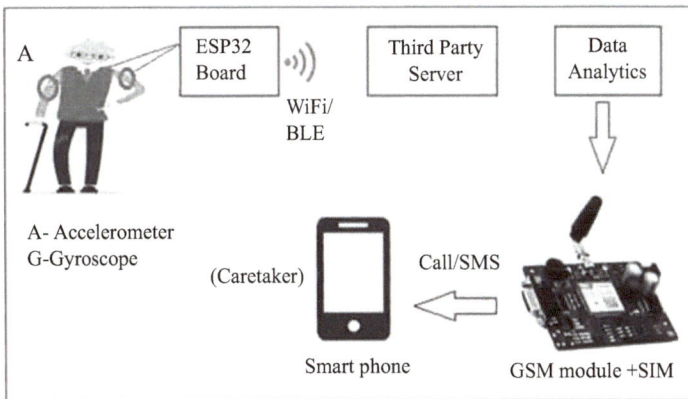

Fig. (8). Overview of fall detection architecture.

The quality of services in healthcare systems can be measured in many ways, including how well they handle data, how quickly they make decisions, how quickly they treat patients, how safe their data is, how well they monitor and control the system in real-time, how well they handle failures, and how good their patients' lives are overall. The goal of this study is to improve healthcare service quality by utilizing IoT and fog computing to optimize both of these factors [34]. Fig. (**9**) shows that the connections made possible by the IoT are extremely useful in a variety of real-world contexts, such as between different kinds of strong contractions, inspecting cameras, and an environmental sensor.

Care that is both more individualized and more comprehensive is what the Internet of Health Things is all about. Because of the wide range of fields in which it has been used, it has captured the interest of many researchers. People who hope to improve their standard of living by using this technology have also taken this tactic [35]. Dynamic electrocardiography, often known as the Holter system, records the heart's electrical activity for a whole day as its foundational premise for long-term heart monitoring. The authors position electrodes in the patient's thorax and then wire them to a separate recording device.

Holter evaluates the effects of various diseases and medications, such as transient ischemic episodes, sinus node disease, Chagas cardiomyopathy, Wolff-Parkinson-White syndrome, post-infarction evaluation, pacemakers, and sinus node disease. Fig. (**10**) shows the overall framework of the heart monitoring system.

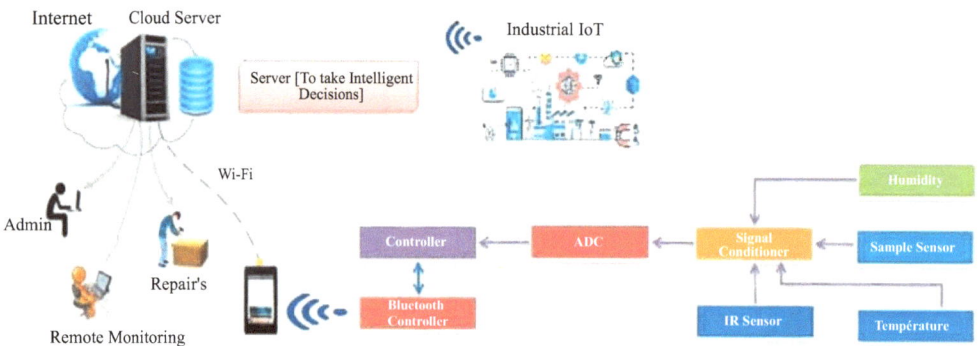

Fig. (9). Architecture diagram of quality healthcare system.

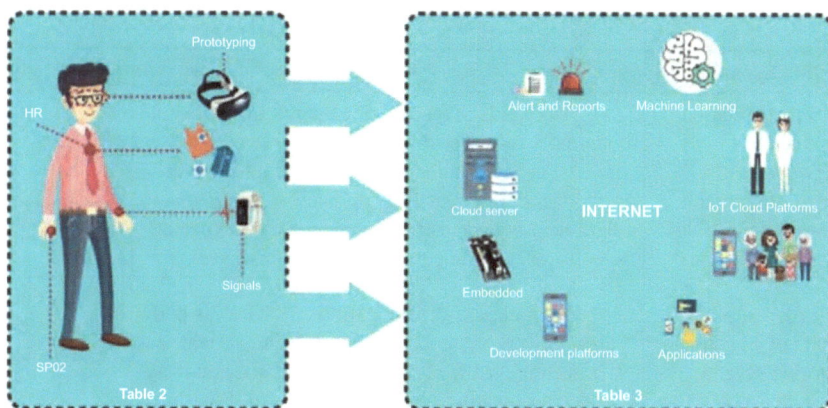

Fig. (10). The overall framework of the heart monitoring system.

CONCLUSION

The Internet of Things (IoT) stands as a cornerstone in the transformation of various industries, propelling them into a new era of connectivity and efficiency. In the realm of transport, IoT-enabled solutions have revolutionized the way we conceive and navigate cities, ultimately paving the way towards the realization of smart cities. Through real-time data acquisition and analysis, IoT empowers transport systems to become more adaptive, sustainable, and responsive to the evolving needs of urban environments. Furthermore, the integration of IoT in healthcare has ushered in a paradigm shift in patient care. The transition from conventional healthcare models to e-healthcare-based systems has facilitated the remote monitoring of patients' conditions, transcending geographical barriers. This breakthrough allows for immediate response and intervention, particularly in emergencies, and ultimately saves invaluable lives. The management of hospitals and the provision of timely, personalized care particularly underscore the significance of IoT in healthcare. Healthcare professionals can remotely track patients' vital signs with real-time monitoring, precisely timing and adapting medical interventions to meet specific demands. This relieves pressure on overworked healthcare institutions while simultaneously improving patient outcomes. Essentially, the IoT is not merely a technological advancement but a transformative force with the potential to reshape our societies for the better. By harnessing the power of real-time data, we unlock unprecedented levels of efficiency, safety, and precision across various domains. We must responsibly and ethically leverage IoT technologies as we progress in this interconnected era, ensuring equitable distribution of benefits for the betterment of humanity as a whole. Table **1** highlights the merits and demerits of the existing work carried out in healthcare.

Table 1. Existing system merits and demerits.

Existing System	Merits	Demerits
MedAi: A smartwatch-based application framework for the prediction of common diseases using machine learning [26]	This system aids in recognizing symptoms that could potentially result in severe illnesses or even fatality, as well as inspiring individuals to adopt a health-conscious way of life. The approach is anticipated to significantly decrease the incidence of unexpected death and undiagnosed terminal diseases.	Requires robust data
Energy consumption improvement of a healthcare monitoring system: application to LoRaWAN [27]	The system depicts the utilization of a wireless body area network system in a medical context, where data is conveyed *via* LoRa technology to a gateway and subsequently sent to the doctor *via* IP-based technology.	The impact of LoRaWAN packet size on device energy efficiency has not been adequately taken into account.
Embedded machine learning using microcontrollers in wearable and ambulatory systems for health and care applications: a review [28, 37]	Microcontrollers possess integrated processing capabilities that facilitate the capture of real-time data and enable local decision-making. This reduces the necessity for constant contact with centralized servers.	Issues regarding hardware heterogeneity, MCU architectures, resource constraints, limited memory management, and software interoperability between devices.
Parkinson's disease management *via* wearable sensors: a systematic review [29]	Wearables have the potential to objectively and continuously monitor movement patterns or physiological variables in laboratory, hospital, and everyday situations, which is particularly useful for monitoring neurological illnesses.	Limited to the specific disease.
Wearable multimodal vital sign monitoring sensor with a fully integrated analog front end [30]	A wearable sensor capable of monitoring the four most critical physiological parameters in patients with cardiovascular disease (CVD) was developed. The parameters of this sensor are derived from the electrocardiogram (ECG) and two-wavelength photoplethysmography (PPG) signals, which are detected by an integrated analog front-end (AFE) integrated circuit (IC).	Limited to the specific disease.

REFERENCES

[1]   Tufan TB, Rhein L, Guler U. Implementation techniques for transcutaneous carbon dioxide monitoring: approaches for wearable smart health applications. IEEE Trans Biomed Eng 2023. [PMID: 37812542]

[2]   Wang J, Viciano-Tudela S, Parra L, Lacuesta R, Lloret J. Evaluation of suitability of low-cost gas sensors for monitoring indoor and outdoor urban areas. IEEE Sens J 2023; 23(18): 20968-75. [http://dx.doi.org/10.1109/JSEN.2023.3301651]

[3]   Incel OD, Bursa SO. On-device deep learning for mobile and wearable sensing applications: a review.

IEEE Sens J 2023.

[4]   Ahamad S. A low-cost methodology for developing personal health monitoring devices to examine psychological states and the impact of exercises. International Journal of Information Technology 2023; 16: 1-3.
[http://dx.doi.org/10.1007/s41870-023-01473-7]

[5]   Mani N, Haridoss P, George B. Smart suspenders with sensors and machine learning for human activity monitoring. IEEE Sens J 2023; 23(9): 10159-67.
[http://dx.doi.org/10.1109/JSEN.2023.3263231]

[6]   JG AK, Aparna V, Varghese A, Sebastian D, Johnson N. Low-Cost IOT Based Healthcare Management System In Remote Areas. In 2022 IEEE 19th India Council International Conference (INDICON). 2022 Nov 24 (pp. 1-6). IEEE.

[7]   Maguire G, Chen H, Schnall R, Xu W, Huang MC. Smoking cessation system for preemptive smoking detection. IEEE Internet Things J 2022; 9(5): 3204-14.
[http://dx.doi.org/10.1109/JIOT.2021.3097728] [PMID: 36059439]

[8]   El Zouka HA, Hosni MM. Secure IoT communications for smart healthcare monitoring system. Internet of Things 2021; 13: 100036.
[http://dx.doi.org/10.1016/j.iot.2019.01.003]

[9]   Kim J, Khan S, Wu P, *et al.* Self-charging wearables for continuous health monitoring. Nano Energy 2021; 79: 105419.
[http://dx.doi.org/10.1016/j.nanoen.2020.105419]

[10]   Kapoor B, Nagpal B, Alharbi M. Secured healthcare monitoring for remote patient using energy-efficient IoT sensors. Comput Electr Eng 2023; 106: 108585.
[http://dx.doi.org/10.1016/j.compeleceng.2023.108585]

[11]   Raja GB, Chakraborty C. Internet of things based effective wearable healthcare monitoring system for remote areas. In Implementation of Smart Healthcare Systems using AI, IoT, and Blockchain, Academic Press, 2023; pp. 193-218.
[http://dx.doi.org/10.1016/B978-0-323-91916-6.00004-7]

[12]   Selvakumar K, Vinodh Kumar E, Sailesh M, *et al.* Realtime PPG based respiration rate estimation for remote health monitoring applications. Biomed Signal Process Control 2022; 77: 103746.
[http://dx.doi.org/10.1016/j.bspc.2022.103746]

[13]   Yan W, Ma C, Cai X, Sun Y, Zhang G, Song W. Self-powered and wireless physiological monitoring system with integrated power supply and sensors. Nano Energy 2023; 108: 108203.
[http://dx.doi.org/10.1016/j.nanoen.2023.108203]

[14]   Komkova MA, Eliseev AA, Poyarkov AA, *et al.* Simultaneous monitoring of sweat lactate content and sweat secretion rate by wearable remote biosensors. Biosens Bioelectron 2022; 202: 113970.
[http://dx.doi.org/10.1016/j.bios.2022.113970] [PMID: 35032921]

[15]   Gnana Sheela K, Rose Varghese A. Machine learning-based health monitoring system. Mater Today Proc 2020; 24: 1788-94.
[http://dx.doi.org/10.1016/j.matpr.2020.03.603]

[16]   Indrakumari R, Poongodi T, Suresh P, Balamurugan B. The growing role of the Internet of Things in healthcare wearables. In Emergence of Pharmaceutical Industry Growth with Industrial IoT Approach, Academic Press, 2020; pp. 163-194.
[http://dx.doi.org/10.1016/B978-0-12-819593-2.00006-6]

[17]   Wu T, Wu F, Qiu C, Redoute JM, Yuce MR. A rigid-flex wearable health monitoring sensor patch for IoT-connected healthcare applications. IEEE Internet Things J 2020; 7(8): 6932-45.
[http://dx.doi.org/10.1109/JIOT.2020.2977164]

[18]   Mamun MAA, Yuce MR. Sensors and systems for wearable environmental monitoring toward IoT-enabled applications: A review. IEEE Sens J 2019; 19(18): 7771-88.

[http://dx.doi.org/10.1109/JSEN.2019.2919352]

[19]   Bharathiraja N, Indhuja RB, Krishnan PA, Anandhan S, Hariprasad S. Real-time fall detection using esp32 and amg8833 thermal sensor: a non-wearable approach for enhanced safety. 2023 Second International Conference on Augmented Intelligence and Sustainable Systems (ICAISS). 1732-6.
[http://dx.doi.org/10.1109/ICAISS58487.2023.10250598]

[20]   Haghi M, Neubert S, Geissler A, *et al.* A flexible and pervasive IoT-based healthcare platform for physiological and environmental parameters monitoring. IEEE Internet Things J 2020; 7(6): 5628-47.
[http://dx.doi.org/10.1109/JIOT.2020.2980432]

[21]   Otta M. Towards a health software supporting platform for wearable devices. Procedia Comput Sci 2022; 210: 112-5.
[http://dx.doi.org/10.1016/j.procs.2022.10.126]

[22]   Virginia Anikwe C, Friday Nweke H, Chukwu Ikegwu A, *et al.* Mobile and wearable sensors for data-driven health monitoring system: State-of-the-art and future prospect. Expert Syst Appl 2022; 202: 117362.
[http://dx.doi.org/10.1016/j.eswa.2022.117362]

[23]   Koene I, Klar V, Viitala R. IoT connected device for vibration analysis and measurement. HardwareX 2020; 7: e00109.
[http://dx.doi.org/10.1016/j.ohx.2020.e00109] [PMID: 35495203]

[24]   Alshamrani M. IoT and artificial intelligence implementations for remote healthcare monitoring systems: A survey. Journal of King Saud University - Computer and Information Sciences 2022; 34(8): 4687-701.
[http://dx.doi.org/10.1016/j.jksuci.2021.06.005]

[25]   Khan MA. An IoT framework for heart disease prediction based on MDCNN classifier. IEEE Access 2020; 8: 34717-27.
[http://dx.doi.org/10.1109/ACCESS.2020.2974687]

[26]   Himi ST, Monalisa NT, Whaiduzzaman MD, Barros A, Uddin MS. MedAi: a smartwatch-based application framework for the prediction of common diseases using machine learning. IEEE Access 2023; 11: 12342-59.
[http://dx.doi.org/10.1109/ACCESS.2023.3236002]

[27]   Taleb H, Nasser A, Andrieux G, Charara N, Cruz EM. Energy consumption improvement of a healthcare monitoring system: application to LoRaWAN. IEEE Sens J 2022; 22(7): 7288-99.
[http://dx.doi.org/10.1109/JSEN.2022.3150716]

[28]   Diab MS, Rodriguez-Villegas E. Embedded machine learning using microcontrollers in wearable and ambulatory systems for health and care applications: a review. IEEE Access 2022; 10: 98450-74.
[http://dx.doi.org/10.1109/ACCESS.2022.3206782]

[29]   Mughal H, Javed AR, Rizwan M, Almadhor AS, Kryvinska N. Parkinson's disease management *via* wearable sensors: a systematic review. IEEE Access 2022; 10: 35219-37.
[http://dx.doi.org/10.1109/ACCESS.2022.3162844]

[30]   Wang Y, Miao F, An Q, Liu Z, Chen C, Li Y. Wearable multimodal vital sign monitoring sensor with fully integrated analog front end. IEEE Sens J 2022; 22(13): 13462-71.
[http://dx.doi.org/10.1109/JSEN.2022.3177205]

[31]   Garg RK, Bhola J, Soni SK. Healthcare monitoring of mountaineers by low power Wireless Sensor Networks. Inform Med Unlocked 2021; 27: 100775.
[http://dx.doi.org/10.1016/j.imu.2021.100775]

[32]   Sharma N, Mangla M, Mohanty SN, *et al.* A smart ontology-based IoT framework for remote patient monitoring. Biomed Signal Process Control 2021; 68: 102717.
[http://dx.doi.org/10.1016/j.bspc.2021.102717]

[33]   Kulurkar P, Kumar Dixit C, Bharathi VC, Monikavishnuvarthini A, Dhakne A, Preethi P. AI-based

elderly fall prediction system using wearable sensors: A smart home-care technology with IoT. Measurement. Sensors (Basel) 2023; 25: 100614.

[34]   Gowda D, Sharma A, Rao BK, *et al.* Industrial quality healthcare services using the Internet of Things and fog computing approach. Measurement. Sensors (Basel) 2022; 24: 100517.

[35]   Santos MAG, Munoz R, Olivares R, Filho PPR, Ser JD, Albuquerque VHC. Online heart monitoring systems on the internet of health things environments: A survey, a reference model and an outlook. Inf Fusion 2020; 53: 222-39.
        [http://dx.doi.org/10.1016/j.inffus.2019.06.004]

[36]   Llu'ıs C, Gomez C, Vidal R. Understanding the impact of packet size on the energy efficiency of LoRaWAN. J Commun Netw (Seoul) 2023; 25: 6.

[37]   Cormac D. Fay, brian corcoran, and dermot diamond. Green IoT Event Detection for Carbon-Emission Monitoring in Sensor Networks. Sensors (Basel). 2024.

CHAPTER 6

Data Science and Data Analytics for Healthcare: Transforming Patient Care Through a Design Thinking Approach to Data Science

M. Kavibharathi[1,*], J. Sumitha[1] and **S. Muthu Vijaya Pandian[2]**

[1] *Department of Computer Science, Dr. SNS Rajalakshmi College of Arts and Science, Tamil Nadu, India*

[2] *Department of EEE, SNS College of Technology, TamilNadu, India*

Abstract: Design thinking is essential for the successful integration of data science in healthcare. The healthcare industry is undergoing a profound transformation driven by the power of data. In this book chapter, we delve into the pivotal role of data science in healthcare, exploring its importance, ethical considerations, and various stages of data collection, pre-processing, analysis, and visualization. With the potential to revolutionize patient care, reduce costs, and drive medical innovations, data in healthcare holds immense promise. The chapter highlights the critical role of data quality, integration, and data visualization in healthcare analytics, emphasizing their impact on patient outcomes and healthcare decision-making. It explores predictive modeling, including supervised learning and model evaluation, showcasing their applications in risk prediction and disease subtyping. Unsupervised learning and anomaly detection are discussed in the context of uncovering hidden patterns and irregularities in healthcare data. Text analytics and natural language processing emerge as essential tools for mining clinical notes and understanding patient sentiment. As healthcare evolves into a data-driven field, data visualization and dashboard design are discussed as tools for conveying complex data in a comprehensible manner. The chapter highlights the importance of design thinking in creating visualizations that are intuitive and easy to interpret for healthcare professionals. The future of healthcare analytics is explored, including AI advancements, precision medicine, and the critical role of telemedicine. Additionally, the chapter addresses ethical and regulatory considerations surrounding data privacy, informed consent, and regulatory compliance. Design thinking principles can guide the development of user-friendly privacy policies and consent forms. This chapter offers a comprehensive perspective on the challenges and opportunities in the field of data science in healthcare, highlighting its potential to revolutionize patient care, improve outcomes, and safeguard the rights and privacy of individuals in a data-driven healthcare landscape.

---

* **Corresponding author M. Kavibharathi:** Department of Computer Science, Dr. SNS Rajalakshmi College of Arts and Science, Tamil Nadu, India; E-mail: kavi.m.bharathi@gmail.com

Sivakumar Rajagopal, Prakasam P., Konguvel E., Shamala Subramaniam, Ali Safaa Sadiq Al Shakarchi & B. Prabadevi (Eds.)

Keywords: Anomaly detection, Big data, Data science, Data integration, Data visualization, Healthcare, Predictive modeling, Precision medicine, Text analytics, Unsupervised learning.

INTRODUCTION

In the modern era, healthcare is on the brink of a data revolution. The advent of data science and data analytics has reshaped the landscape of the healthcare industry, offering unprecedented opportunities to enhance patient care, optimize operational efficiency, and propel medical research to new heights. In this book chapter, we embark on a journey into the realm of data science and data analytics in healthcare, exploring the profound impact these technologies have on the delivery of healthcare services, clinical decision-making, and the overall well-being of patients.

Data has become the lifeblood of the healthcare sector, flowing from diverse sources such as electronic health records (EHRs), medical devices, genomic data, and patient-generated data. The sheer volume and complexity of this healthcare data necessitate advanced analytical techniques to unlock its potential. Proper management and analysis of this data have the power to uncover valuable insights, predict disease outbreaks, personalize treatment plans, and improve the overall quality of healthcare services.

However, harnessing the potential of data in healthcare is not without its challenges. The ethical considerations surrounding data privacy and security are paramount, as healthcare data is sensitive and personal. Regulatory compliance, informed consent, and data protection are critical aspects that require careful attention. Additionally, the integration and interoperability of disparate data sources pose technical challenges, demanding innovative solutions for seamless data flow within healthcare systems.

This chapter takes a comprehensive journey through the key facets of data science and data analytics in healthcare. We explore the importance of data quality, data integration, and data visualization in the context of healthcare analytics. We delve into predictive modeling techniques that empower healthcare professionals with the ability to foresee critical health events. The chapter also addresses the power of unsupervised learning and anomaly detection in identifying hidden patterns and irregularities within healthcare data.

Text analytics and natural language processing emerge as invaluable tools for mining clinical notes and understanding patient sentiments, enabling the development of personalized healthcare plans and improved patient outcomes. The role of data visualization and dashboard design in conveying complex healthcare data in an understandable manner is discussed, highlighting their significance in aiding healthcare decision-makers.

The future of healthcare analytics is examined through the lens of AI advancements, precision medicine, and the growing influence of telemedicine. We also delve into the ethical and regulatory considerations that underpin data science in healthcare, safeguarding patient privacy, ensuring informed consent, and adhering to regulatory compliance. As we progress through this chapter, we will gain a deeper understanding of the transformative potential of data science and data analytics in healthcare.

DATA SCIENCE IN HEALTHCARE

Importance of Data in Healthcare

Data plays a crucial role in healthcare as it has the potential to improve patient care, lower costs, and revolutionize medical therapies. The healthcare industry generates massive amounts of data from various sources such as hospital records, medical examinations, and biomedical research [1, 2]. Proper management and analysis of this big data are essential to derive meaningful information and improve public health [3, 4]. Data analytics is becoming an escalating tool in healthcare systems, allowing for descriptive, diagnostic, predictive, and prescriptive analysis [5]. By using data efficiently, healthcare organizations can monitor performance, prevent hospitalizations, combat opioid abuse, improve antimicrobial stewardship, and reduce pharmaceutical spending. However, there are challenges in data acquisition, integration, and usability that need to be addressed for effective implementation [6 - 9]. Implementing better data management and integration can bridge gaps in care, improve data analysis, and contribute to a healthier population. In Fig. (1) , which can be seen in the following section, a flowchart vividly illustrates the intricate data flow within healthcare systems, underlining the interdependence of diverse data sources and the processes of analysis.

Ethical Considerations

Ethical considerations of data in healthcare are crucial due to the sensitive nature of health information and the potential risks to individuals' rights and opportunities. Reasonable security standards are needed to protect electronic health records (EHRs) [10]. Healthcare informatics professionals should be informed of their rights, duties, and responsibilities, and have guidelines and ethical tutoring to prevent conflict or misconduct in handling patient information [11]. The availability of diverse sources of health data and the advancements in data science raise ethical and regulatory challenges in the use of biomedical big data [12]. The use of informatics devices and software in healthcare facilitates communication and information flow but also increases the risk of information misuse [13]. The generation and analysis of health data offer significant opportunities for knowledge generation, medical practice improvement, and innovation, but ethical considerations must be taken into account [14].

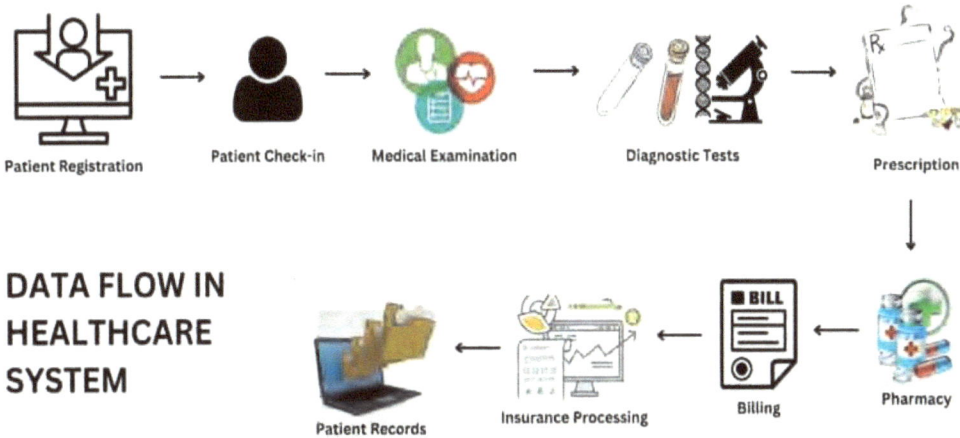

Fig. (1). Flowchart depicting data flow in healthcare systems.

DATA COLLECTION AND PRE-PROCESSING

Data Sources

Data sources in data science in healthcare include time-series health check-up data [15], medical big data collected from various sources such as disease and drug registries, electronic health records, claims and billing data, and census data [16], a wide range of health data including computerized physician order entry, electronic medical records, clinical notes, medical images, genomic data, and clinical decision support systems [17], patient data collected using electronic devices like smart contact lenses [18], and information available through computer abstract hospital databases [19]. These data sources provide valuable information for conducting analyses, developing personalized simulation models, and improving healthcare outcomes.

Data Quality

Data quality is a critical aspect of data science in healthcare. The use of data cataloging tools allows for the collection and organization of health information [16]. However, the large amount of data does not always guarantee quality, especially in the medical field where the consequences of poor data quality can be detrimental to patients. To ensure data quality, systematic and rigorous processes are needed, including qualitative or statistical control processes and indicators to identify data gaps or anomalies. In precision healthcare, data science plays a key role in analyzing health-related big data and constructing personalized simulation models to improve individual health conditions [10]. The QVida+ Project utilizes data science algorithms to estimate patients' quality of life without constant questionnaire filling. Measuring data quality in electronic health records is crucial for research purposes and can benefit patient care and quality of life.

Data Integration

Data integration plays a crucial role in data science in healthcare. It involves combining multiple types of data from different sources into a single infrastructure, allowing for efficient and effective healthcare analytics. The healthcare sector faces challenges in integrating diverse medical data from heterogeneous sources, but data integration enriches the data and enhances its value. It also paves the way for predicting diseases or outbreaks. Various data integration technologies and tools have been developed for the healthcare domain, and future research directions in the integration of big healthcare data are being explored. Data integration is considered a key component in achieving integrated care, as it allows for the combination of different types of data and facilitates access, editing, and contribution to electronic health records. It is essential for improving data analysis, bridging gaps in care, and leveraging technology for patients' general well-being. In the diagram provided in Fig. (**2**) , you will see a visual representation of the data collection methods we have discussed in this section. This illustrative depiction offers a clear perspective on the diversity of data sources and the intricacies of their integration, highlighting the complexity of data within healthcare analytics.

EXPLORATORY DATA ANALYSIS IN HEALTHCARE

Descriptive Statistics

Descriptive statistics play a crucial role in data science in the field of healthcare. They are used to analyze and summarize data, providing insights into various aspects of health research and patient care. Descriptive statistics help in understanding the distribution and variability of data, as well as measuring central tendency and dispersion [20]. They are important at every stage of a research study, from design issues to reporting of results [21 - 24]. In the rapidly developing field of medicine and healthcare service, data analysis is essential for improving workflow management, patient care, and predicting the development of diseases [25]. Data science enables the analysis of health- related big data, allowing for personalized healthcare plans and the construction of simulation models for individual health conditions.

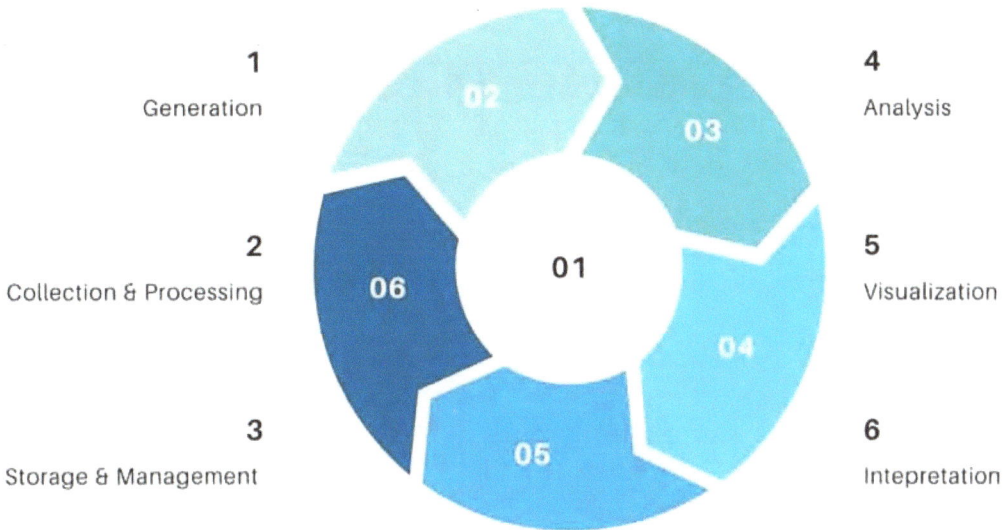

1		4
Generation	02 / 03	Analysis
2		5
Collection & Processing	06 / 01	Visualization
3		6
Storage & Management	05 / 04	Intepretation

Fig. (2). Data collection methods.

Data Visualization

Data visualization plays a crucial role in data science in healthcare. It helps in meaningful and reliable data evaluation, allowing doctors, healthcare staff, and medical specialists to generate reports, analyze trends, and identify patterns and correlations. Visualization of data points and interactive dashboards enable quick analysis of large datasets, saving time and potentially saving lives. The healthcare industry is increasingly becoming data-driven, and advances in computing have made it possible to collect and analyze vast amounts of digital health data. Leveraging data visualization in healthcare improves patient outcomes, enhances planning and policy, and helps in critical decision-making. It is an efficient and effective approach to communicating information to patients, delivering clinical insights, and optimizing healthcare systems [26 - 27], which is clearly shown in Fig. (3).

PREDICTIVE MODELING IN HEALTHCARE

Supervised Learning

Supervised learning is a fundamental concept in data science within the healthcare domain. It involves leveraging historical data to establish relationships and make predictions on new, unseen data. Traditional supervised learning techniques, including logistic regression and Cox proportional hazard models, have been fixtures in the field of medicine for many years and are valued for their interpretability. However, the modern era of machine learning has introduced algorithms that prioritize making predictions as accurately as possible, often at the expense of interpretability and represented in Fig. (**4**). Supervised learning has been instrumental in the development of risk prediction models for specific diseases, while unsupervised learning has been applied to unearth previously unknown subtypes of diseases [28, 29].

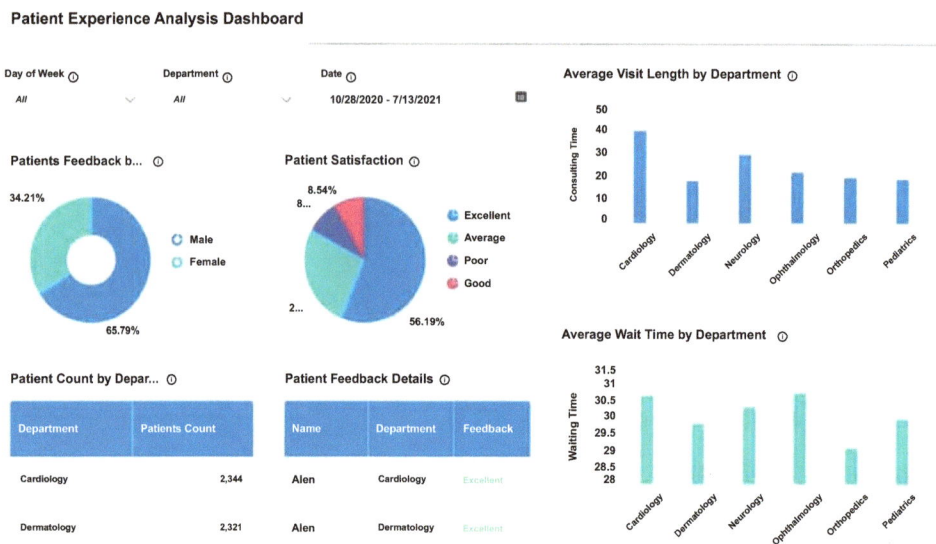

Fig. (3). Data visualization example.

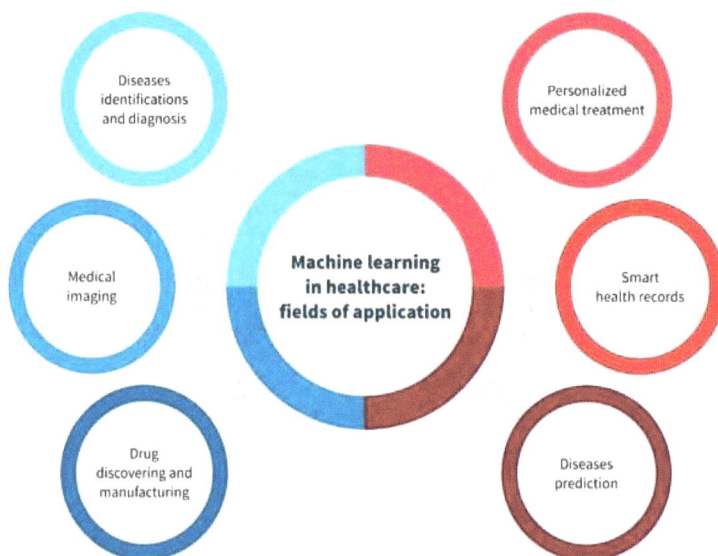

Fig. (4). ML in healthcare.

Model Evaluation

Model evaluation stands as a pivotal aspect of data science in healthcare, demanding the creation and utilization of reliable models [30]. Decision-analytic models serve as common tools for evaluating the value of new healthcare technologies and informing decisions related to resource allocation. Nevertheless, concerns regarding decision-analytic modeling processes persist. One substantial concern pertains to the effects of different structural choices for models on the comparability and accuracy of economic evaluations [31]. The development of disease-specific reference models has the potential to address this concern. Furthermore, data science plays a vital role in precision healthcare by facilitating the analysis of extensive health-related datasets. Personalized simulation models of health conditions can be constructed and utilized to modify individual behaviors, thus contributing to precision healthcare. In conclusion, the establishment and application of dependable models, coupled with the analysis of large health-related datasets, are indispensable components of effective model evaluation in data science in healthcare.

Table 1. Model performance table of diabetes risk prediction.

Metric	Value
Area under the curve (AUC)	0.85
Sensitivity (True positive rate)	0.78
Specificity (True negative rate)	0.90
Accuracy	0.86
Precision (Positive predictive value)	0.82
F1 score	0.80
Matthews correlation coefficient	0.65
Brier score	0.12
Log-loss	0.36

CLUSTERING AND UNSUPERVISED LEARNING

Unsupervised Learning

Unsupervised learning in data science has been applied in healthcare to discover unknown disease subtypes. Machine learning techniques, such as unsupervised learning, have the advantage of adaptability and flexibility compared to traditional biostatistical methods, making them deployable for tasks such as risk stratification, diagnosis, classification, and survival prediction [32]. The digitization of healthcare data has led to an increase in processes and services in the healthcare sector, and machine learning is the key enabling technology to extract insights from this vast amount of data. Unsupervised learning algorithms can analyze and assess large amounts of complex healthcare information, providing opportunities for data-driven insights and improving the quality of care delivery [33]. By leveraging unsupervised learning, healthcare professionals can gain valuable knowledge from diverse and fragmented healthcare data sources, leading to more robust results across populations [34] which is shown in Fig. (5).

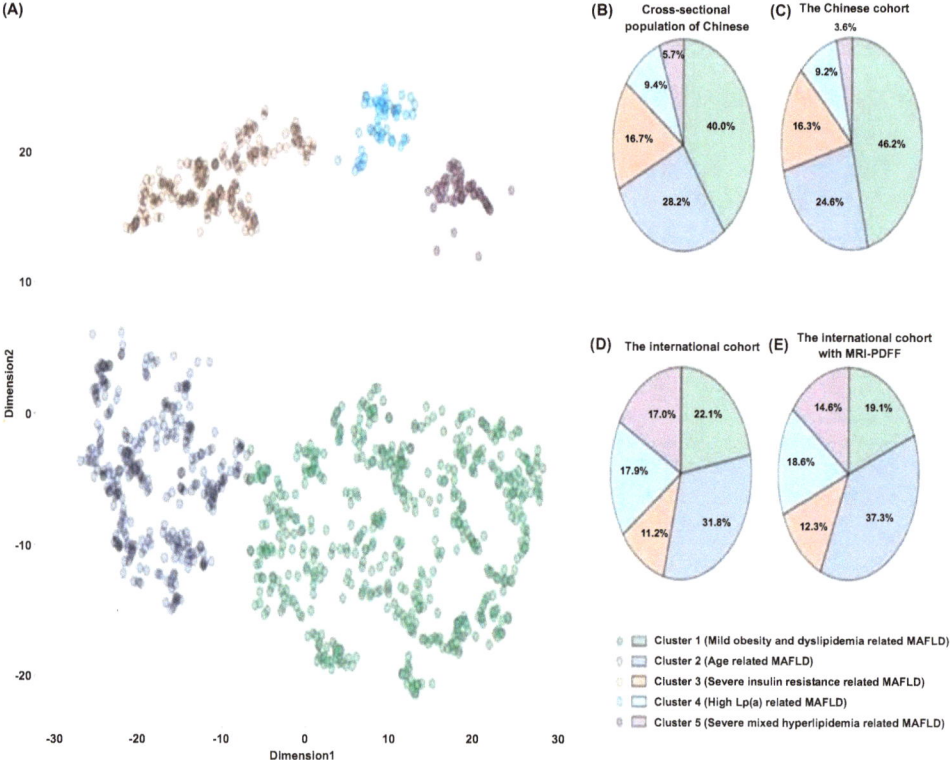

Fig. (5). Cluster visualizations.

Anomaly Detection

Anomaly detection in healthcare data is an important problem with various applications such as detecting anomalous readings, patient health conditions, health insurance fraud, and fault detection in mechanical components. Several state-of-the-art anomaly detection algorithms have been compared, and the isolation-based method iForest has shown better overall performance in terms of AUC and runtime [35]. Traditional detection techniques for big data are complex, so fuzzy logic-based anomaly detection has been introduced to achieve efficient data analysis with greater accuracy and reduced training time [36]. A big data-based medical data anomaly detection method has been proposed, which utilizes technologies such as relationship network analysis, data standardization, and data update to detect abnormal card swiping behaviors and help standardize the use of medical insurance cards [38]. A new powerful method of image anomaly detection has been introduced, which outperforms state-of-the-art approaches in complex medical image analysis tasks [39]. An improved approach for anomaly detection in healthcare has been proposed, which takes into account personal status based on latent factors and utilizes the Hidden Markov Model (HMM) and Hotelling's theory for state estimation and statistical detection of abnormal data. An example of how Anomaly Detection Results are made is shown in Fig. (6).

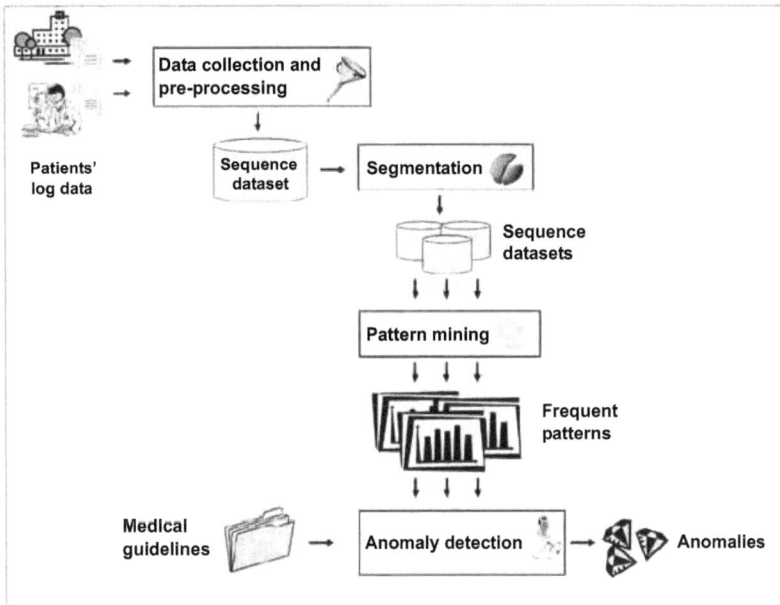

Fig. (6). Anomaly detection results.

TEXT ANALYTICS AND NLP IN HEALTHCARE

Clinical Notes

Clinical notes play a crucial role in data science in healthcare. They provide abundant information about patient health and can be used for various purposes such as personalized health care plans, pattern discovery, and predictive modeling. Clinical notes are often recorded in free-text format during patient visits and contain valuable insights that may not be captured by structured information alone. Machine learning models can be trained on clinical notes to improve readmission prediction and in-hospital mortality prediction. However, the quality of medical notes is a concern, and it is important to select the most valuable information from the notes to achieve accurate predictions. Overall, clinical notes are a valuable source of information for data science in healthcare [40].

Sentiment Analysis

Sentiment analysis in data science plays a crucial role in the healthcare industry. It involves classifying and categorizing user opinions based on their expressions of feelings towards specific data [3]. Social media platforms like Twitter are widely used for sentiment analysis in healthcare. Twitter data can be collected in real time using tools like Apache Flume [50]. The analysis of sentiments on social media platforms like Twitter can provide insights into the public perception of healthcare workers [5]. Sentiment analysis can also be applied to medical datasets to understand patients' perspectives and improve healthcare services [4]. It can help in identifying patients' concerns and devising strategies for their resolution [6]. Sentiment analysis in healthcare has the potential to enhance treatment standards and improve patient outcomes. It can also be used to analyze attitudes and feelings of the global society towards specific healthcare goods, people, or thoughts. An example of a sentimental analysis of COVID-19 is shown in Fig. (7) for a better understanding.

DATA VISUALIZATION AND DASHBOARDS

Dashboard Design

Dashboard design in data science in healthcare involves transforming fragmented data into meaningful information that can be easily accessed and understood. Dashboards can support decision-making and inform quality improvement activities in healthcare services. Designing effective dashboards is a challenging task due to the complex nature of healthcare organizations and the distinct needs of end users. The theoretical underpinnings of healthcare dashboards are poorly characterized, and there is a need for a clearer understanding of how dashboards are developed, implemented, and evaluated. Human-centered design approaches can be used to develop data dashboards in the healthcare sector, taking into account stakeholder requirements, expert reviews, user evaluations, and usability evaluations [24]. A framework for designing healthcare dashboards through technical architecture has been proposed, which includes principles and guidelines for multilayered system architecture. Privacy protection is also an important aspect of dashboard design in healthcare, with the use of privacy dashboards and other data protection methods being suggested. Fig. (8) portrays the dashboard of healthcare.

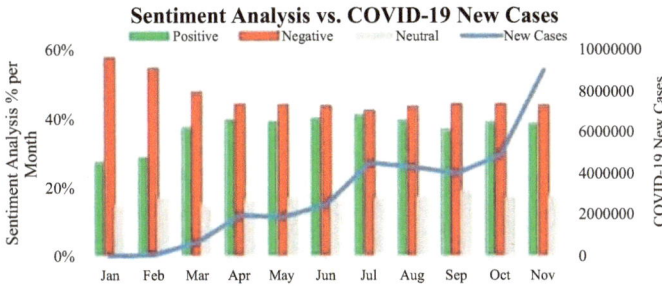

Fig. (7). Sentiment analysis chart.

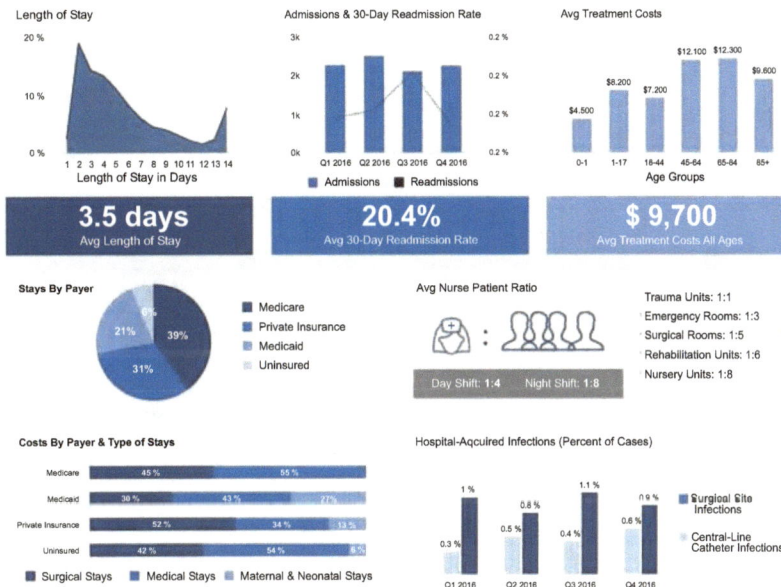

Fig. (8). Example of healthcare dashboard.

Real-time Monitoring

Real-time monitoring in data science plays a crucial role in healthcare. Continuous data monitoring through digital devices and wearables has shown significant benefits in improving patient outcomes and supporting healthcare decision-making. IoT-based patient monitoring systems, powered by the Internet of Things, have been developed to provide real-time data on a patient's physiological state, allowing medical personnel to monitor patients remotely and provide better care [41]. These systems utilize wearable sensor nodes to monitor metrics such as heart rate, temperature, and blood pressure, with the data being stored in the cloud for analysis [42]. Real-time data analysis techniques, particularly machine learning methods, have been widely investigated in health monitoring systems, showing promising results in health diagnosis, early symptom detection, and disease prediction [43]. However, the choice of data analysis techniques may vary depending on the specific health domain, and further research is needed to identify the most suitable methods for each context. And Fig. (**9**) gives an example of real-time monitoring.

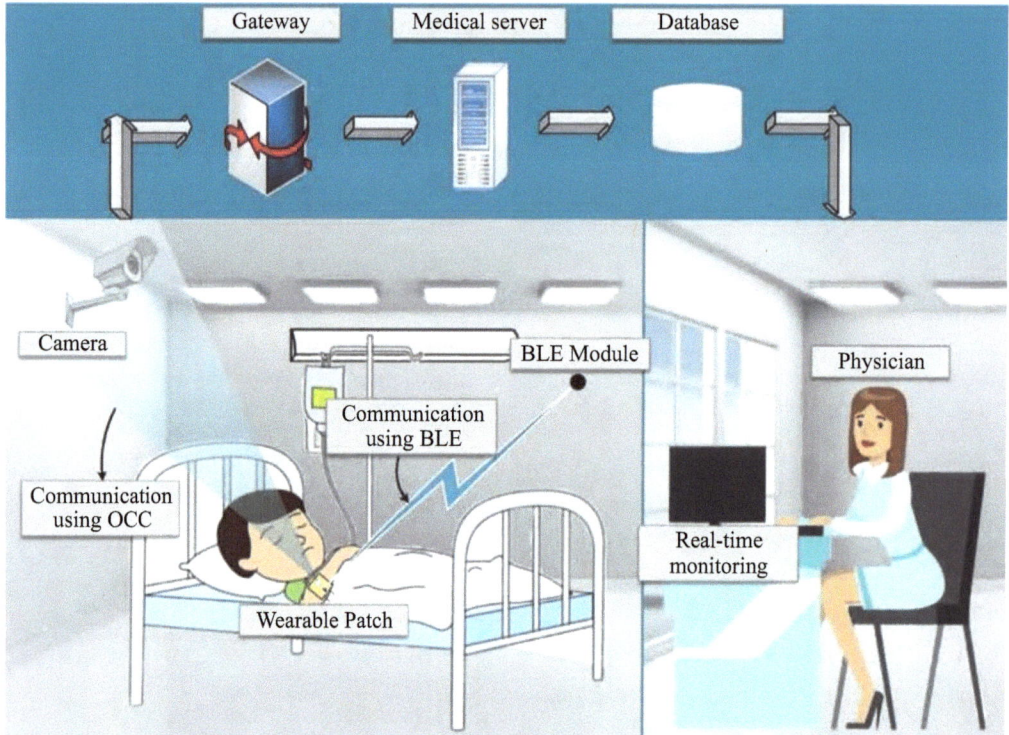

Fig. (9). Visual example of real-time monitoring.

FUTURE TRENDS IN HEALTHCARE ANALYTICS

AI Advancements

AI advancements in data science have had a significant impact on healthcare. AI-powered techniques are being used to improve various aspects of medical services, including management tasks, diagnostic procedures, image processing, machine digitalization, and patient care. The use of AI in healthcare has the potential to enhance care quality, lower costs, and improve patient outcomes [42]. AI is being utilized to automate existing healthcare tasks, increase access to care, and augment healthcare capabilities [43]. However, there are concerns regarding the safety and security of electronic health records and patient information, which need to be addressed. The efficient usage of AI in healthcare can lead to improved decision-making, predictive modelling, time consumption, and resource management. Overall, AI advancements in data science have the potential to revolutionize the healthcare industry by leveraging big data analytics and improving various aspects of medical services. The current AI trends is shown in the Fig. (**10**).

Fig. (10). Current AI trends in healthcare industry.

Precision Medicine

Precision medicine in healthcare involves leveraging data science and advanced analytics to enhance our understanding of patient data and improve diagnosis and treatment. By integrating artificial intelligence (AI) and machine learning (ML) technologies, precision medicine can uncover associations between genomic makeup and health, identify disease biomarkers, and accelerate drug discovery [30]. The analysis of big data, including omics data and electronic health records (EHR), is crucial for precision medicine. However, challenges such as data heterogeneity and unintentional biases need to be addressed Data governance plays a vital role in ensuring the success of precision medicine by protecting data subjects' interests and facilitating data integration, collection, and access. Overall, precision medicine in data science has the potential to revolutionize healthcare by providing personalized and effective treatments based on individual patient characteristics and data analysis [44].

Telemedicine

Telemedicine is a vital topic in healthcare informatics, especially in the context of the COVID-19 pandemic. It involves the remote delivery of medical services using telecommunication technology. While telemedicine offers numerous benefits, such as facilitating long-distance consultations between doctors and patients, it also poses cybersecurity risks that could lead to data leakage and misuse [45]. Data security measures are crucial in telemedicine systems, particularly in the transmission of sensitive medical data over the internet. Various encryption techniques, such as ciphertext-policy attribute- based encryption and Secure Better Portable Graphics (SBPG) architecture, have been developed to protect data during transmission Additionally, the integration of Internet of Things (IoT) technology in telemedicine systems requires attention to data security, with data encryption being the most frequently used method [17]. Overall, telemedicine in data science plays a significant role in improving healthcare services by leveraging technology and ensuring the secure transmission of medical information.

ETHICAL AND REGULATORY CONSIDERATIONS

Data Privacy

Data privacy is a crucial concern in data science in healthcare, and it is essential to ensure the protection of sensitive health data. As the use of data analytics in healthcare continues to expand, effective privacy protection measures are critical to prevent issues such as fraud, identity theft, and privacy breaches. Researchers have proposed privacy-preserving data collection protocols that anonymize healthcare data without relying on third-party involvement or private channels for data transmission. The relationship between health and privacy is a pivotal issue, involving two fundamental rights that are vulnerable to technological advancements. To address these challenges, a dataset condensation approach has emerged as a promising way to share healthcare data for AI research, allowing for de-identification of individual-level information while preserving the original deep-learning utilities.

Informed Consent

Ensuring informed consent in data science in healthcare is a complex issue that needs careful consideration. The collection, storage, and curation of human genomic data for biomedical research have raised ethical concerns about data sharing and the need for consent [10]. Traditional informed consent, where researchers fully explain study goals to subjects, is challenging in the context of modern analytics and big data, where unexpected correlations may be discovered. Alternative consent models, such as blanket consent, broad consent, dynamic consent, and meta-consent, have been proposed to address these challenges. These models may involve a degree of uncertainty for participants, as the future uses of their data are often unknown at the time of collection. The Revised Common Rule introduces a form of "broad consent," allowing varied forms of data use and reuse, but it may lower the standards for other types of information collection.

The potential re-identification of de-identified biospecimens also raises concerns about privacy and data security. Ensuring informed consent in data science in healthcare requires a careful balance between the benefits of research and the protection of participants' rights and privacy.

Regulatory Compliance

Regulatory compliance is a critical consideration in data science in healthcare to ensure patient safety, maintain quality standards, and foster trust between healthcare providers and patients. The growing volume of digitally stored information in health institutions necessitates the development of rules to govern its use and understanding without losing important data. Compliance with regulatory requirements is essential for improving quality and informing policy and practice in health and social care services advances in technology, such as mobile apps and wearables, have raised data protection compliance issues, including consent, transparency, research use, and data security. Healthcare database analyses have been recognized as valuable supplements to randomized clinical trials for generating evidence on the effectiveness, harm, and value of medical products in routine care. To enable successful regulatory decision-making based on healthcare database analyses, meaningful, valid, expedited, and transparent evidence is required.

CONCLUSION

In the final analysis, the integration of data science and data analytics into healthcare represents a transformative force with the potential to revolutionize patient care, drive medical innovations, and enhance the overall healthcare landscape. As we conclude this chapter, it is evident that data-driven insights are shaping the future of healthcare in profound ways. The importance of data quality, data integration, and data visualization cannot be overstated. Healthcare professionals and decision-makers are increasingly relying on these tools to derive actionable insights from the vast troves of data generated within the healthcare ecosystem. From predicting disease outbreaks to tailoring treatment plans for individual patients, these technologies empower healthcare providers to make informed decisions that have a direct impact on patient well-being.

The world of predictive modeling, powered by advanced machine learning and AI techniques, offers a glimpse into the future of proactive healthcare. Early detection, risk prediction, and personalized healthcare plans are becoming more attainable than ever, offering the promise of improved patient outcomes and cost-effective care delivery. Unsupervised learning and anomaly detection are becoming indispensable for identifying hidden patterns and irregularities within healthcare data. These techniques hold the potential to uncover subtle signals that may otherwise go unnoticed, contributing to a deeper understanding of health trends and individual patient needs.

Text analytics and natural language processing have opened up new frontiers in mining clinical notes and understanding patient sentiments. These technologies enable healthcare providers to better tailor their care plans to the needs and preferences of individual patients, ultimately improving patient satisfaction and engagement. Data visualization and dashboard design bridge the gap between complex data and actionable insights. As healthcare becomes increasingly data-driven, these tools play a vital role in ensuring that healthcare professionals and decision-makers can understand and act upon the information at their disposal.

Looking to the future, the ongoing advancements in AI and precision medicine promise to bring even greater transformations to the healthcare industry. The era of telemedicine is here, offering accessible and efficient healthcare delivery that transcends geographic boundaries. However, it also brings with it the challenge of data security, which requires ongoing vigilance.

On the ethical and regulatory front, the importance of data privacy, informed consent, and regulatory compliance cannot be underestimated. Striking the right balance between utilizing data for improving healthcare and safeguarding patient rights and privacy is an ongoing challenge that requires a multifaceted approach.

In conclusion, data science and data analytics are redefining the healthcare landscape, offering a promising path toward better patient care, more effective healthcare services, and groundbreaking medical discoveries. As we navigate this data-driven future, it is crucial to remain cognizant of the ethical and regulatory considerations that accompany this transformation, ensuring that healthcare remains a field rooted in the best interests of patients and the broader community. With continued innovation, responsible use of data, and a commitment to patient-centered care, the future of healthcare shines ever brighter.

REFERENCES

[1]   Guoping Liu , Lee KY, Jordan HF. TDM and TWDM de Bruijn networks and ShuffleNets for optical communications. IEEE Trans Comput 1997; 46(6): 695-701.
        [http://dx.doi.org/10.1109/12.600827]

[2]   Dalianis H, Henriksson A, Kvist M, Velupillai S, Weegar R. Health bank-a workbench for data science applications in healthcare. CAiSE Industry Track 2015; 1381: 1-8.

[3]   Chen T, Keravnou-Papailiou E, Antoniou G. Medical analytics for healthcare intelligence – Recent advances and future directions. Artif Intell Med 2021; 112: 102009.
        [http://dx.doi.org/10.1016/j.artmed.2021.102009] [PMID: 33581829]

[4]   Khandelwal A. Big data analytics in healthcare: an overview. SSRN. 4539651. 2023; p.
        [http://dx.doi.org/10.2139/ssrn.4539651]

[5]   Sarker IH. Data science and analytics: an overview from data-driven smart computing, decision-making and applications perspective. SN Comput Sci 2021; 2(5): 377.
[http://dx.doi.org/10.1007/s42979-021-00765-8] [PMID: 34278328]

[6]   Liang Y, Kelemen A. Big Data science and its applications in health and medical research: Challenges and opportunities. J Biom Biostat 2016; 7(3): 2.
[http://dx.doi.org/10.4172/2155-6180.1000307]

[7]   Kumar S, Singh M. Big data analytics for healthcare industry: impact, applications, and tools. Big data mining and analytics. 2018; 2(1): 48-57.

[8]   Davenport TH, Harris JG. Competing on analytics: the new science of Winning. Language. 2007; 15 (217p): 24

[9]   Rejeb A, Rejeb K, Treiblmaier H, *et al.* The Internet of Things (IoT) in healthcare: Taking stock and moving forward. Internet of Things. 2023; 14: 100721.

[10]   Merikangas KR, McClair VL. Epidemiology of substance use disorders. Hum Genet 2012; 131(6): 779-89.
[http://dx.doi.org/10.1007/s00439-012-1168-0] [PMID: 22543841]

[11]   Boulos MN, Brewer AC, Karimkhani C, Buller DB, Dellavalle RP. Mobile medical and health apps: state of the art, concerns, regulatory control and certification. Online J Public Health Inform 2014; 5(3): 229.
[PMID: 24683442]

[12]   Obermeyer Z, Emanuel EJ. Predicting the future—big data, machine learning, and clinical medicine. N Engl J Med 2016; 375(13): 1216-9.
[http://dx.doi.org/10.1056/NEJMp1606181] [PMID: 27682033]

[13]   Schnipper JL, Liang CL, Ndumele CD, Pendergrass ML. Effects of a computerized order set on the inpatient management of hyperglycemia: a cluster-randomized controlled trial. Endocr Pract 2010; 16(2): 209-18.
[http://dx.doi.org/10.4158/EP09262.OR] [PMID: 20061280]

[14]   Di Ieva A, Russo C, Liu S, *et al.* Application of deep learning for automatic segmentation of brain tumors on magnetic resonance imaging: a heuristic approach in the clinical scenario. Neuroradiology 2021; 63(8): 1253-62.
[http://dx.doi.org/10.1007/s00234-021-02649-3] [PMID: 33501512]

[15]   Esteva A, Kuprel B, Novoa RA, *et al.* Dermatologist-level classification of skin cancer with deep neural networks. nature. 2017 Feb; 542(7639): 115-8. E. J. Topol, High-performance medicine: the convergence of human and artificial intelligence. Nat Med 2019; 25(1): 44-56.
[PMID: 30617339]

[16]   Johnson KW, Torres Soto J, Glicksberg BS, *et al.* Artificial intelligence in cardiology. J Am Coll Cardiol 2018; 71(23): 2668-79.
[http://dx.doi.org/10.1016/j.jacc.2018.03.521] [PMID: 29880128]

[17]   Rajkomar A, Oren E, Chen K, *et al.* Deep learning for cardiovascular medicine: A practicalprimer. Eur Heart J 40(25): 2058-73.

[18]   Beam AL, Kohane IS. Big data and machine learning in health care. JAMA 2018; 319(13): 1317-8.
[http://dx.doi.org/10.1001/jama.2017.18391] [PMID: 29532063]

[19]   Horng S, Sontag DA, Halpern Y, Jernite Y, Shapiro NI, Nathanson LA. Creating an automated trigger for sepsis clinical decision support at emergency department triage using machine learning. PLoS One 2017; 12(4): e0174708.
[http://dx.doi.org/10.1371/journal.pone.0174708] [PMID: 28384212]

[20]   Kononenko I. Machine learning for medical diagnosis: history, state of the art and perspective. Artif Intell Med 2001; 23(1): 89-109.

[http://dx.doi.org/10.1016/S0933-3657(01)00077-X] [PMID: 11470218]

[21]   Chen X, Lin Z, Ma H. Artificial intelligence in healthcare: anticipating challenges in ethical, legal, and social impact. J Healthc Eng 2018.

[22]   Collins GS, Reitsma JB, Altman DG, Moons KGM. Transparent reporting of a multivariable prediction model for individual prognosis or diagnosis (TRIPOD): the TRIPOD statement. Circulation 2015; 131(2): 211-9.
[http://dx.doi.org/10.1161/CIRCULATIONAHA.114.014508] [PMID: 25561516]

[23]   Dilsizian SE, Siegel EL. Artificial intelligence in medicine and cardiac imaging: harnessing big data and advanced computing to provide personalized medical diagnosis and treatment. Curr Cardiol Rep 2014; 16(1): 441.
[http://dx.doi.org/10.1007/s11886-013-0441-8] [PMID: 24338557]

[24]   Berkowitz DA, Brown K, Morrison S, *et al.* Improving Low-acuity Patient Flow in a Pediatric Emergency Department: A System Redesign. Pediatr Qual Saf 2018; 3(6): e122.
[http://dx.doi.org/10.1097/pq9.0000000000000122] [PMID: 31334454]

[25]   Cadamuro J. Rise of the machines: The inevitable evolution of medicine and medical laboratories intertwining with artificial intelligence—A narrative review. Diagnostics (Basel) 2021; 11(8): 1399.
[http://dx.doi.org/10.3390/diagnostics11081399] [PMID: 34441333]

[26]   Friesen M, Hamel C, McLeod R. A mHealth application for chronic wound care: findings of a user trial. Int J Environ Res Public Health 2013; 10(11): 6199-214.
[http://dx.doi.org/10.3390/ijerph10116199] [PMID: 24256739]

[27]   Nykänen P. Decision support systems from a health informatics perspective. Tampere University Press 2000.

[28]   Friedman CP, Wyatt J. Evaluation methods in biomedical informatics. Springer Science & Business Media 2005.

[29]   Gulati V, Raheja N. PCSVD: A hybrid feature extraction technique based on principal component analysis and singular value decomposition. Journal of Autonomous Intelligence 2023; 6: 2.
[http://dx.doi.org/10.32629/jai.v6i2.586]

[30]   Deo RC. Machine learning in medicine. Circulation 2015; 132(20): 1920-30.
[http://dx.doi.org/10.1161/CIRCULATIONAHA.115.001593] [PMID: 26572668]

[31]   Jabeur N, Yasar A, Mohamad Y, Melchiori M. Guest editorial: special issue on data science approaches and applications. Comput Inf 2022; 41(1): 1-11.
[http://dx.doi.org/10.31577/cai_2022_1_1]

[32]   Barnes S, Hamrock E, Toerper M, Siddiqui S, Levin S. Real-time prediction of inpatient length of stay for discharge prioritization. J Am Med Inform Assoc 2016; 23(e1): e2-e10.
[http://dx.doi.org/10.1093/jamia/ocv106] [PMID: 26253131]

[33]   Breiman L. The two cultures. Quality control and applied statistics. 2003; 48(1): 81-2.

[34]   Frank L, Basch E, Selby JV. The PCORI perspective on patient-centered outcomes research. JAMA 2014; 312(15): 1513-4.
[http://dx.doi.org/10.1001/jama.2014.11100] [PMID: 25167382]

[35]   Johnson AEW, Ghassemi MM, Nemati S, Niehaus KE, Clifton D, Clifford GD. Machine learning and decision support in critical care. Proc IEEE 2016; 104(2): 444-66.
[http://dx.doi.org/10.1109/JPROC.2015.2501978] [PMID: 27765959]

[36]   Li X, Tian D, Li W, *et al.* Artificial intelligence-assisted reduction in patients' waiting time for outpatient process: a retrospective cohort study. BMC Health Serv Res 2021; 21(1): 237.
[http://dx.doi.org/10.1186/s12913-021-06248-z] [PMID: 33731096]

[37]   Banerjee A, Chen S, Fatemifar G, *et al.* Machine learning for subtype definition and risk prediction in heart failure, acute coronary syndromes and atrial fibrillation: systematic review of validity and

clinical utility. BMC Med 2021; 19(1): 85.
[http://dx.doi.org/10.1186/s12916-021-01940-7] [PMID: 33820530]

[38]   Motwani M, Dey D, Berman DS, *et al.* Machine learning for prediction of all-cause mortality in patients with suspected coronary artery disease: a 5-year multicentre prospective registry analysis. Eur Heart J 2017; 38(7): 500-7.
[PMID: 27252451]

[39]   Yoon J, Alaa A, Hu S, Schaar M. ForecastICU: a prognostic decision support system for timely prediction of intensive care unit admission. InInternational Conference on Machine Learning. 1680-9.

[40]   Olsen L, Aisner D, McGinnis JM. The learning healthcare system: workshop summary.

[41]   Shafique K, Khawaja BA, Sabir F, Qazi S, Mustaqim M. Internet of things (IoT) for next- generation smart systems: A review of current challenges, future trends and prospects for emerging 5G-IoT scenarios. IEEE Access 2020; 8: 23022-40.
[http://dx.doi.org/10.1109/ACCESS.2020.2970118]

[42]   Y LeCun Y, Bengio Y, Hinton G. Deep learning. nature. 2015 May 28; 521(7553): 436-44.

[43]   Ahun E, Demir A, Yiğit Y, *et al.* Perceptions and concerns of emergency medicine practitioners about artificial intelligence in emergency triage management during the pandemic: a national survey-based study. Front Public Health 2023; 11: 1285390.
[http://dx.doi.org/10.3389/fpubh.2023.1285390] [PMID: 37965502]

[44]   Rejeb A, Rejeb K, Treiblmaier H, *et al.* The Internet of Things (IoT) in healthcare: Taking stock and moving forward. Science Direct. Volume 11, July 2023.

[45]   Khaled H. Application of internet thing in healthcare domain. Journal of Umm Al-Qura University for Engineering and Architecture. 2023.

CHAPTER 7

The Internet of Things for Healthcare: uses, Particular Cases, and Difficulties

K.P. Parthiban[1], S. Muthu Vijaya Pandian[2,*], M. Muthukrishnaveni[3] and M. Kavibharathi[4]

[1] *Department of EEE, VSB College of Engineering and Technical Campus, Coimbatore, India*

[2] *Department of EEE, SNS College of Technology, Tamil Nadu, India*

[3] *Department of Physics, Sri Ramakrishna Institute of Technology, Coimbatore, India*

[4] *Department of Computer Science, Dr. SNS Rajalakshmi College of Arts and Science, Coimbatore, India*

Abstract: Internet of Things (IoT) integration in healthcare has completely changed patient care and administration. The revolutionary effects of IoT technology on the healthcare industry are examined in this abstract. The Internet of Things (IoT) enables real-time monitoring of patient's vital signs through the seamless connectivity of medical equipment and sensors, guaranteeing prompt intervention and customized treatment programs. Healthcare providers may obtain vital information through remote patient monitoring, which improves diagnostic precision and maximizes resource use. Furthermore, IoT-powered innovative healthcare systems provide preventative care by continually gathering and evaluating patient data, which makes it possible to identify health problems early. Notwithstanding these developments, issues like data security and interoperability still need to be resolved if IoT in healthcare is to reach its full potential. The significant effects of IoT on healthcare delivery are highlighted in this abstract, highlighting the necessity of a comprehensive strategy to handle both possibilities and problems in this quickly changing environment.

Keywords: IoT technology, (IoT) Internet of Things, Medical equipment, Smart healthcare system, Sensors.

INTRODUCTION

The "Internet of Things" (IoT) refers to various applications, technologies, protocols, and initiatives. It is fundamentally a network of objects that are linked to the Internet. These include IoT-enabled physical items and IoT-enabled equipment. Things and data are the foundation and core of what IoT is and what it

* **Corresponding author S. Muthu Vijaya Pandian:** Department of EEE, SNS College of Technology, Tamil Nadu, India; E-mails: mvpeeehod@gmail.com, muthu.s.ihub@snsgroups.com

Sivakumar Rajagopal, Prakasam P., Konguvel E., Shamala Subramaniam, Ali Safaa Sadiq Al Shakarchi & B. Prabadevi (Eds.)

enables. IoT assets and devices have electronic parts and software for data collection, organization, and sharing. The phrase "Internet of Things" was coined by Kevin Ashton. He researched radio frequency identification (RFID) in the late 1990s.

This technique enables information to be stored on tiny radio frequency tags attached to various objects and read from a distance [1]. It allows, for example, following the flow of products, enhancing the supply management system, and even preventing theft. It is a little sticker or customized label in a plastic casing. These RFID tags are now often used in the trade sector. Additionally, Kevin Ashton used the phrase "Internet of Things" while describing the fundamental concept of his creation. He previously predicted that everything on the Internet of Things (IoT) will have a digital equivalent that will serve as its virtual image. The scope of applications for RFID technology is expanding right now. This technique is commonly used to automate industrial operations, particularly when sophisticated manufacturing of automobiles and appliances (refrigerators and washing machines) is involved.

Certain libraries, like the Vatican Library, which has more than two million copies of books in its collection, have adopted RFID to expedite inventory and book searches, automate book deliveries, and prevent theft. This technique is being used or implemented by more than 700 of the biggest libraries in the world [2, 3]. In several nations worldwide, new passports also come with RFID tags. These forms of identification are known as biometrics or electronic passports, and they have a chip with the same data as the printed version. This technology is "taking root" in medicine more and more.

RFID bracelets, for instance, are used in maternity facilities to link the baby's identity to the mother. They are accustomed to monitoring the movements of patients who require continual care in traditional hospitals. It was recently proposed that a wireless sensor network be used to track and manage things by coupling a tracker to a heart rate monitor [4]. Today's gadgets connect with smartphones, social networks, cloud computing, big data analytics, and GPS devices to enable the contemporary IoT.

Since the 2000s, as the number of devices linked to the Internet rose quickly, the direction of IoT has been actively developing. The vast quantity of big data utilized by the Internet of Things raises privacy concerns for its consumers. Preserving users' rights is the main goal of developing and implementing the General Data Protection Regulation (GDPR). Alexia Kounoudes and associates examined the challenges of implementing GDPR in the Internet of Things. To properly investigate the issue of user privacy, the writers carried out a thorough literary analysis [5].

The Organization for Economic Cooperation and Development's group of twenty (G20) Artificial Intelligence (AI) Guidelines and the General Principles for Human G20, as well as the European Commission's Coordination Plan and Ethical Recommendations on AI Reliability, should all be considered when working with the Internet of Things. Specifically, human-centered values and justice, sustainable growth and prosperity, transparency and clarity, dependability, protection, security, and accountability are five interconnected principles that should be followed [6 - 8].

For IoT to function continuously and with high quality, it needs a specialized environment that has platforms for controlling the network, devices, and apps. This environment must also include various "smart" devices that are directly connected and have sensors, network access, and information transfer capabilities. This system cannot function without at least one of these parts. Governments, corporations, mobile and Internet service providers, and even regular citizens will need to work closely together to fully realise the Internet of Things promise.

This chapter looks at the IoT foundations inside the healthcare system, emphasizing the use of IoT technology for rapidly developing customized health. It discusses the newest and most advanced IoT-derived methods and well-known health cases. This study also focuses on the financial, ethical, and technical barriers to creating a more advanced healthcare system that can identify and diagnose illnesses early on. These cutting-edge health systems might be useful to healthcare practitioners, giving them access to the relevant patient information at the right moment. As a result, quickly and effectively taking care of medical issues.

This chapter covers IoT technology's function and uses in the healthcare industry. It then presents a few chosen medical scenarios that illustrate an IoT-driven healthcare system and addresses the possible obstacles to its widespread adoption.

IoT and Healthcare

IoT applications in healthcare are among the most noble. Physicians may use the Internet to aid patients through IoT. Using portable Internet of things-based health monitoring devices may greatly decrease patient-physician distance. You may assess each patient individually using IoT, determine the best course of action based on their health situation, and choose how best to treat them. Physicians may remotely monitor and provide real-time care for their patients by using portable sensors.

However, a constant Internet connection is necessary for real-time metrics. Despite its rapid development in this field, several medical companies have not yet fully adopted IoT [9]. There are still several challenges in creating suitable Internet apps for conventional medicine. The Internet of Things is expected to draw more medical researchers in the upcoming years due to a notable rise in medical research projects. Medical practitioners must gather a lot of big data, analyze it, and interpret it to make individualized, well-informed judgments. It all requires a lot of time and work.

IoT-based new technologies have the potential to expedite and streamline this process. As electronic health registration becomes more widespread, there is a noticeable increase in the volume of digitalized medical data. Viewing and evaluating all of this data takes a long time. Additionally, it is necessary to teach medical professionals about AI-based technology, which is closely related to the Internet of Things [10, 48].

By coordinating digital technologies like AI and the Internet of Things, doctors may more effectively customize therapy to patients' requirements.

These technologies enable the handling of considerably larger amounts of data for storing and analysis, allowing for careful monitoring of the development of a certain process or illness. Positive advances in healthcare administration can be achieved by deftly fusing the potential of new techniques for diagnosis, collection, and analysis with real-world personal experience [11]. Fig. (**1**) presents the idea of IoT in the medical field.

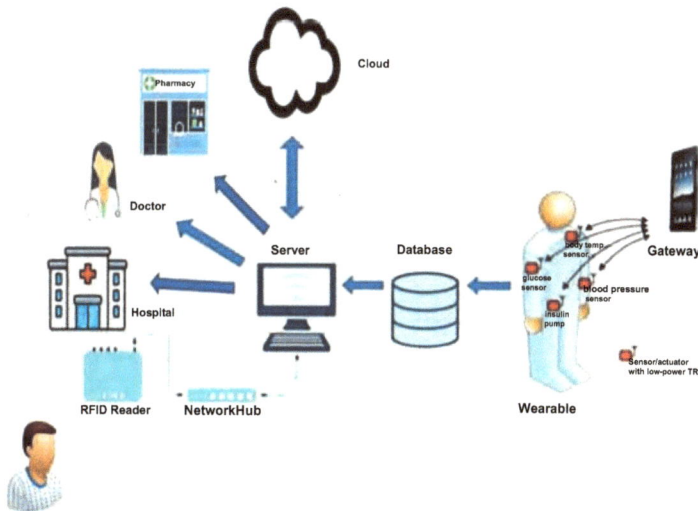

Fig. (1). IoT in the Healthcare.

The Internet of Things eventually brings wearable and portable technologies that are network-enabled. These devices can detect, trigger, synergize, and link with similar media over the Internet. The Internet of Things profoundly alters data distribution, consumption, and creation. While ordinary people regularly utilize these systems to monitor their physical states, exercise, vital signs, food, and sleep, IoT devices occasionally acquire and process ecological data that impacts an individual's health. Ultimately, this interoperability has ushered in a new age of medical alternative manufacturing.

Applications of IoT in Healthcare

IoT in healthcare has the potential to significantly improve patient management, clinical practice, and research. It has several uses in the insurance and business sectors [12 - 15]. Four guiding concepts underpin the IoT's contribution in each scenario. The first concept is data collecting, made possible by a network of linked devices like cameras, monitors, detectors, sensors, and equators.

The second principle concerns data conversion. Analogue data from sensors and other relevant equipment must be converted to digital data to be processed further. The third principle covers data storage, often accomplished by a cloud-based system. The fourth element is using sophisticated analytics modalities to analyze data and ultimately give consumers the knowledge they need to make decisions [15 - 17].

Most healthcare systems use the previously described concepts, from manually written patient records to networked laboratory databases. Their ability to have instantaneous effects from IoT-based choices and continuous data flow set them apart in the context of IoT. Wearable technology makes up the majority of the IoT infrastructure used with patients.

Depending on the patient's history and the parameters that need to be monitored, wearables may include glucose level monitors, blood pressure, pulse/heart rate, and oxygen saturation monitors. These gadgets can provide individualised care in the event of a sudden deterioration or an urgent ailment. They can also be reminders if linked to appointment and referral systems, calorie-counting software, or physical activity [13, 18, 19].

From a physician's perspective, IoT provides a real-time link to their patients, colleagues, and clinic or laboratory. Notification of an arrhythmia harming a patient can be sent to a cardiologist, and information regarding hypoglycemia endangering a patient can be sent to a diabetologist. Patients can receive prompt medical advice and assistance in both situations. Doctors can evaluate patients' adherence. It can also involve device monitoring and the potential consequences (such as elevated blood pressure) that may arise from patients' noncompliance with therapy.

According to this, pillbox opening frequency may be tracked by counting how often they were opened daily. According to available evidence, data from IoT devices may help doctors choose the best course of action for their patient's care and management. This makes a significant contribution to the field of customized healthcare. Future treatment-to-outcome research may be built around this larger-scale big data [20, 21].

Hospitals and research centers foster larger-scale IoT applications. This occurs because of the heavy workload and wide range of data that must be handled there, these institutions' accountability, and the financing they receive. In addition to the previously discussed health monitoring of inpatients and outpatients, hospitals and labs may utilize IoT to secure equipment like oxygen pumps, wheelchairs, defibrillators, and nebulizers.

Additionally, research institutions can continuously and laboriously monitor the progress of experimental activity, the deployment of equipment, and the availability of resources [22, 23]. In numerous cases, communication and sensor devices eventually develop into multifaceted information technology solutions. Researchers and physicians have developed innovative healthcare solutions with the help of expanding IoT technology.

The importance of IoT-related health research in providing preventive care and services at lower costs and higher quality is considerable. The Internet of Things is gaining traction as a cutting-edge research topic across several academic and corporate domains, particularly medicine. Remarkably, given the explosive growth of smartphones and wearables, IoT-derived methods transform healthcare from a conventional hub-centered system to an even more individualised one.

Individualized medical services are offered *via* e-health to satisfy people's healthcare needs. In the big data age, IoT is a significant advancement that supports a number of timely technological software programs to optimize services. IoT data analytics is being used by the medical system as a consumer data source to obtain more information, identify illnesses early, and establish critical conditions for improved quality of life.

Finally, there is a rapidly growing need for the prompt implementation of an enhanced healthcare system. IoT devices can quickly gather and share data with other cloud-based platforms, making it possible to gather, store, and analyze enormous amounts of data. IoT devices might be used to remotely track the local environment or computerise corporate processes. IoT applications in healthcare promise to improve patient quality of life, lower costs, and expand access to treatment.

IoT applications can help improve the functioning of insurance companies and the healthcare sector. IoT may be the foundation for data storage, product assessment, medical evaluation, and quicker compensation services. The main area of IoT use in healthcare is depicted in Fig. (**2**). However, data protection laws and the interference of financial interests in managing such information might provide significant legal challenges for such applications [13, 24].

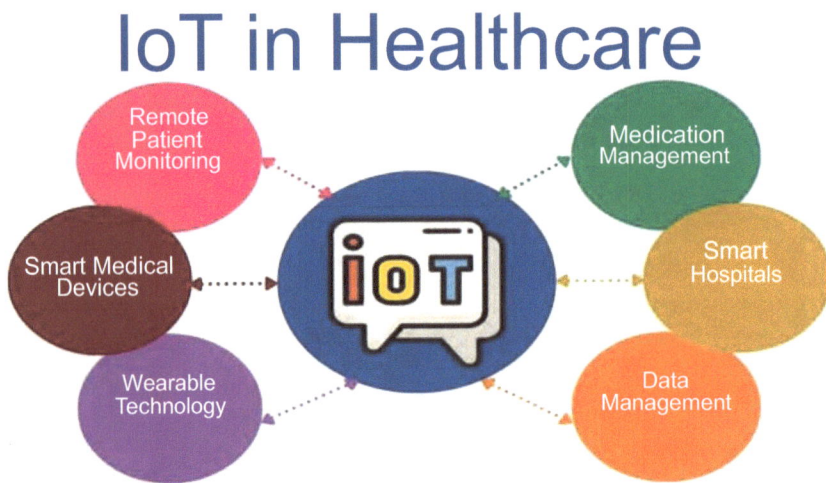

Fig. (2). Application of IoT in Healthcare.

Selected Cases of Using IoT in Healthcare

These days, wearable technology and smartphone applications offer symptom monitoring, health education, exercise, collaborative sickness management, and cogent care. Analytics software programs have the potential to greatly improve data interpretation and reduce the amount of time required to put the generated data back together. Big data perspectives would influence the electronic growth of commercial processes, time management, and the medical field. With the world's population becoming older, improving knowledge and interpretation of health and wellbeing data is imperative, reducing chronic and diet-related illnesses, improving mental capacities, and promoting healthy lifestyles is imperative.

While enumerating every IoT healthcare application is impossible, we will present a summary of the most well-known ones. The scientific literature and some commercial resources clearly show that IoT is anticipated to play a significant role in the management of exaggerated acute conditions, drug delivery, and adherence monitoring, patient-driven self-assessment procedures, cancer care, and mental health in the near future.

Clinical practice has previously tried wearables for cancer care (CC). A randomized clinical trial was reported at the American Society of Clinical Oncology annual meeting in 2018. The study focused on head and neck cancer patients tracked *via* symptom-tracking software that sent regular and urgent information to their doctors, as well as a blood pressure cuff and weight scale with Bluetooth capabilities. The research had about 400 patients, and those who utilized the IoT-based system reported less severe symptoms than the control group, which underwent weekly physical assessments [20, 21, 25].

Diabetes is a model illness for evaluating treatment compliance and self-monitoring in various settings, such as oral medication, insulin injections, blood pressure monitoring, and glucose testing. Numerous current gadgets based on the Internet of Things can perform continuous glucose monitoring. While type 1 Diabetes Mellitus (T1D) patients primarily require continuous monitoring and rapid management, mounting evidence indicates that more frequent or continuous monitoring may help patients with type 2 Diabetes (T2D) avoid problems [26, 27].

Smart insulin pens are useful for evaluating patients with diabetes mellitus and their adherence to their treatment regimens (DM). While the current gadgets are designed for insulin injections, pillboxes might benefit from comparable devices. These days, wearables like these are linked to smartphone applications and are often evaluated by doctors. By using these modalities in an Internet of Things setting, doctors might be alerted earlier when patients are not receiving therapy and take appropriate action [28, 29]. The use of closed-loop, or automated, insulin delivery devices in T1D therapy has long been anticipated.

Potential problems with regulation and administration have impeded the Use of these devices in clinical settings. Several patient and physician advocacy initiatives have already been noted, acknowledging that IoT may play a major role in overcoming these challenges. An automated and Internet of Things (IoT) protected closed-loop system can be crucial for T1D patients who are at risk of developing diabetic ketoacidosis, even if there are a few procedures that must be followed [30, 31].

Asthma, as opposed to DM, is a chronic illness characterized by an exaggerated pattern that presents a promising opportunity for IoT-based healthcare. For hundreds of millions of people worldwide, it is a substantial hardship. Most patients are youthful, energetic individuals looking for a steady quality of life. Early diagnosis and treatment of an impending exacerbation are crucial, and this may be achieved through IoT devices that measure saturation or warn about common allergens present. In the same context, IoT-based inhalers may give patients' doctors trustworthy data on adherence and the patient's capacity to operate the device appropriately [14, 32].

Of course, both mental health issues and asthma have a chronic component. In addition to the monitoring choices outlined above, IoT can improve patient assistance programs. When paired with AI modalities, IoT can offer helpful chatbots for various applications, from diagnosing suicidal thoughts to treating patients with dementia or moderate cognitive impairment regularly through cognitive rehabilitation [16].

Management of asthma exaggerations, drug delivery and adherence monitoring, cancer care, patient-driven self-assessment procedures, and supportive care for degenerative mental health conditions are just a few of the examples of IoT implementation in healthcare that we have highlighted in this section (Fig. **3**). These methods demonstrate how, if further and ideally accepted, IoT can change healthcare practice, patient management, and research.

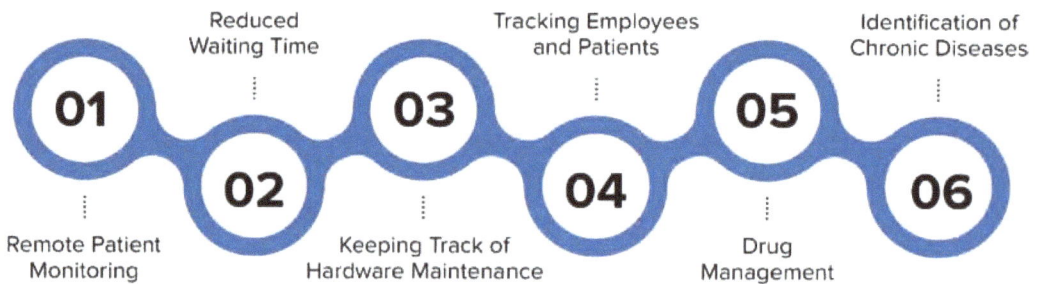

Fig. (3). Selected cases of IoT implementation in healthcare.

Challenges of IoT in Healthcare

Thanks to IoT, healthcare is predicted to undergo a revolution. But a revolution cannot occur without fighting and hardship. The tension between morality and technology determines how innovation is integrated. The difficulties posed by IoT in healthcare can be categorized, as shown in Fig. (**4**), as technological, financial, and ethical [28, 33, 34].

Technical Challenges

The fact that IoT is still not a part of daily life raises technical issues. Most nations lack access to fifth-generation (5G) wireless technology and Internet of Things services. Patients aside, the great majority of healthcare providers, researchers, and practitioners are not fully aware of what IoT entails and what applications of IoT are outside of healthcare in daily life. As said, 5G can be regarded as the initial technical difficulty with IoT deployment in healthcare [21, 35].

Installing many antennae is necessary for 5G implementation, yet it is expensive, time-consuming, and has been linked to health risks. Even if there is not enough data to support this assertion, more studies must be done to certify the security of widespread 5G technology deployment. Taking on the worldwide consequences of 5G, convincing legislators, and altering (with their help) Opinion research can take just as much time, if not more, than safety investigations. The potential advantages of IoT applications in healthcare might be a compelling case in support of the 5G rollout [16, 23].

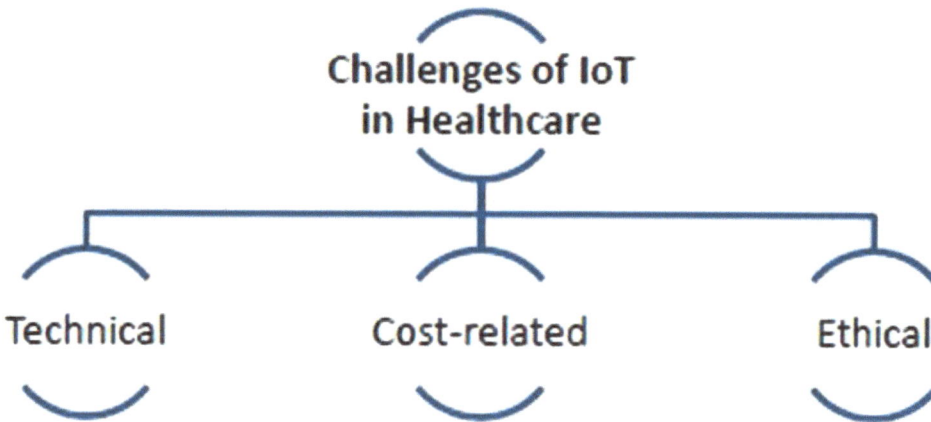

Challenges of IoT in Healthcare

Technical   **Cost-related**   **Ethical**

Fig. (4). IoT in healthcare challenges.

Data integration presents the next technological difficulty. Various devices imply multiple data sources. For technological and budgetary reasons, the healthcare industry's wearables and data collection devices cannot readily be converted to a single data-gathering pattern. Currently, manufacturers cannot agree on communication protocols and standards [21, 36].

Patients with the same illness can utilize several wearables that target the illness specifically or the patient's vital signs generally. Individuals with diabetes, for instance, could utilize several glucose and vital sign monitoring devices in addition to a different shut-off insulin pump. This explains at least three distinct types of information from the identical patient. Although these data will be handled in the end, this will take extra time, which can be problematic in urgent medical situations. Management has also been shown to be problematic when dealing with a large number of persons [13, 37].

Financial Challenges

Regarding the topic's financial implications, we plan to discuss potential cost-efficacy obstacles in more detail. Financially speaking, IoT falls within the category of applications for remote health. The International Data Corporation claims that (IDC), e10.41 is the current European funding for remote health monitoring. Billion then increased to more than e12.4 billion. Such a budget might seem advantageous concerning the development of IoT-based healthcare. Still, the intricacies of IoT implementation have already turned off a lot of potential investors. Origins of third-party service companies' involvement in financial instability ensure the caliber of IoT and the related infrastructure for connectivity.

Empirical findings show that without proof and experience, public and private healthcare providers would not be prepared to support creating an IoT healthcare network, from other nations or healthcare systems [38, 39]. More specifically, with the IoT in 2014, the size of the healthcare market was expected to be $60 billion. It is anticipated to have attained a $136 billion net wealth by 2021 [40]. Notable is the fact that it is anticipated that the IoT's Compound Annual Growth Rate (CAGR) in healthcare will attain or surpass

12.5 per cent during the projection. If this increasing trend continues, it will be maintained, depending on the ability of healthcare providers who are contracted out and suppliers to attain and sustain a sufficient degree of comprehension and collaboration [41].

If implemented, IoT can lower healthcare costs. Traditionally, healthcare expenses are separated into direct and indirect costs. The former includes the costs incurred by healthcare providers, whereas the latter comprises costs incurred by healthcare recipients, including missed work, unpaid medical expenses, and the involvement of family members or other caregivers in how they're handled. Considering that no IoT healthcare service has been implemented in any substantial healthcare system, there is no proof of its cost-effectiveness [12, 42].

Economists have previously emphasized how the Internet of Things might help establish a healthcare paradigm that is both affordable and effective. Supply chain management, rigorous quality control, proactive asset management, inventory management, and product packaging optimization chain management are acknowledged as a need for the financial success of the Internet of Things in medical treatment. However, now, these ideas appear to be more related to industry than healthcare-centered. Close cooperation will be necessary for them to adjust to clinical practice and profound comprehension between physicians, healthcare management, and economists [15, 16].

Ethical Challenges

The data management and care paradigm is the root cause of the ethical debate surrounding IoT in healthcare. Informational privacy, data sharing, and autonomy are the primary contentious issues when it comes to the administration of sensitive health data. Difficulties with data ownership, permission, and unclear value also exist. Within the care paradigm, the decontextualization, dehumanization, and isolation of doctor-patient contact health and wellbeing, as well as the dangers of receiving treatment from unqualified personnel, are alarming in relation to ethics [36, 43].

Ethical considerations may impact regulations about IoT in healthcare. The Universal Declaration of Human Rights and other regional and global norms should be fundamental to any relevant law. The General Data Protection Regulation (GDPR) applies in the EU. In this context, additionally, it anticipates that policymaking may provide moral roadblocks to IoT adoption in the medical field [18, 47]. While the GDPR only relates to the European Union, it significantly impacts IoT research in the healthcare industry.

Understanding ethical responsibility requires looking at a few real-world examples in the context of IoT for healthcare. Sensors tracking people at work and home will become commonplace; consumers could even forget they have them. However, the same inconspicuous sensors will always keep an eye on its consumers' personal life, such as the typical variations in heart rate during a heated argument or joyful moment. Sensors that detect and analyze sound waves may also be able to "overhear" private talks. For the benefit of their well-being, even if the users agree to this, this surveillance might infringe upon the privacy of their friends, family, and coworkers [12, 28].

Researchers have already developed solutions that promise a selective memory of Internet of Things-connected sensors. To our understanding, it might be challenging to set boundaries between clinically significant private and personal data. Daily family disputes or discussions about questionable behaviour that are kept concealed may indicate latent hypertension or arrhythmia.

If the sensors are unable to determine what is significant, consult a doctor or data scientist. Such a selection entails a fundamental invasion of privacy [14, 44].

However, a significant ethical responsibility is associated with using sensor-selected data alone to make potentially dangerous judgments [45]. What happens if an IoT-mediated monitoring system triggers an increased anticoagulant dose that causes significant gastrointestinal bleeding? One may contend that researching the usage of sensors will supply the required evidence-based recommendations. Nonetheless, Patients' lives and well-being cannot be in danger; they have to bear this risk in any other case.

Another component of IoT healthcare services that should not be overlooked is cybersecurity. Cloud-mediated services are necessary for data processing and storage [36]. Even though all the moral issues surrounding healthcare and outsourcing services that will have access to this data are taken care of, hacking is still a serious risk. Potential employees are treated unfairly by insurance firms and HR departments. Workers may breach medical history and biometric data. In the same situation, any other organization or person may make financial or other claims to avoid interpreting reasonable data [35, 46].

DISCUSSION

The acceptance and execution of choices based on the gathering and analysis of vast amounts of data are made possible by IoT technology. Furthermore, managing such massive volumes of data that are dynamically humans has created AI technology, which is evolving. One of the technological revolution's main advantages is the widespread IoT implementation process. IoT will rank among humankind's greatest inventions. The tremendous increase in the last two years has led to the popularity of IoT. It is well recognized that while making significant decisions, taking into account all the vast quantities of data about several topics and items tangential to the procedures about which a choice is made.

Essentially, the choice needs to be captured as closely in real time as possible. Due to certain physiological and mental constraints, people are not always able to make snap decisions and well-informed choices. IoT and AI technologies are, therefore, relevant. The medical industry may gain a great deal from the sensible application of these technologies. One arena where IoT is becoming increasingly commonplace is contemporary healthcare. IoT directly impacts people's lives and highlights the significance of medicine as a field of study in contemporary society.

Among the most promising systems are medical ones. These are important fields that use a lot of numerical data. The substantial information flow makes it difficult to transform it into a specific final product and consume it utilizing cutting-edge technology and interactive information. Based on research studies, most health industry executives think IoT will spark a revolution in the medical field in the upcoming years. The following three areas will primarily involve remote patient health monitoring, preventing chronic illness exacerbations, and information gathering.

The IoT's fastest-growing market is health. Scientists predict that there will be ten times more linked medical gadgets in the next ten years. According to several analytics, there will be 92.1 million wearable sensors for medical guidance in the next two years in 2025. There were just 2.4 million of them. In addition, various gadgets and intelligent systems are meant to help, not replace, medical professionals like physicians and nurses their output. Using Internet technology, doctors may provide remote medical care, which is particularly significant when the epidemiological regime is starting to deteriorate.

IoT makes it possible to identify the best way to treat each patient individually and to carry out his state of well-being. Compute IoT may also make medical care practices easier by automating data collection in healthcare facilities, improving the performance of medical staff, enabling more precise illness diagnosis, and keeping an eye on patients' conditions and the disease's progression in real time.

Furthermore, IoT can enhance the efficiency of illness prediction and prevention. That might happen with the logical and appropriate use of AI and the Internet of Things; the number of mistakes in medical reducing practice, to a large extent, can help save more people. These days, many discussions are sparked by the introduction and wide use of AI in many areas of human endeavours.

The Internet of Things is facing dangers and obstacles in its technological growth. Among the primary systemic obstacles is the comprehension of the benefits of IoT. As a result, something is lacking or insufficient in its plan for development. Significant risks and obstacles include political, technical, financial, and economic; educational and motivational; regulated by law; security, privacy, technology compatibility, and standardization. These obstacles need further research to ascertain the course of action for overcoming them as they might give rise to various risks while using the Internet of Things.

CONCLUSION

Modern technologies, such as devices that regularly monitor timely health-related data or health biometrics, are quickly becoming beneficial in the health arena. With the widespread availability of cell phones and Internet technology, healthcare mobile applications are used by patients and clinicians to manage health. Integrating IoT methods of big data is essential in the field of health.

IoT in the healthcare industry is transforming efficient healthcare, building a foundation for contact across various health sectors, offering online assistance at each turn, and enabling the quick adaptation of contemporary medicine to the needs within a period. Healthcare practitioners might use these cutting-edge healthcare technologies to deliver timely, accurate patient data.

As a result, medical decisions may be made quickly and effectively. However, the appearance of AI in many areas of human endeavors has seen a rise in the general application of AI in this period. Various discussions show that there are various dangers and obstacles to technological advancement. One of the primary systemicobstacles is a lack of knowledge about the ideals of IoT and, thus, its development strategy's deficiency or lack.

More research is necessary to defeat these obstacles and decide the course of action for addressing them. This chapter has offered novel perspectives on IoT technology's function and specific uses in the healthcare industry.

Additionally, a few medical scenarios illustrated an IoT-driven healthcare system and the potential difficulties in implementing it in the healthcare industry.

Lastly, more studies on AI, IoT, or any other technologies connected to the different facets of their creation, application, and use are recommended as the path for healthcare in the future.

REFERENCES

[1]   Greengard S. The internet of things. Cambridge, MA: MIT Press 2015.
        [http://dx.doi.org/10.7551/mitpress/10277.001.0001]

[2]   Nayak R. Radio frequency identification (RFID) technology and application in the fashion and textile supply chain. CRC Press, 2019.

[3]   Solanki MR. Application of RFID technology in libraries and role of librarians Indian J. Agric. Libr. Inf. Serv 2019; p. 35.

[4]   Li S, Xu LD, Zhao S. The internet of things: a survey. Inf Syst Front 2015; 17(2): 243-59.
        [http://dx.doi.org/10.1007/s10796-014-9492-7]

[5]   Kounoudes AD, Kapitsaki GM. A mapping of IoT user-centric privacy preserving approaches to the GDPR. Internet of Things 2020; 11: 100179.
        [http://dx.doi.org/10.1016/j.iot.2020.100179]

[6]   Hleg AI. Ethics guidelines for trustworthy AI. B-1049 Brussels. 2019.

[7]   Ministers T. digital economy ministers. G20 Ministerial Statement on Trade and Digital Economy. 2019.

[8]   European Commission. Coordinated plan on artificial intelligence. 2018.

[9]   Sadoughi F, Behmanesh A, Sayfouri N. Internet of things in medicine: A systematic mapping study. J Biomed Inform 2020; 103: 103383.
        [http://dx.doi.org/10.1016/j.jbi.2020.103383] [PMID: 32044417]

[10]   Paranjape K, Schinkel M, Nanayakkara P. Short keynote paper: Mainstreaming personalized healthcare–transforming healthcare through a new era of artificial intelligence. IEEE J Biomed Health Inform 2020; 24(7): 1.
        [http://dx.doi.org/10.1109/JBHI.2020.2970807] [PMID: 32054591]

[11]   Ergen O, Belcastro KD. AI Driven advanced Internet of Things (Iotx2): The future seems irreversibly connected in medicine. Anatol J Cardiol 2019; 22(2) (Suppl. 2): 15-7.
        [http://dx.doi.org/10.14744/AnatolJCardiol.2019.73466] [PMID: 31670717]

[12]   Bandyopadhyay D, Sen J. Internet of things: Applications and challenges in technology and standardization. Wirel Pers Commun 2011; 58(1): 49-69.
[http://dx.doi.org/10.1007/s11277-011-0288-5]

[13]   Gopal G, Suter-Crazzolara C, Toldo L, Eberhardt W. Digital transformation in healthcare–present and future information technologies architectures. Clin Chem Lab Med 2019; 57(3): 328-35.
[http://dx.doi.org/10.1515/cclm-2018-0658] [PMID: 30530878]

[14]   Mittelstadt B. Ethics of the health-related internet of things: a narrative review. Ethics Inf Technol 2017; 19(3): 157-75.
[http://dx.doi.org/10.1007/s10676-017-9426-4]

[15]   Psiha MM, Vlamos P. IoT applications with 5G connectivity in medical tourism sector management: third-party service scenarios. InGeNeDis 2016. Geriatrics 2017; 141-54. [Springer International Publishing].

[16]   Latif S, Qadir J, Farooq S, Imran M. How will 5G wireless (and concomitant technologies) revolutionize healthcare? Future Internet 2017; 9(4): 93.
[http://dx.doi.org/10.3390/fi9040093]

[17]   O'brolcháin F, De Colle S, Gordijn B. The ethics of smart stadia: a Croke Park project stakeholder analysis. Sci Eng Ethics 2019; 25(3): 737-69.
[http://dx.doi.org/10.1007/s11948-018-0033-5] [PMID: 29497969]

[18]   Gope P, Hwang T. BSN-Care: A secure IoT-based modern healthcare system using body sensor networks. IEEE Sens J 2016; 16(5): 1368-76.
[http://dx.doi.org/10.1109/JSEN.2015.2502401]

[19]   Özdemir V, Hekim N. Birth of industry 5.0: Making sense of big data with artificial intelligence, "the Internet of things" and next-generation technology policy. OMICS 2018; 22(1): 65-76.
[http://dx.doi.org/10.1089/omi.2017.0194] [PMID: 29293405]

[20]   Li D. 5G and intelligence medicine—how the next generation of wireless technology will reconstruct healthcare? Precis Clin Med 2019; 2(4): 205-8.
[http://dx.doi.org/10.1093/pcmedi/pbz020] [PMID: 31886033]

[21]   Russell CL. 5 G wireless telecommunications expansion: Public health and environmental implications. Environ Res 2018; 165: 484-95.
[http://dx.doi.org/10.1016/j.envres.2018.01.016] [PMID: 29655646]

[22]   Chai PR, Zhang H, Jambaulikar GD, *et al.* An internet of things buttons to measure and respond to restroom cleanliness in a hospital setting: descriptive study. J Med Internet Res 2019; 21(6): e13588.
[http://dx.doi.org/10.2196/13588] [PMID: 31219046]

[23]   Stefano GB, Kream RM. The micro-hospital: 5G telemedicine-based care. Medical science monitors basic research. 2018; 24: 103.

[24]   Joyia GJ, Liaqat RM, Farooq A, Rehman S. The Internet of Medical Things (IoMT): Applications, benefits, and future challenges in the healthcare domain. J Commun 2017; 12(4): 240-7.

[25]   Schüz J, Espina C, Villain P, *et al.* European Code against Cancer 4th Edition: 12 ways to reduce your cancer risk. Cancer Epidemiol 2015; 39 (Suppl. 1): S1-S10.
[http://dx.doi.org/10.1016/j.canep.2015.05.009] [PMID: 26164654]

[26]   Gruson D. Ethics and artificial intelligence in healthcare, towards positive regulation. Soins; la Revue de Reference Infirmiere. 2019. 1; 64(832): 54-7.

[27]   Nikus K, Lähteenmäki J, Lehto P, Eskola M. The role of continuous monitoring in a 24/7 telecardiology consultation service—a feasibility study. J Electrocardiol 2009; 42(6): 473-80.
[http://dx.doi.org/10.1016/j.jelectrocard.2009.07.005] [PMID: 19698956]

[28]   Baker SB, Xiang W, Atkinson I. Internet of Things for Smart Healthcare: Technologies, Challenges, and Opportunities. IEEE Access 2017; 5: 26521-44.

[http://dx.doi.org/10.1109/ACCESS.2017.2775180]

[29] Tyagi S, Agarwal A, Maheshwari P. A conceptual framework for IoT-based healthcare system using cloud computing. In 2016 6th International Conference-Cloud System and Big Data Engineering (Confluence) 2016; 503-7.
[http://dx.doi.org/10.1109/CONFLUENCE.2016.7508172]

[30] Sharma M, Singh G, Singh R. An advanced conceptual diagnostic healthcare framework for diabetes and cardiovascular disorders. arXiv preprint arXiv:190110530 2019.

[31] Zhang P, Schmidt D, White J, Mulvaney S. Towards precision behavioral medicine with IoT: Iterative design and optimization of a self-management tool for type 1 diabetes. 2018 IEEE International Conference on Healthcare Informatics (ICHI). 64-74.
[http://dx.doi.org/10.1109/ICHI.2018.00015]

[32] Gomes BTP, Muniz LCM, Da Silva e Silva FJ, *et al.* A middleware with comprehensive quality of context support for the Internet of things applications. Sensors (Basel) 2017; 17(12): 2853.
[http://dx.doi.org/10.3390/s17122853] [PMID: 29292791]

[33] Baldini G, Botterman M, Neisse R, Tallacchini M. Ethical design in the Internet of things. Sci Eng Ethics 2018; 24(3): 905-25.
[http://dx.doi.org/10.1007/s11948-016-9754-5] [PMID: 26797878]

[34] Parmentier F. Données de santé et intelligence artificielle: une vision géostratégique.
[http://dx.doi.org/10.1016/j.soin.2019.06.013]

[35] Yang J, Luo J, Lin F, Wang J. Content-sensing based resource allocation for delay-sensitive VR video uploading in 5G H-CRAN. Sensors (Basel) 2019; 19(3): 697.
[http://dx.doi.org/10.3390/s19030697] [PMID: 30744050]

[36] Panwar N, Sharma S, Singh AK. A survey on 5G: The next generation of mobile communication. Phys Commun 2016; 18: 64-84.
[http://dx.doi.org/10.1016/j.phycom.2015.10.006]

[37] Garcia-Morchon O, Falck T, Wehrle K. Sensor network security for pervasive e-health. Secur Commun Netw 2011; 4(11): 1257-73.
[http://dx.doi.org/10.1002/sec.247]

[38] Backman W, Bendel D, Rakhit R. The telecardiology revolution: improving the management of cardiac disease in primary care. J R Soc Med 2010; 103(11): 442-6.
[http://dx.doi.org/10.1258/jrsm.2010.100301] [PMID: 20959351]

[39] Li S, Li M, Xu H, Zhou X. Searchable encryption scheme for personalized privacy in IoT-based big data. Sensors (Basel) 2019; 19(5): 1059.
[http://dx.doi.org/10.3390/s19051059] [PMID: 30832294]

[40] Rayan RA, Tsagkaris C, Iryna RB. The Internet of Things for healthcare: applications, selected cases and challenges. IoT in Healthcare and Ambient Assisted Living 2021; pp. 1-5.
[http://dx.doi.org/10.1201/9781003140443]

[41] Mohd A. Internet of Things (IoT) Healthcare Market by Component Global Opportunity Analysis and Industry Forecast, 2014-2021. Allied Market Research 2016.

[42] Cirillo F, Wu FJ, Solmaz G, Kovacs E. Embracing the future internet of things. Sensors (Basel) 2019; 19(2): 351.
[http://dx.doi.org/10.3390/s19020351] [PMID: 30654571]

[43] Bowes A, Dawson A, Bell D. Ethical implications of lifestyle monitoring data in ageing research. Inf Commun Soc 2012; 15(1): 5-22.
[http://dx.doi.org/10.1080/1369118X.2010.530673]

[44] Agrafioti F, Bui FM, Hatzinakos D. Medical biometrics in mobile health monitoring. Secur Commun Netw 2011; 4(5): 525-39.

[http://dx.doi.org/10.1002/sec.227]

[45] Brey P. Freedom and privacy in ambient intelligence. Ethics Inf Technol 2005; 7(3): 157-66.
[http://dx.doi.org/10.1007/s10676-006-0005-3]

[46] Coeckelbergh M. E-care as craftsmanship: virtuous work, skilled engagement, and information technology in health care. Med Health Care Philos 2013; 16(4): 807-16.
[http://dx.doi.org/10.1007/s11019-013-9463-7] [PMID: 23338289]

[47] Rejeb Abderahman, Rejeb Karim, Treiblmaier Horst, *et al.* The Internet of Things (IoT) in healthcare: Taking stock and moving forward. Science Direct. 2023; 11.

[48] Khaled H. Application of internet thing in healthcare domain. Journal of Umm Al-Qura University for Engineering and Architecture. 2023.

## CHAPTER 8

The 5G Revolution in Healthcare: Shaping the Future of Medicine

Natraj N.A.[1,*]**, Prasad J.**[2]**, Bhuvaneswari M.**[3] **and Suriya K.**[4]

[1] *Symbiosis Institute of Digital and Telecom Management, Symbiosis International (Deemed University), Pune, India*

[2] *Department of Electronics and Communication Engineering, KPR Institute of Engineering And Technology, Coimbatore, India*

[3] *Department of Mechatronics Engineering, Sri Krishna College of Engineering and Technology, Coimbatore, India*

[4] *Department of Electronics and Communication Engineering, SNS College of Technology, Coimbatore, India*

Abstract: The next generation of cellular network technology, 5G, is poised to transform the medical industry completely. 5G has the potential to bring novel and forward-thinking approaches to the diagnosis, treatment, and management of diseases thanks to its fast speed, low latency, and large capacity. The proposed article digs into the incredible impact that 5G technology has had on the healthcare field, ushering in a new era that will be distinguished by unprecedented connectedness and innovation. Initially, the evolution of wireless networks is examined, laying the historical groundwork for the paradigm-shifting arrival of 5G is explored. Then, the fundamental features of 5G, focusing on its high speed, low latency, and reliable characteristics, which, when combined, push the limits of what is currently thought to be possible in the healthcare sector, are explored. We then analyse the synergy of 5G in different disruptive technologies like AI and IoT in the proliferation of healthcare care technology. The necessity of data security and the privacy of 5g in healthcare technology are analysed further. Our goal is to provide an illuminating look into the world of medicine in the not-too-distant future when 5G has the potential to transform the way healthcare is delivered, eliminate barriers based on location, and project medical innovation to new heights. This article lets readers understand how 5G will revolutionise the healthcare industry by making it more user-friendly, cost-effective, and efficient.

Keywords: 5g, AI, Disruptive technologies, Health care, IoT, Telemedicine.

---

* **Corresponding author Natraj N.A.:** Symbiosis Institute of Digital and Telecom Management, Symbiosis International (Deemed University), Pune, India; E-mail: natraj@sidtm.edu.in

Sivakumar Rajagopal, Prakasam P., Konguvel E., Shamala Subramaniam, Ali Safaa Sadiq Al Shakarchi & B. Prabadevi (Eds.)

INTRODUCTION

The most recent advancement in cellular network technology is 5G, which stands for the fifth generation. It is aimed to deliver more dependable connections, have lower latency, and quicker speeds than those provided by earlier generations. 5G may completely change many businesses, including the medical, transportation, and manufacturing sectors. 5G can achieve higher performance by utilising various emerging technologies. The Millimetre-wave spectrum, often the mmWave spectrum, is among the most significant. The frequency band known as the mmWave spectrum is far higher than those utilised by earlier cellular network generations [1]. Because of this, 5G can carry more data at once and achieve higher download rates.

The Core Concepts Behind 5G

It is intended that 5G technology will overcome the constraints of its forerunners, such as 4G (LTE) and 3G, by delivering unrivalled speeds, reducing the amount of latency experienced, and increasing the capacity for connectivity. These characteristics, taken together, have a transformative effect on the digital world, opening up doors to opportunities in various fields, including medicine, transportation, manufacturing, and the entertainment industry [1].

Millimetre Wave (mmWave) Spectrum: The Key to Increasing Transmission Speed

Utilisation of millimetre wave (mmWave) spectrum is an essential pillar supporting the development of 5G technology. This spectrum runs at higher frequencies than the bands used by earlier cellular networks. Because of this, 5G can carry enormous amounts of data at speeds that have never been seen before. 5G can reach download and upload speeds of up to 20 gigabits per second (Gbps) by utilising mmWave technology. This incredible speed is nearly 100 times quicker than 4G, which will profoundly change how we interact with digital content. However, it is essential to keep in mind that the mmWave technology does come with some drawbacks. The combination of its shorter wavelength and higher frequency results in less coverage when compared to bands that operate at lower frequencies. Consequently, mmWave technology is optimal for use in urban areas, dense city centres, and sites with a high demand for data.

The capacity multiplier is known as Massive MIMO

Massive Multiple Input, Multiple Output, or MIMO for short, is yet another ground-breaking technology that is essential to 5G. Massive MIMO is a method for sending and receiving data concurrently that requires the installation of massive arrays of antennas at base stations and on user devices. Because of this enormous breakthrough in antenna technology, 5G's capacity to accommodate a large number of users and devices at the same time has been significantly improved. The capability of Massive MIMO to generate numerous data streams contributes significantly to an increase in the effectiveness of the network. The dependability of connections and the amount of data that can be sent within a given frequency range are improved due to this improvement in spectral efficiency. This technology has far-reaching ramifications for applications like the Internet of Things, which require constant communication for many devices. Fig. (**1**) shows the mmWave band in 5G and Fig. (**2**) shows the implementation of Massive MIMO in 5G.

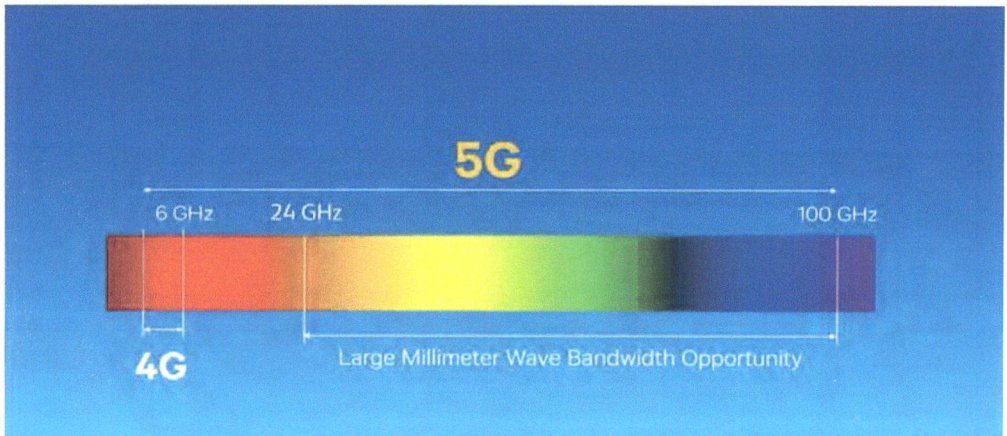

Fig. (**1**). mmWave band in 5G.

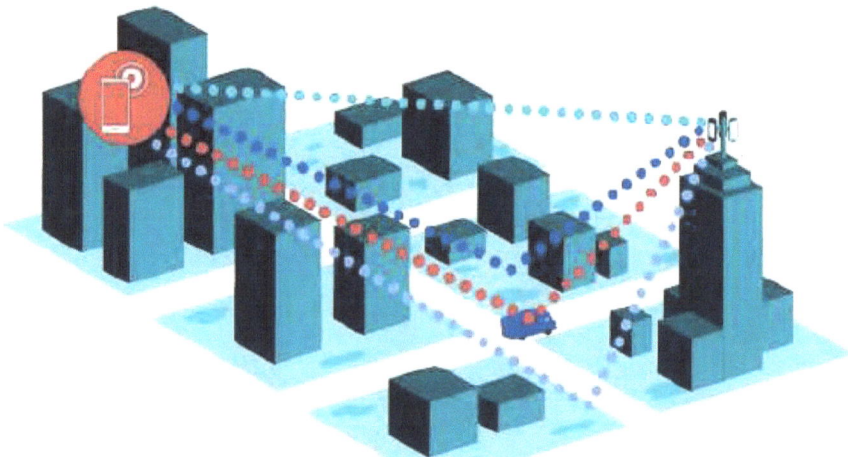

Fig. (**2**). Massive MIMO implementation in 5G.

The Reduction of Latency Is a Game-Changer

One of the most game-changing characteristics of 5G is its lower latency, which refers to the delay in data transfer. The latency of 5G connections can be as low as one millisecond or even lower. This response time, which is very close to instantaneous, is essential for applications that require real-time interactions. Some examples of such applications are remote surgery, autonomous vehicles, and augmented reality experiences. 5G is poised to revolutionise how we interact with technology by reducing latency, making it possible for previously inconceivable applications to become a reality. Fig. (**3**) gives the overview of latency resuction measures in 5G.

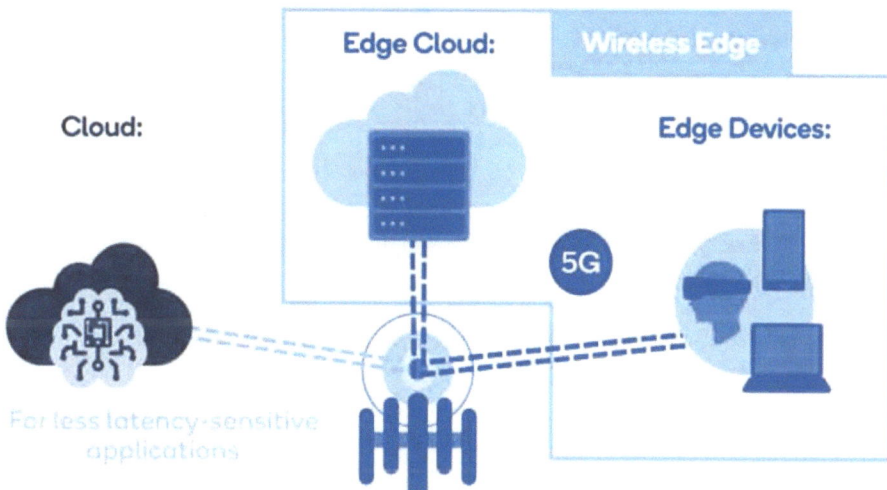

Fig. (**3**). Overview of latency reduction measures in 5G.

The Revolutionary Effects that 5G Will Have on Different Industries

The introduction of 5G technology will usher in a new age marked by increased creativity and productivity across various business sectors. 5G is poised to revolutionise how businesses function and redefine the limits of what is possible because it promises lightning-fast speeds, ultra-low latency, and unsurpassed dependability. In this article, we will investigate how 5G alters the landscape of various industries, including healthcare, transportation, manufacturing, and entertainment.

The Rise of Digital Technology in Healthcare

A paradigm change towards more accessibility, precision, and positive outcomes for patients is what the 5G technology promises to bring about in the field of healthcare. Telemedicine has become an integral part of modern medical practice because of technological advances that allow for the transmission of enormous amounts of data at breakneck speeds. Patients can now consult with medical specialists from the convenience of their homes thanks to high-definition video conferencing and data sharing in real-time. This helps to bridge geographical gaps and increases patients' access to care [2]. Another essential component of 5G's influence on the healthcare industry is the expansion of remote patient monitoring. It is possible for wearable devices equipped with 5G connectivity to collect and transmit vital signs continuously. This gives medical professionals access to real-time data, enabling them to make better-informed decisions. Because of this, it is beneficial for patients with chronic diseases because it makes it possible to take preventative action and minimises the frequency of required in-person visits. One of the most impressive applications of 5G in the medical field is likely to be in the field of remote surgical procedures. Surgeons can now execute procedures on patients located thousands of kilometres away with a level of precision that has never been seen before, thanks to the near-instantaneous response speeds and reliability of 5G. This invention has the potential to revolutionise the field of surgery, making specialised treatment available to locations that are now underserved and improving the overall quality of the delivery of healthcare [3, 8].

Transportation's Quantum Leap

It is impossible to emphasise how significant the impact of 5G will have on the transportation sector. On the way to completely autonomous vehicles, lower latency and smoother communication are necessary. When 5G networks are established, autonomous vehicles can communicate effectively with surrounding infrastructure, pedestrians, and other vehicles in real time. This will ensure that transportation is both safe and efficient. Our streets will be transformed into safer and more efficient thoroughfares due to this game-changing innovation, which will reduce the number of accidents and improve traffic flow. In addition to supporting autonomous vehicles, transportation systems that are enabled with 5G can increase general mobility, traffic flow, and pollution reduction. In case of traffic lights, for instance, the timing of their activation can be dynamically adjusted in response to real-time traffic data, thereby lowering both congestion and fuel usage. Public transport could also benefit from 5G, providing passengers a more connected and streamlined travelling experience.

Productivity Improvements in the Manufacturing Industry

The advent of 5G technology has ushered in a new era of increased productivity and decreased expenses within the field of manufacturing. Implementing technology known as Massive Multiple Input, Multiple Output, or MIMO, in conjunction with the reduced latency offered by 5G, makes it possible to monitor and manage machines in real time. This enables industrial processes to be fine-tuned with greater precision, leading to a gain in efficiency and reduced downtime. In addition, 5G makes it possible for robots and self-driving machinery to coordinate their actions within industrial facilities seamlessly. These robots can work with human workers, increasing production and worker safety. Maintenance activities can also be simplified since sensors connected to 5G networks can offer real-time data on equipment conditions. This enables predictive maintenance, reducing the frequency and severity of breakdowns, which can be expensive [4].

Reconceptualizing Entertainment

The high speeds and low latency of 5G networks represent a significant step forward for the entertainment industry. Experiences in augmented reality and virtual reality are taken to new heights, providing consumers with content that is both immersive and participatory. The options are virtually limitless, ranging from gaming to educational simulations. The combination of high-definition streaming in 4K and 8K and minimum buffering allows viewers to experience content in its finest possible quality with no interruptions. The gaming industry, in particular, stands to gain a great deal from implementing 5G technology. Cloud gaming services can deliver gaming experiences on mobile devices that are on par with consoles' offerings, removing sophisticated hardware requirements. This makes gaming more accessible to a larger population and democratises it simultaneously.

In summary, the effect that 5G technology will have on many industries will be nothing short of revolutionary. Telemedicine, remote monitoring, and even remote surgery become more accessible and efficient due to this development, paving the way for a digital revolution in the healthcare industry. In the realm of transportation, 5G promises to make roads safer and more efficient, while in the realm of industry, it opens the door to new heights of efficiency and automation. The entertainment industry is being revolutionised, and now consumers can participate in activities formerly considered science fiction. We may predict a future in which connectivity and innovation interact in ways that have never been seen before, reshaping industries and improving the quality of life for individuals and society. As 5G networks continue to develop and mature, we can look forward to this future.

5G TECHNOLOGY IN HEALTHCARE

5G wireless technology ushers in a new era of connectivity and innovation. 5G will transform several industries, including healthcare, by providing quicker, more reliable wireless connections. 5G is a massive improvement over 4G (LTE) and 3G. It has high capacity, ultra-low latency, and remarkable speed. These traits will transform the digital landscape, allowing industries to maximise sophisticated connections. 5G could revolutionise healthcare delivery, patient care, and research. Healthcare personnel may quickly access patient records, high-resolution medical photos, and essential data with its lightning-fast data transfer capabilities, reduced wait times, faster diagnosis, and more efficient healthcare delivery. Healthcare uses 5G for telemedicine, remote patient monitoring, augmented reality in medical teaching, and IoT-connected medical devices. Wearable gadgets with 5G connectivity monitor vital indicators continuously, enabling proactive healthcare and reducing hospital stays [5]. With its ability to support many devices simultaneously, 5G improves the healthcare IoT ecosystem, from wearable fitness trackers to advanced medical equipment. As 5G networks evolve, healthcare will benefit from increased accessibility, patient-centric care, and medical research. Technology like 5G will make the future healthier and more connected. Fig. (**4**) shows the different applications of 5G in healthcare sector.

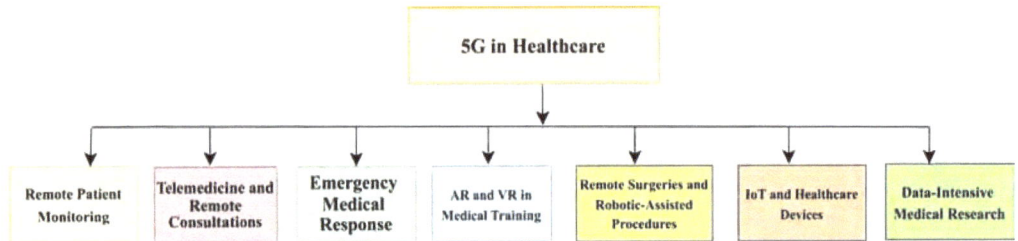

Fig. (4). Overview of 5G Applications in Healthcare.

The advent of 5G technology holds immense potential to revolutionise various domains within the healthcare sector, encompassing patient care and medical research. This is mainly attributed to its unparalleled swiftness, minimal delay, and extensive bandwidth. The most promising areas where 5G is set to impact profoundly are explained below.

Telemedicine and Remote Consultations Revolutionized by 5G: Bridging Healthcare Gaps

Telemedicine and other forms of remote consultation are two areas where 5G promises to impact the healthcare industry significantly. Previously hampered by slow connections, 5G networks' high throughput, low latency, and other desirable characteristics are poised to usher in a new era of telemedicine [6]. This revolutionary combination of 5G and telemedicine has the potential to revolutionise healthcare delivery by increasing reach and improving access to specialists.

Virtual Doctor Visits in High Definition

High-definition video calls are one of the most noticeable improvements to telemedicine made possible by 5G. 5G wireless technology has far lower latency and faster data transfer rates than previous generations. This results in natural, high-quality video and audio communications between doctors and patients, recreating an atmosphere not unlike an in-person consultation. Patients can now communicate clearly and timely with doctors and nurses, regardless of location. The distance between patient and doctor is no longer an issue thanks to these telemedicine consultations. Patients in underserved areas or more remote rural areas can get their medical needs met without leaving their homes. By decreasing the number of patients who must use physical hospitals, the healthcare system can allocate resources more effectively to those who need them most.

Brighter and More Reachable Future for All

Beyond the ease of remote doctor visits, 5G-enabled telemedicine will have far-reaching consequences. The future healthcare system will be more user-friendly, client-focused, and cost-effective. Even in the most remote and underdeveloped areas, people can access high-quality medical care and specialists' quick intervention and direction. Access to healthcare is simplified for the elderly and those with mobility impairments, giving them more independence in managing their health despite their circumstances. The use of telemedicine to reduce healthcare inequities is further strengthened by 5G technology. It enables near-instantaneous transfer of data, which enables remote monitoring of patients with chronic diseases, tracking of their progress, and proactive intervention. By reducing inpatient stays and ER visits, this method improves patient outcomes and lowers overall healthcare expenses. However, as we embrace the revolutionary promise of 5G-powered telemedicine, we must confront some difficulties. It is still essential to ensure that people living in rural and outlying locations have equal access to 5G networks. Strong privacy and data security precautions for patients are mandatory. The licencing, reimbursement, and care standards for telehealth services, among other aspects of the industry, are all in flux and require updated regulatory frameworks.

To summarize, 5G has opened the door to a new era of telemedicine and remote consultations, overcoming limitations in connectivity and increasing access to medical treatment. Distance is no longer a barrier to great care thanks to the healthcare ecosystem made possible by the confluence of high-definition virtual visits and the remarkable capabilities of 5G. Telemedicine, which stands to revolutionise healthcare delivery, accessibility, and the patient experience as 5G networks continue to develop and grow, is already an integral part of the healthcare landscape. This movement is about more than just cutting-edge innovation; it is about making everyone's life better and creating a brighter, more equitable future.

Remote Patient Monitoring Revolutionized by 5G: A Proactive Approach to Healthcare

The introduction of 5G technology into the medical field has heralded the beginning of a revolutionary shift in how patients are cared for, most notably in remote patient monitoring. Wearable devices equipped with 5G connections transform the healthcare landscape by continuously collecting and reporting vital signs. These gadgets take advantage of the unrivalled real-time data transfer capabilities of 5G. Individuals managing chronic conditions profit the most from this proactive strategy since it enables healthcare providers to monitor patients with an accuracy and timeliness that has never been seen before. Fig. (**5**) shows the remote monitoring for health in 5G.

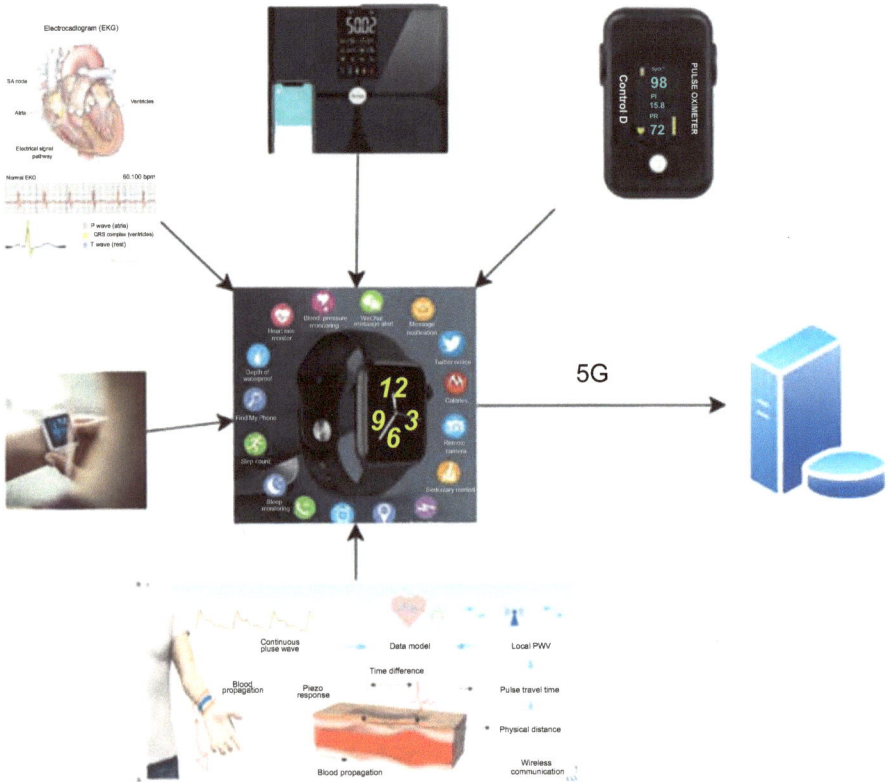

Fig. (5). Remote monitoring for healthcare in 5G.

Monitoring in Real Time Utilising Wearable Technology

The practice of remotely monitoring patients has been hailed for some time as a way to improve the standard of care provided to patients while also enabling them to continue living independently and in an environment that is more comfortable for them at home. However, constraints in connectivity have frequently prevented it from functioning to its full potential. These restrictions are crumbling in the face of the arrival of 5G technology. Wearable devices equipped with 5G connectivity can monitor a wide variety of vital indications, such as a person's heart rate, blood pressure, glucose levels, and more. These gadgets supply a never-ending stream of data that is updated in real time and sent immediately to healthcare providers and electronic health records. Because this data flow is so fluid, medical personnel can get insights regarding a patient's health status and overall well-being that are unavailable to them otherwise.

Early Detection and Preventative Measures: A Proactive Approach

The capability of 5G-enabled remote patient monitoring to provide proactive intervention and early detection of health issues is the most attractive component of this type of monitoring. Patients who suffer from disorders that last for an extended period, such as diabetes, hypertension, or cardiovascular disease, stand to gain much from continuous monitoring. Healthcare professionals can define thresholds and warnings within the monitoring system, enabling early actions if vital signs vary from normal ranges. For instance, if a patient's glucose levels reveal an alarming surge, the healthcare provider can receive an immediate alert and begin conversing with the patient to assess the situation. This can be done to determine whether or not the patient requires medical attention. This early detection and response can prevent a health concern from worsening, reducing the need for care in an emergency setting or hospitalisation of the patient. Patients are provided with timely information and assistance from their healthcare team, which enables them to assume responsibility for their healthcare.

Improving the Outcomes of Patient Care While Cutting Costs of Healthcare

The influence of remote patient monitoring made possible by 5G will transcend beyond the scope of an individual patient's experience. It can improve overall healthcare outcomes while at the same time reducing the strain placed on the healthcare system. Medical professionals can make real-time adjustments to treatment plans, medication dosages, or recommendations for lifestyle changes when patients with chronic diseases are monitored remotely, and trends in their health data are identified. This individualised approach to treatment not only improves patient outcomes but also helps the healthcare system allocate its resources more effectively, which contributes to the overall system's overall cost savings. In addition, using 5G for remote patient monitoring can result in considerable financial savings. When health problems are handled head-on, hospital admissions and emergency department trips decrease, resulting in lower healthcare expenses. Additionally, it decreases the stress placed on healthcare institutions, which is especially helpful during times of high demand like the COVID-19 pandemic.

Implementing 5G into the healthcare industry is causing a revolution in remote patient monitoring, transforming it into a preventative and data-driven approach to healthcare delivery. Wearable devices that are connected with 5G connectivity provide continuous, real-time data transmission, which enables healthcare personnel to monitor patients who have chronic diseases remotely and to react swiftly when it is necessary to do so. This game-changing technology not only boosts patient outcomes and improves their quality of life but also makes the healthcare system more efficient and reduces costs, ultimately redefining the standard of care for patients managing chronic health issues.

Emergency Medical Response using 5G

In emergency medical care, the passage of time is frequently the decisive factor determining whether or not a patient will survive [7, 5]. This essential component of medical treatment has been completely reimagined as a result of the advent of 5G technology, which offers unrivalled rates of speed and dependability and has the potential to determine the difference between a successful intervention and a fatal conclusion. The ultra-low latency of 5G ensures that vital information, such as patient data and real-time video feeds, can be transmitted instantaneously to first responders and medical professionals. This transforms the time it takes emergency medical personnel to respond to a situation and enables quicker decision-making and potentially life-saving interventions.

Regarding emergency medical scenarios, the ultra-low latency that 5G provides is nothing short of a game-changer. It is beneficial to send data in real-time, whether a paramedic reacting to an accident, a remote professional supervising a complicated surgery, or a medical team organising a crucial rescue operation. With 5G, crucial information such as a patient's medical history, diagnostic imaging, and real-time video feeds from the scene of an emergency can be transmitted to medical experts nearly instantaneously.

5G ensures that first responders and medical professionals have access to the information they require precisely when they require it, which is a significant benefit in situations where every second counts. Paramedics can obtain information from specialists in real time, which enables them to make crucial decisions regarding treatment protocols, the administration of medication, or life-saving treatments. Additionally, remote consultations with specialists are made more accessible, enabling fast expert views on arduous instances. This is possible even if the specialist is far from the emergency location [8, 6].

This capability is critical in situations like accidents, cardiac arrests, or traumatic injuries because prompt intervention can make all the difference in these types of situations. 5G's contribution to emergency medical response extends to disaster management, making it possible to coordinate resources and rapidly deploy specialised teams more effectively. This is made possible by 5G's ability to transmit data at higher speeds. As a direct consequence, it is possible to save lives, and the level of care provided in times of emergency is considerably improved.

In a nutshell, the 5G technology represents a game-changing force in emergency medical response. Because of its extremely low latency, it guarantees that vital information will flow in real-time. It also enables medical professionals to make snap decisions based on accurate and up-to-date data, which can be the difference between life and death. 5G ensures that emergency medical response is quicker, more effective, and ultimately more likely to lead to excellent patient outcomes. This is true whether it is a paramedic on the front lines or a professional delivering distant guidance. In urgent medical care, where every second is essential, 5G technology is vital to a better and more promising future.

5G Unleashes Augmented and Virtual Reality (AR/VR) in Medical Training

Augmented reality (AR) and virtual reality (VR) are gaining the spotlight in the new realm of opportunities that have become available due to the advent of 5G technology in medical education and training. Immersive training experiences are now possible thanks to the high speed and low latency of 5G networks, which has changed how healthcare professionals learn, practise, and collaborate. Applications that use augmented and virtual reality are becoming an essential part of medical training. It is to the benefit of surgeons, medical students, and experts alike.

Immersive Surgical Training using AR and VR in the Healthcare Industry

One of the most potent applications of AR and VR in the healthcare industry is immersive surgical training. Virtual reality (VR) simulations offer a risk-free environment in which surgeons can practise performing complex surgeries. Surgeons can use these simulations before going into the operating theatre to hone their abilities, experiment with novel procedures, and boost their confidence by replicating real-life surgical circumstances with surprising precision. This not only raises the level of expertise of working surgeons but also hastens the education of those interested in becoming surgeons. As a result, the learning curve linked with actual operations is lowered.

Lessons on Anatomy That are Interactive

AR and VR technologies in medical education can benefit medical students enormously. Students have the unique opportunity to investigate the human body in all its three-dimensional glory through interactive anatomy classes in virtual reality. They can perform hands-on, immersive activities such as dissecting virtual cadavers, viewing complex anatomical structures from various perspectives, and gaining a profound grasp of human physiology. This method not only improves one's ability to remember information but also helps one develop a deeper understanding of the complexity of the human body.

Real-Time Cooperation Across Borders

The collaborative character of contemporary medicine is also considerably boosted by 5G-enabled augmented and virtual reality (AR and VR). Real-time cooperation sessions are available for specialists and other medical professionals, regardless of their location. This has far-reaching ramifications, enabling specialists worldwide to electronically meet and discuss difficult situations, exchange their views, and cooperatively design treatments [9]. AR and VR make it possible for healthcare professionals to work together more efficiently and effectively, whether collaborating on treating a rare medical condition, a complex surgical procedure, or interdisciplinary consultations.

Integration of AR and VR into medical training contributes to the creation of more knowledgeable and skilled healthcare professionals. Incorporating AR and VR into medical training contributes to the creation of more knowledgeable and skilled healthcare professionals. Surgeons are better prepared to undertake complex surgeries, medical students receive an immersive and exciting education, and specialists can tap into the experience of other experts worldwide to improve their clinical decision-making [10]. In the long run, this will result in more excellent patient care, safer surgical procedures, and overall improvements in health. To summarize, the introduction of 5G technology has allowed augmented reality and virtual reality applications to be used in medical education.

5G for Data Intensive Medical Research in Healthcare

The collection and analysis of data is the engine that drives forward movement in the field of medical research. The introduction of 5G technology represents a revolutionary step forward in medical research, which is becoming increasingly reliant on massive datasets, high-resolution imaging, and complex simulations in the modern era. 5G is ushering in a new era of data-intensive medical research, thereby revolutionising how researchers communicate, analyse data, and speed up the pace of medical breakthroughs. 5G's unrivalled capacity to transport massive volumes of data at astonishing speeds makes it one of the most impressive technological advances in recent history.

The Benefits of Conducting Medical Studies That Rely Heavily on Data

Small-scale trials have given way to complicated, data-intensive endeavours in the field of medical research. These endeavours encompass genetics, drug development, epidemiological studies, and other fields. Accessing, sharing, and analysing massive datasets promptly is necessary if researchers successfully create novel treatments, identify genetic predispositions, and solve the mysteries behind diseases [11]. This is where 5G comes into play, completely altering how medical research is conducted.

Collaboration on a Global Scale and Analysis of Data in Real-time

The capacity of 5G to stimulate worldwide collaboration and real-time data analysis is one of the most significant contributions it will make to medical research. Researchers worldwide can now communicate easily, sharing and analysing enormous datasets in real time. This strategy of working together expedites the discovery process by permitting the pooling of a wide variety of knowledge and resources to conduct more thorough and up-to-date research. One example of a work that requires a large amount of data is the analysis of massive genetic datasets in the field of genomics, which aims to uncover disease markers or possible therapeutic targets. These datasets may be analysed more quickly and effectively by geneticists and researchers thanks to 5G, which may result in early illness diagnosis and treatments that are better tailored to individual patients. The real-time nature of 5G enables epidemiologists to more quickly and precisely monitor disease outbreaks, follow the transmission of illnesses, and make educated decisions amid public health emergencies [12].

Facilitating the Running of Complicated Simulations

Complex simulations are used to model disease processes, drug interactions, and treatment results as part of data-intensive medical research. This is one of the many aspects of medical research. These simulations call for a significant amount of computer power and depend on continuous data streaming. Because 5G can handle data-intensive operations with agility, it gives researchers the capacity to perform simulations more efficiently and precisely, which ultimately leads to a more excellent knowledge of diseases and the creation of more effective treatments.

Innovative Steps Towards Advancing Healthcare

It is impossible to overestimate the importance of 5G in expediting medical advances, mainly when medical research delves deeper into the complexity of diseases and the nuances of the human body. It plays a critical role in fostering innovation, enhancing the precision of research results, and facilitating our progress towards a better knowledge of medical diseases. 5G is at the vanguard of some pioneering and potentially game-changing discoveries in the healthcare field. These discoveries include elucidating the genetic basis of diseases, rapidly developing and testing novel medications, and swiftly responding to newly emerging health concerns [13].

5G technology enables researchers to collaborate globally, analyse large datasets in real time, and precisely carry out sophisticated simulations. This technology not only speeds up the process of discovery, but it also has the potential to completely transform the healthcare industry by paving the way for significant advances in genetics, drug development, epidemiology, and other related fields. The future of medical research is expected to be data-driven, linked, and characterised by quickly translating information into concrete healthcare solutions as 5G networks continue to develop and evolve.

CHALLENGES AND CONSIDERATIONS OF 5G IN HEALTHCARE INDUSTRY

The transformative impact of 5G technology on the healthcare sector is indisputable, with its potential advantages evident. Nevertheless, incorporating 5G technology in the healthcare sector presents a series of obstacles and factors that require careful deliberation to guarantee its effective implementation and extensive influence.

Network Infrastructure and Accessibility

Incorporating 5G technology into the healthcare sector has considerable potential, albeit accompanied by a notable obstacle: a resilient network architecture requirement. The efficacy of 5G technology in the healthcare sector is contingent upon the widespread implementation of small cells and the augmentation of 5G network coverage. This is crucial to guarantee consistent accessibility to this revolutionary technology for healthcare facilities, distant clinics, and marginalised areas [14]. The lack of extensive network coverage may hinder the realisation of the advantages of 5G technology, such as instantaneous remote consultations and data-intensive medical research, hence limiting accessibility for a significant portion of the population.

The Significance of Network Infrastructure

The network infrastructure is the fundamental framework for implementing and operating 5G technologies. The system comprises a complex arrangement of interconnected cellular units, base stations, and fibre-optic links, facilitating the expeditious transfer of information with exceptionally minimal delay. Establishing a robust infrastructure in the healthcare sector is of utmost importance as it permits efficient and rapid data transfer. This infrastructure is essential for enabling real-time remote consultations, supporting wearable devices, and powering innovative medical applications [15].

The Deployment of Small Cells

The fundamental capabilities of 5G rely on small cells, which are cellular radio access nodes characterised by low power and short range. The deployment of these technologies is crucial to expand the coverage of 5G networks in densely populated urban areas and isolated and rural locations. Small cells facilitate 5G connection for healthcare institutions, including remote clinics and healthcare facilities. This facilitates providing advanced healthcare services, such as telemedicine and remote monitoring, to patients residing in underprivileged regions.

Enhancing Accessibility through Broadened Coverage

The expansion of 5G coverage encompasses more than merely enhancing healthcare service accessibility; it also serves as a means to mitigate healthcare inequities. Numerous marginalised communities and remote areas need more dependable internet access, with the challenge of limited availability of high-speed 5G networks. It is imperative to address the digital gap in order to enable equitable access to the healthcare improvements facilitated by 5G, irrespective of individuals' geographical location or economic circumstances.

Maximising the Capabilities of 5G Technology

Establishing comprehensive network infrastructure and expanding coverage plays a crucial role in fully harnessing the capabilities of 5G technology within the healthcare sector [16]. The effectiveness and efficiency of real-time remote consultations, remote patient monitoring, data-intensive medical research, and remote procedures are contingent upon the dependability and speed of 5G networks. The potential of 5G to revolutionise healthcare may be constrained if there is inadequate network infrastructure and limited accessibility.

To summarize, the formidable obstacles associated with deploying network infrastructure and expanding coverage are indispensable to actualising the potential benefits of 5G technology in the healthcare sector. By effectively tackling these obstacles, healthcare stakeholders may ensure the widespread dissemination of 5G's advantages throughout the healthcare ecosystem, thereby fostering improvements in patient care, facilitating advancements in medical research, and augmenting the overall quality of healthcare services [17].

5G-Connected Healthcare Security and Privacy Issues

Data security and privacy become critical as healthcare implements 5G technology to transform patient care and medical research. Rapid patient data transfer and storage in a 5G ecosystem raises serious challenges that require constant attention. Robust security processes are necessary to protect patient data and reduce data breach risks.

Critical Safeguard: Encryption

In 5G-connected healthcare, encryption is essential. Healthcare providers can protect sensitive data from hackers by encrypting it during transmission and storage. This protects patient privacy and prevents data theft without decryption keys.

Controlling Access and Authentication

Data security requires authentication procedures, too. Multi-factor authentication restricts patient data access to medical professionals. Access restrictions restrict healthcare information to those with specific roles and responsibilities. This granular management reduces unauthorised access data breaches.

Storage Data Encryption

Healthcare data is secure in EHRs and data repositories. These systems should encrypt patient data at rest with solid algorithms. Security audits and penetration testing can find weaknesses and maintain encryption effectiveness.

Regulatory Compliance

Compliance with healthcare data protection rules, including HIPAA in the US, is essential. These requirements require healthcare organisations to protect patient data with adequate security measures. If not, healthcare facilities risk legal trouble and reputation damage.

Equitable Access to 5G Technology in Healthcare: Bridging the Divide

As the healthcare industry embraces 5G technology's disruptive promise, equal access for everybody, regardless of location, economic status, or background, becomes crucial. 5G can reduce connectivity and healthcare inequities, which have long been problematic [18]. Prioritising equitable 5G technology access can prevent healthcare inequities from worsening.

The Healthcare Digital Divide

Healthcare has been plagued by the digital divide, which limits access to high-speed internet and technology. Due to poor connectivity, remote areas, rural areas, and economically disadvantaged people sometimes need help accessing healthcare, telemedicine, and health information. Depending on deployment and accessibility, 5G technology might worsen or bridge these discrepancies.

Bridging 5G Gaps

It is vital to take a diversified approach to secure healthcare parity in the era of 5G technology. First and foremost, the expansion of 5G infrastructure needs to reach unserved and rural areas. This requires deploying small cells and base stations to extend network coverage to remote places. Providing low-cost 5G services and equipment across all socioeconomic levels is also extremely important. This can be accomplished through subsidies, incentives, and strategic community partnerships to cut access prices [19]. In addition, providing digital literacy education and training on healthcare services connected to 5G networks is an essential component of equitable access. This is in recognition of the fact that a large number of people are likely to require assistance to make good use of these resources. Integrating telemedicine projects is crucial since it makes it possible for medical personnel to access patients regardless of their location or connectivity status, hence eliminating geographical barriers to care. In conclusion, creating community relationships with local governments, advocacy groups, and community organisations is a collaborative strategy to identify and remove access hurdles experienced by marginalised communities, paving the path towards a more equitable healthcare landscape. This may be accomplished by working together to identify and address these barriers.

The Way Forward

Equitable access to 5G healthcare technology is ethical, improves results, and reduces inequities. Healthcare stakeholders can work together to ensure that 5G technology creates a more equitable healthcare landscape where everyone can benefit from its advancements by prioritising infrastructure expansion, affordability, digital literacy, telemedicine initiatives, and community partnerships.

In summary, the use of 5G technology in the medical field holds the potential to bring about revolutionary improvements. Still, it has its share of difficulties and essential considerations. To create a 5G healthcare ecosystem that is both effective and ethical, it is necessary to address network infrastructure deficiencies, protect patient data, modify regulatory and reimbursement frameworks, provide equal access, and promote interoperability [20]. By working together to find solutions to these problems, stakeholders in the healthcare industry can unlock the full potential of the fifth-generation wireless standard (5G) to boost patient care, advance medical research, and propel innovation in healthcare.

THE FUTURE OF 5G IN HEALTHCARE: UNLEASHING THE POWER OF CONVERGENCE

There is a strong correlation between the development of 5G technology and the future of healthcare, which holds the promise of revolutionary shifts that will affect not only patient care but also medical research and the entire healthcare ecosystem. When we look to the future, we see several significant trends and emerging technologies, the most notable of which are artificial intelligence (AI) and machine learning (ML), that are set to complement and amplify the capabilities of 5G in the healthcare industry, so ushering in a new era of innovation and precision.

Real-time Healthcare Delivery *via* Remote Devices

The extraordinary speed of 5G and its incredibly low latency will make it possible to provide real-time remote healthcare on a scale never before seen. Telemedicine will evolve to become more interactive and responsive, enabling medical professionals to conduct virtual consultations, diagnose complex illnesses, and monitor patients with an accuracy that was previously unattainable. The diagnostic capabilities of remote healthcare services will be significantly improved using AI-driven testing and analysis technologies.

Diagnostics and Decision Support Powered by Artificial Intelligence

The combination of 5G and AI is expected to usher in a new era of diagnostics and decision assistance in the medical field. AI algorithms enable real-time medical data analysis, paving the way for accurate and prompt diagnosis. For instance, AI-driven image recognition systems can assist radiologists in spotting anomalies in medical images [20, 10]. The ability of machine learning algorithms to forecast illness patterns and identify at-risk groups enables proactive healthcare measures to be considered.

Computing in the Periphery for Real-Time Insights

Edge computing is made feasible by 5G, enabling data processing to occur closer to the source of the data, lowering latency and making it easier to gain immediate insights. In medicine, this denotes that wearable devices and sensors can process and send vital signs and health data in real-time, making it possible for preventative interventions for patients who suffer from chronic illnesses. When deployed at the edge, AI algorithms have the potential to enable continuous monitoring as well as early detection of health problems.

Individualised Medical Treatment and the Research and Development of New Drugs

The development of personalised medicine will be pushed forward more quickly by 5G, AI, and ML. The ability to analyse genetic data and effortlessly communicate it enables the development of treatment regimens that are specifically suited to an individual's genetic makeup. Simulations and data analysis that are driven by artificial intelligence will also assist medication discovery and development. This might ultimately lead to more tailored therapies and faster drug approvals.

Improvements in Research and Collaborative Efforts

The cooperative aspect of medical research will be facilitated more by 5G and AI technologies. Researchers worldwide can work together in real time, sharing and analysing big datasets, which speeds up the process of making new medical discoveries. Tools that AI drives can be used to aid in the process of sorting through massive amounts of research data, locating patterns, and providing insights that can lead to scientific discoveries.

In a nutshell, the future of 5G in healthcare will be a convergence of technological advancement and innovative medical practices. As 5G networks mature and AI and ML advance, the healthcare industry will experience a paradigm shift. Real-time remote healthcare delivery, AI-powered diagnostics, edge computing, personalized medicine, and enhanced research collaboration are just a glimpse of the possibilities. By harnessing these advancements, healthcare can become more accessible, precise, and efficient, ultimately improving patient care and advancing medical science.

CONCLUSION

5G is poised to transform several industries in novel ways as a beacon of innovation in the rapidly changing technological landscape. With numerous benefits over its predecessors, including lightning-fast speeds, low latency, reliable connectivity, and the capacity to support vast numbers of devices, 5G has the potential to transform communication and entire economic sectors completely. The revolutionary potential it brings has mainly spurred interest in its application in healthcare. The impact of 5G on healthcare is extensive, with promises of better patient care, remote surgery capability, telehealth consultation ease, and improved wearable sensor performance for real-time health monitoring. It can cross geographic boundaries, providing medical knowledge to underserved areas and opening new channels for precise and accessible treatment. However, there are obstacles to realising 5G's potential. A substantial financial barrier is presented by the expense of implementing 5G infrastructure, which includes the deployment of tiny cells and the expansion of network coverage to rural places. Additionally, the availability and distribution of spectrum frequencies for 5G use demand rigorous organisation and planning. To successfully implement 5G in healthcare, issues with data security, privacy, regulatory frameworks, and equitable access to technology must also be resolved. Despite these difficulties, 5G's potential is still evident. Its ability to reinvent businesses and foster innovation highlights its importance in our constantly connected society. Obstacles can be overcome with continued research, partnerships, and clever approaches, enabling 5G to reach its full potential and significantly impact the healthcare industry and many other industries. As we negotiate the rapidly changing world of 5G technology, one thing is sure: its transformational power is poised to impact the future in ways we have only just begun to envisage.

REFERENCES

[1]   Devi DH, Duraisamy K, Armghan A, *et al.* 5g technology in healthcare and wearable devices: A review. Sensors (Basel) 2023; 23(5): 2519.
      [http://dx.doi.org/10.3390/s23052519] [PMID: 36904721]

[2]   Bhattacharya S. The impact of 5g technologies on healthcare. Indian J Surg 2023; 85(3): 531-5.
      [http://dx.doi.org/10.1007/s12262-022-03514-0]

[3]   Natraj NA, Mitra S, Hallur GG. An investigative study on internet of things in healthcare In Handbook of Research on Machine Learning-Enabled IoT for Smart Applications Across Industries. IGI Global 2023; pp. 116-26.

[4]   Joshi S, Tiwari V, Agarwal B. 5G in healthcare: a survey. In 5G Wireless Communication System in Healthcare Informatics, CRC Press, 2023; pp. 1-10.

[5]   Dhaliwal A. Adopting 5G-enabled e-healthcare for collaborative pandemic management. International Journal of e-Collaboration (IJeC). 2023 Jan 1;19(1):1-8.

[6]   Singh S. 5G enabled network technology trends for smart healthcare systems. 5G Wireless Communication System in Healthcare Informatics. 2023 May 9:29-43.

[7]   Gnanasankaran N, Subashini B, Sundaravadivazhagan B. Amalgamation of deep learning in healthcare systems. In Deep Learning for Healthcare Decision Making, River Publishers, 2023; 10: pp. 1-23.
[http://dx.doi.org/10.1201/9781003373261-1]

[8]   Mijwil MM, Aggarwal K, Doshi R, Hiran KK, Sundaravadivazhagan B. Deep learning techniques for COVID-19 detection based on chest X-ray and CT-scan images: a short review and future perspective. Asian J Appl Sci 2022; 10: 3.
[http://dx.doi.org/10.24203/ajas.v10i3.6998]

[9]   Gautam A, Chirputkar A, Pathak P. Opportunities and challenges in applying artificial intelligence-based technologies in the healthcare industry. 2022 International Interdisciplinary Humanitarian Conference for Sustainability (IIHC) 2022; 1521-4.
[http://dx.doi.org/10.1109/IIHC55949.2022.10059767]

[10]   Saxena P, Prabhu S. Framework for predicting suicidal attempts using healthcare data and artificial intelligence. 2023 International Conference on Innovative Data Communication Technologies and Application (ICIDCA) 2023; 1085-9.
[http://dx.doi.org/10.1109/ICIDCA56705.2023.10099967]

[11]   Chatterjee S, Kulkarni P. Healthcare consumer behaviour: the impact of digital transformation of healthcare on consumer. CARDIOMETRY 2021; (20): 135-44.
[http://dx.doi.org/10.18137/cardiometry.2021.20.134143]

[12]   AlQahtani SA. An evaluation of e-health service performance through the integration of 5G IoT, fog, and cloud computing. Sensors (Basel) 2023; 23(11): 5006.
[http://dx.doi.org/10.3390/s23115006] [PMID: 37299731]

[13]   Natraj NA, Bhavani S. A hybrid power efficient localization scheme for wireless sensor networks. Int J Appl Eng Res 2014; 9(23): 23219-33.

[14]   Attaran M. The impact of 5G on the evolution of intelligent automation and industry digitization. J Ambient Intell Humaniz Comput 2023; 14(5): 5977-93.
[http://dx.doi.org/10.1007/s12652-020-02521-x] [PMID: 33643481]

[15]   Adu-Manu KS, Koranteng GA, Brown SN. 5G enabling technologies: revolutionizing transport, environment, and health.

[16]   Wilson S, Thangamani M, Konguvel E. Detection of breast tumour and speckle noise removal using bilateral filter and bivariate shrinkage. Int J Comput Appl 2015; 116: 3.
[http://dx.doi.org/10.5120/20318-2383]

[17]   Natraj NA, Rathish CR, Gopinath S, Sindhuja P, Madhumitha MS. Technologies and impact of 5g In IoT and artificial intelligence. Empirical Aspects of Advancements in Science, Engineering and Technologies.:66.

[18]   Elango K, Muniandi K. A low-cost wearable remote healthcare monitoring system. Role of Edge Analytics in Sustainable Smart City Development: Challenges and Solutions. 2020; 15:219-42.
[http://dx.doi.org/10.1002/9781119681328.ch11]

[19]   Natraj NA, Kamatchi Sundari V, Ananthi K, Rathika S, Indira G, Rathish CR. Security enhancement of fog nodes in IoT networks using the IBF scheme. International Conference on Image Processing and Capsule Networks 2022; 119-29.
[http://dx.doi.org/10.1007/978-3-031-12413-6_10]

[20]   Krishna KM, Madhan MG, Ashok P. Performance predictions of VCSEL based cascaded fiber-FSO RoF system for 5G applications. Optik (Stuttg) 2022; 257: 168740.
[http://dx.doi.org/10.1016/j.ijleo.2022.168740]

CHAPTER 9

Generative Adversarial Networks in Medical Imaging: Recent Advances and Future Prospects

Harshit Poddar[1] and **Sivakumar Rajagopal**[2,*]

[1] *School of Electronics Engineering (SENSE) Vellore Institute of Technology, Vellore, India*

[2] *Department of Sensor and Biomedical Technology, School of Electronics Engineering, Vellore Institute of Technology, Vellore, India*

Abstract: Generative Adversarial Networks (GANs) represent a significant breakthrough in the realms of machine learning and deep learning, providing novel solutions to the constraints of conventional generative models. This article explores the transformative uses of GANs in the domain of medical imaging, specifically focusing on super-resolution applications in Magnetic Resonance Imaging (MRI), generation of synthetic images for skin lesion categorization, and overall improvement in diagnostic accuracy. The fundamental structure of GANs, comprising a Generator and a Discriminator engaged in adversarial training, facilitates the creation of high-fidelity synthetic medical images. These developments play a crucial role in fortifying machine learning models through the amalgamation of synthetic data with authentic medical datasets, thereby enhancing the precision of diagnostic algorithms and the standard of healthcare provision. Notable innovations encompass the Fused Attentive GAN (FA-GAN) for enhanced MRI clarity and the employment of Pix2Pix GANs for precise brain imaging. Moreover, GAN-centric techniques for the classification of skin lesions, leveraging the ISIC dataset, have showcased substantial enhancements in diagnostic efficacy. Despite their considerable potential, the incorporation of GANs in the healthcare domain necessitates careful navigation of key ethical considerations like patient confidentiality and bias alleviation. It is vital to underscore the need for robust assessment metrics beyond visual accuracy to ensure the clinical applicability of GAN-generated data. This manuscript underscores the continual progressions and the imperative requirement for ethical governance in the utilization of GANs, which hold the potential to transform personalized healthcare, expedite pharmaceutical discoveries, and enrich telemedicine, representing a significant stride forward in medical research and patient welfare.

Keywords: GANs (Generative Adversarial Networks), Medical Imaging, MRI (Magnetic Resonance Imaging), Super-resolution, Skin lesion classification image synthesis.

* **Corresponding author Sivakumar Rajagopal:** Department of Sensor and Biomedical Technology, School of Electronics Engineering, Vellore Institute of Technology, Vellore, India; E-mail: rsivakumar@vit.ac.in

Sivakumar Rajagopal, Prakasam P., Konguvel E., Shamala Subramaniam, Ali Safaa Sadiq Al Shakarchi & B. Prabadevi (Eds.)

INTRODUCTION

Generative Adversarial Networks (GANs) are an assortment of artificial neural networks that are utilized in the domains of machine learning and deep learning. They were first introduced in 2014 by Ian Goodfellow and his colleagues as a solution to the limitations encountered by previous generative models, which faced difficulties in generating data that accurately depicted the intricacy and statistical properties of real-world data [1]. Traditional methodologies, such as manual feature engineering and probabilistic models like Gaussian Mixture Models, proved inadequate in producing diverse and realistic data. The realm of medical imaging stands as a fundamental pillar of contemporary healthcare, furnishing crucial insights into the internal structures and functions of the human body. Modalities like Magnetic Resonance Imaging (MRI), Computed Tomography (CT) scans, and diverse forms of microscopy serve as indispensable instruments for healthcare practitioners. Nevertheless, these technologies come with inherent constraints such as image resolution, clarity, and the availability of meticulously labeled datasets. Herein lies the transformative potential of GANs, which offer innovative solutions to enrich the quality and applicability of medical images.

To confront these challenges, the invention of GANs occurred, presenting a more adjustable and potent approach to generative modeling [2]. The core structure of GANs, as depicted in Fig. (**1**), consists of a dual-network architecture with a generator and a discriminator that participate in adversarial training. The generator, which functions as a transfer function, employs input noise to generate data that closely aligns with the desired data distribution [2 - 6], while the discriminator is tasked with differentiating between authentic data and synthetic data. The triumph of a GAN model is determined by achieving an overall loss of 0.5, as indicated in Equation (1).

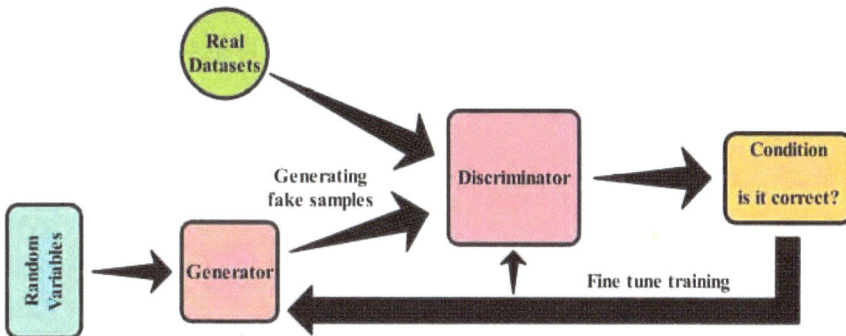

Fig. (1). The fundamental structure of Generative Adversarial Networks (GANs).

GANs have a significant impact on the healthcare sector due to their profound applications. Particularly, they demonstrate immense potential in the generation of synthetic medical images, which are essential in the training and assessment of algorithms used for medical image processing [6 - 10, 12 - 19]. The integration of synthetic data into real medical datasets through GANs enhances the resilience of machine learning algorithms, thereby enhancing the accuracy of diagnoses and the overall quality of healthcare. Furthermore, GANs play a pivotal role in the development of predictive models that can identify subtle patterns in patient data, enabling early detection and prognosis of diseases [12].

A compelling application of GANs in medical imaging materializes in the refinement of MRI images through super-resolution methodologies. Traditional MRI technology, while informative, often grapples with limitations in image resolution. High-resolution images play a pivotal role in precise diagnostic and therapeutic determinations, especially in intricate anatomical sites like the brain. Tailored for super-resolution tasks, GANs have exhibited the capacity to substantially amplify the clarity and intricacy of MRI images. Techniques such as the Fused Attentive GAN (FA-GAN) harness the adversarial training process to generate images with heightened resolution and diminished noise, thereby facilitating more precise and confident diagnoses. Another notable application of GANs emerges in the creation of synthetic images for skin lesion classification. Skin cancer, encompassing melanoma, stands as a critical domain where early and precise diagnosis can profoundly influence patient outcomes. Machine learning models trained on extensive datasets of skin lesion images are increasingly leveraged to aid in diagnosis. Nonetheless, acquiring adequate labeled data poses a significant hurdle. GANs tackle this challenge by producing realistic synthetic images to supplement existing datasets. By augmenting the diversity and volume of training data, GANs contribute to enhancing the accuracy and resilience of diagnostic algorithms. For instance, through the utilization of the International Skin Imaging Collaboration (ISIC) dataset, GAN-generated images have played a pivotal role in refining the performance of machine-learning models dedicated to classifying skin lesions.

$$\min_{G} \max_{D} V(D,G) V(D,G) = \mathbb{E}_{x \sim pdata(x)}[\log D(x)] + \mathbb{E}_{z \sim p_z(z)}[\log(1 - D(G(z)))] \quad (1)$$

G = Generator

D = Discriminator

$P_{data}(x)$ = distribution of real data

P(z) = distribution of generator

x = sample from $P_{data}(x)$

z = sample from P(z)

D(x) = Discriminator network

G(z) = Generator network

Continuing our examination of the applications of Generative Adversarial Networks (GANs), we observe their notable impact on the pharmaceutical field. Various categories of GANs, as depicted in Fig. (**2**), are specifically designed to tackle distinct tasks and overcome various challenges. Instances of such GANs encompass Vanilla GANs, cGANs, WGANs, CycleGANs, StyleGANs, SRGANs, LAPGAN, and DCGAN [4, 10].

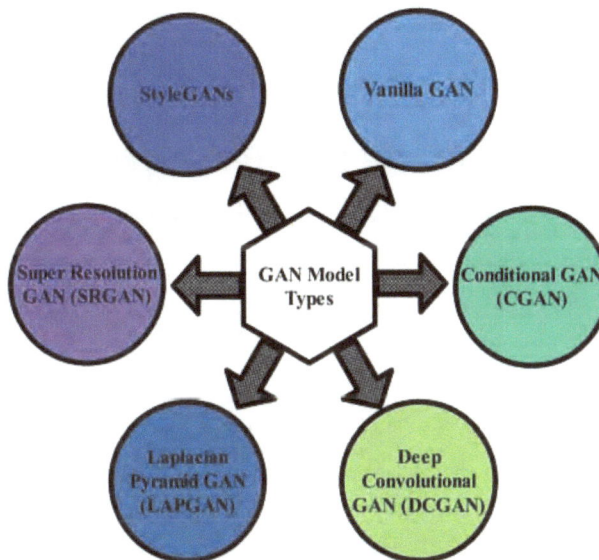

Fig. (2). Different forms of Generative Adversarial Networks (GANs).

RECENT DEVELOPMENTS

Generative Adversarial Networks (GANs) have emerged as a transformative phenomenon in the realm of medical imaging, providing groundbreaking solutions to enduring obstacles across diverse applications [20 - 25]. In this exhaustive exploration, we delve into a series of interconnected themes elucidated in scholarly articles and research papers. These themes not only underscore the potential of GANs but also accentuate the necessity for ethical considerations, rigorous evaluation metrics, and continuous research endeavors to harness the full potential of GANs while effectively addressing their challenges.

The commencement of our expedition is marked by the inception of a Fused Attentive Generative Adversarial Network (FA-GAN) as a formidable instrument for the amplification of MRI image resolution (Fig. **3**). Jiang *et al.* [1] introduce the FA-GAN paradigm, whereby HR references are employed during the training process to direct the optimization of model parameters. This methodology holds the potential to augment the caliber of magnetic resonance images (MRI), thereby bestowing outcomes of greater excellence in comparison to alternative techniques, as illustrated by the utilization of Fréchet Inception Distance (FID) scores [1].

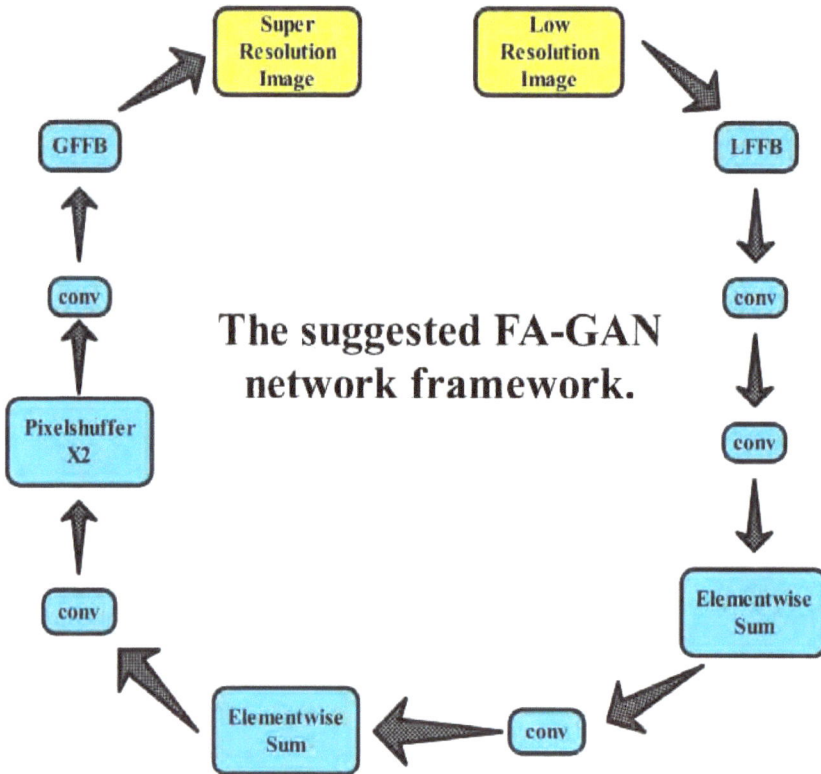

Fig. (3). The structure of the FA-GAN network that has been suggested is composed of the following components: the local feature fusion block, abbreviated as LFFB, and the global feature fusion block, abbreviated as GFFB [1].

The utilization of this particular model is confined to LR images exclusively, which are employed as the primary input to produce SR images. The process can be accurately described as follows:

$$I_{LR} = f(I_{HR}) \qquad (2)$$

$$I_{HR} = g(I_{LR}) = f_1(I_{LR}) + R \qquad (3)$$

where I_{LR}, I_{HR} $\in \mathbb{R}m \times n$ represent LR and HR, respectively, and MRI images of size $m \times n$ and f: I_{HR} $\in \mathbb{R}m \times n$ $\rightarrow$ I_{LR} $\in \mathbb{R}m \times n$ denotes the down-sampling process which creates a LR counterpart from an HR image [1].

The primary findings of the paper suggest that the FA-GAN method integrates HR references to guide the optimization of model parameters during the training process. Furthermore, Table **1** presents a comparison between the model proposed by Jiang *et al.* [1] and other existing models, with the bold and italic values representing the proposed model. Moreover, the GAN learning is made stable through the implementation of spectral normalization (SN) in the discriminator network. The LFFB comprises three-pass networks that employ convolution kernels of different sizes, enabling the extraction of image features at various scales.

Table 1. The average FID (4×) was computed for different methods [1].

Method FID	SRGAN	CA-SR-GAN	SA-SR-GAN	FA-GAN
Cardiac	35.78 ± 2.58	26.65 ± 2.49	24.23 ± 2.32	18.97 ± 2.13
Brain	20.38 ± 3.06	17.22 ± 3.19	16.57 ± 2.86	12.43 ± 2.36
Knee	43.87 ± 4.21	39.89 ± 4.34	38.43 ± 3.71	33.59 ± 3.79
MMWHS	35.32 ± 3.47	34.77 ± 3.59	32.55 ± 2.96	28.15 ± 2.98

The GFFB consists of three elements: the channel attention module, the self-attention module, and the fusion operation. These components enhance important features of the MRI image, resulting in a super-resolution image that closely resembles the original image [9, 10]. To selectively highlight informative characteristics and suppress less useful ones, the channel attention mechanism utilizes a one-dimensional vector derived from global data. The paper proposes a novel technique for generating high-resolution MRI images using an attentive generative adversarial network [26]. This approach demonstrates its capability to produce high-quality images that closely resemble the original image.

Moving forward, Bissoto *et al.* [2] direct their attention towards the utilization of Generative Adversarial Networks (GANs) in the analysis of skin lesions. Their investigation highlights the potential of synthetic images as a substitute for real ones and the advantages this substitution can bring to test datasets that exist outside the distribution [27]. While placing significant emphasis on the favorable outcomes of GAN-based augmentation, the authors also emphasize the significance of ethical considerations stemming from the inherent expenses and risks associated with the implementation of GANs in medical applications. This underscores the requirement for a balanced approach in harnessing the capabilities of GANs in the realm of healthcare [2].

Table 2. Optimal epochs and FID values for each generative model, guiding training dataset checkpoint selection based on FID and GAN training duration [2].

GAN Arch.	Epochs	FID
SPADE	300	16.62
Pix2pixHD	400	19.27
PGAN	890	39.57
StyleGAN2	565	15.98

The construction of the reference training set of genuine images was based on the training set of the ISIC 2019 challenge. This particular training set was partitioned into a training set consisting of 14,805 samples and a validation set comprising 1,931 samples, which were employed in all of the conducted experiments [2]. Conversely, the test set was composed of 3,863 samples.

In conjunction with the previously mentioned training and validation sets, the authors also integrated the clinical attribute of semantic masks of the ISIC Challenge 2018 Task 2. However, it is important to note that these masks were only accessible for 2,594 images, making up approximately one-sixth of the training set. To carry out the experiments, the researchers employed five gold-standard datasets, specifically derm7pt and dermofit, while making adjustments to eliminate certain classes or duplicates [2]. To ensure the reliability of their findings, the authors replicated each experiment ten times. This involved varying the selection of authentic images from the training set and repeatedly conducting fine-tuning of the target model. In the realm of brain imaging, Aljohani *et al.* [3] have introduced a profound pix2pix GAN methodology. Their investigation reveals the efficacy of their approach in generating accurate synthetic medical images when compared to other methods. The Pix2Pix framework, encompassing the generator and discriminator components, assumes a pivotal role within this methodology [28, 29]. The generator establishes a correlation between the source and target images, whereas the discriminator distinguishes between genuine and counterfeit images. This underscores the adaptability of GANs in tackling specific challenges within medical imaging, such as brain imaging [3].

The comprehensive system known as the Pix2Pix framework consists of two main components: a generator and a discriminator. The foundation of this framework lies in the principles of the Conditional Generative Adversarial Network (CGAN) [28]. The generator's primary purpose is to establish a connection between the source image x and a randomly generated noisy image z, ultimately resulting in the creation of the target image y. This can be described as a mapping process. On the other hand, the discriminator plays a crucial role in distinguishing between genuine and counterfeit images, particularly concerning y when given x. The objective function of Pix2Pix can be represented by equation (4) as indicated in Fig. (**4**) [3].

$$L_{Pix2Pix}(G,D) = \mathbb{E}_{x,yPdata(x,y)}[\log D(x,y)] + \mathbb{E}_{xPdata(x),zPz(z)}[\log(1 - D(x,G(x,z)))] \quad (4)$$

1: **for** *iteration* =number of training... **do**
2:    **for** *k* steps... **do**
3:        Sample minibatch of m samples $y^1,...,y^m$ from data distribution $p_{data}(y)$.
4:        Sample minibatch of m noise samples $z^1,...,z^m$ from noise prior $p_g(z)$.
5:        Sample minibatch of m samples $x^1,...,x^m$ from data generating distribution $p_{data}(x)$.
6:        Update the discriminator by ascending its stochastic gradient:
$\nabla_{\theta d} \frac{1}{m} \sum_{i=1}^{m} [log D(x^i,y^i)] + log(1 - D(x^i,G(x^i,z^i)))]$
7:        **end for**
8:    Sample minibatch of m noise samples $z^1,...,z^m$ from noise prior $p_g(z)$.
9:    Update the generator by descending its stochastic gradient: $\nabla_{\theta g} \frac{1}{m} \sum_{i=1}^{m} log(1 - D(x^i,G(x^i,z^i)))$
10: **end for**

Fig. (4). The pseudo-code representing the training process of Pix2Pix GAN is provided in the literature. It is important to note that the variable "k" denotes the quantity of network updates that occur during each iteration of the training process [3].

Table 3. The results of the three models in the dataset of brain slice images were evaluated based on the metrics of mean absolute error (MAE), peak signal-to-noise ratio (PSNR), and structural similarity index measure (SSIM) [3].

Method	MAE	PSNR	SSIM
Cycle GAN	0.059	59.56	0.0344
Pix2Pix GAN	0.057	60.51	0.354
Deep Pix2Pix GAN	0.038	61.32	0.391

The authors put forward a plethora of potential applications for their approach, encompassing both the training of machine learning models and the mitigation of the transmission of substantial quantities of medical images. However, they recognize the presence of challenges and limitations when using synthetic medical images, such as the need to ensure that these images accurately depict real-world medical conditions and the possibility of biased training data. The algorithm functions in the following manner [3].

The initial step involves the creation of a dataset consisting of paired images, where each pair consists of an actual medical image and a corresponding synthetic image. Various techniques, such as image rotation, scaling, and translation, are employed to generate the synthetic images. Subsequently, the Pix2Pix GAN is trained on this image dataset using a combination of adversarial loss, which promotes the production of realistic images, and L1 loss, which encourages the production of images that resemble those in the dataset [3].

Upon completion of the training process, the GAN is capable of generating new synthetic medical images from a single input image. The generator uses the input image to produce a synthetic image, which is then evaluated for its realism by the discriminator. Both the generator and discriminator are trained simultaneously in an adversarial manner, where the generator aims to create increasingly lifelike images, while the discriminator strives to distinguish between genuine and synthetic images [3]. This process continues until the generator achieves the creation of synthetic images that are indistinguishable from authentic medical images.

Qin *et al.* [4] postulated a technique for generating images in the process of categorizing skin lesions by utilizing Generative Adversarial Networks (GANs). Their methodology, predicated upon the International Skin Imaging Collaboration (ISIC) 2018 dataset, highlights the potential of GANs in augmenting the accuracy and efficacy of skin lesion categorization. The precision and accuracy of their model emphasize the impact GANs can exert on the realm of healthcare, particularly in the precise identification and classification of skin lesions. Table **4** showcases the outcomes of the model proposed by Qin *et al.* [4], with the noteworthy results being highlighted in bold.

Table 4. Model classification outcomes [4].

Class. Model	Accuracy	Sensitivity	Precision	AP
CNN	0.936	0.499	0.631	0.599
ResNEt50	0.936	0.544	0.563	0.535
Tranf.- ResNet50	0.944	0.714	0.718	0.820
Tansf.-ResNet50+	0.952	0.743	0.769	0.831

Abbr. Transfer-ResNet50+ refers to the utilization of Transfer-ResNet50 in conjunction with the application of data augmentation techniques.

The investigation carried out by Qin *et al.* [4], employs an open dermoscopy dataset that has been made accessible by the International Skin Imaging Collaboration (ISIC) 2018 classification challenge, which is also known as the ISIC 2018 dataset. This dataset consists of 10,015 dermatoscopic images of skin lesions that have been collected from Austria and Australia. The images are 24-bit RGB images with a consistent size of 600 x 450 pixels and a resolution of 96 dots per inch. The ISIC 2018 dataset encompasses seven diagnostic categories that cover more than 95% of skin lesion types. These categories are explicitly defined within the dataset, and concise explanations of each category are provided. The dataset exhibits notable differences within the same category and significant similarities across different categories [4]. Furthermore, it is unbalanced, with melanocytic nevi accounting for the majority of images while dermatofibroma represents a small portion. The images within the dataset vary in terms of image quality, perspectives, and lighting conditions, and may contain artifacts such as hair or marks.

The proposed technique yielded an accuracy rate of 0.952, with a sensitivity of 0.743 and specificity of 0.966, when compared to alternative methods. The authors suggest that this approach holds the potential to improve the accuracy of skin lesion diagnosis in clinical settings. However, further research is necessary to enhance the process of image synthesis and broaden the range of generated images. Li *et al.* [5] expand the scope of their study by delving into the broader landscape of deep learning methods in the field of medical image superresolution. They draw attention to the challenges faced in medical imaging, such as the limitation of spatial resolution, and investigate a variety of super-resolution techniques that encompass approaches based on both optimization and learning. The authors provide insight into the potential applications of deep learning in diverse medical imaging modalities, including but not limited to CT, MRI, retinal vascular fundus images, microscopy, and endoscopy [5].

Skandarani and colleagues [6] conducted a comprehensive examination of various GAN architectures and their applications in three distinct medical imaging modalities: cardiac cine-MRI, liver CT, and RGB retina images. The datasets differ in terms of size, image modalities, and organ shapes, thereby encompassing a diverse assortment of data for training and evaluation purposes [30]. The investigation highlights the adaptability of GANs in tackling the intricacies of medical data, with a particular focus on image synthesis. FID scores are utilized to quantitatively assess the quality of synthesized images, emphasizing the visual realism achieved by GANs. Nonetheless, the study also underscores the limitations in reproducing the complete complexity of medical datasets, thereby necessitating further research and advancements in generative models [6].

The findings of the study indicate that GANs can generate medical images that exhibit a realistic appearance according to FID standards, successfully deceiving trained experts in a visual Turing test. However, when it comes to segmentation results, none of the GAN models were able to replicate the comprehensive richness observed in actual medical datasets. This observation highlights the limitations associated with traditional metrics employed to evaluate GANs and proposes that task-based evaluations offer a more robust alternative. The research concludes that the effectiveness of GANs as a source of medical imaging data is unreliable, even if the generated images closely resemble real data. Consequently, the study emphasizes the necessity for further investigation into GANs that take into account the intricacies inherent in medical data to develop improved generative models.

Table **5** is an extensive resource that offers a comprehensive overview of the various datasets and targets that necessitate the implementation of diverse types of Generative Adversarial Network (GAN) models for distinct objectives.

Table 5. Various datasets and targets necessitate the utilization of diverse types of GAN models for different purposes [2].

Ref_{Year}	Type of GAN	Model Target	Target	Dataset
[7]$_{2021}$	translation-based	U-Net	covid lung CT	Radiopaedia
[8]$_{2020}$	DCGAN	custom CNN	skin lesions	ISIC 2017
[9]$_{2020}$	translation-based	U-Net, DenseUNet	opt. cover. Tomogr	RETOUCH
[10]$_{2020}$	PGAN	VGG19	Histopathology	TCGA, CARE
[11]$_{2020}$	pix2pix-based	MobileNet	skin lesions	private (49920 clinical images)
[12]$_{2019}$	CycleGAN	U-Net	abdomen CT	Kidney NIH, (Liver, Spleen)

The underlying theme that unifies these subjects is the transformative potential of GANs in the healthcare and medical imaging domains. These investigations serve as exemplars of the advancements achieved in generating medical images of high quality, classifying skin lesions, and developing super-resolution applications. The ability of GANs to improve the accuracy, efficiency, and quality of medical imaging is readily apparent. However, they also highlight the importance of addressing ethical considerations in data generation, employing rigorous evaluation metrics that transcend mere visual realism, and pursuing ongoing research to enhance the efficacy of generative models that can aptly capture the nuanced characteristics inherent in medical data. The article presents a proposition by D. Mahapatra *et al.* regarding a Geometry-Aware Shape Generative Adversarial Network (GeoGAN), which is capable of generating realistic images of retinal OCT scans while preserving the connections between geometry and shape. By incorporating geometric relation-based augmentation, this approach enhances the segmentation performance of retinal OCT images [9].

The GeoGAN approach introduces diversity and accuracy in image generation by integrating shape loss, adversarial loss, and uncertainty sampling. It effectively captures the geometric relation between different layers and enhances the quality and accuracy of the generated images. The proposed method surpasses current segmentation methods on the public RETOUCH dataset, which comprises retinal OCT images acquired through various procedures. Ablation studies and visual analysis demonstrate the advantages of incorporating geometry and diversity in the segmentation process. In comparison to other methods, the GeoGAN approach exhibits greater agreement among ophthalmologists, underscoring the significance of modeling geometric relationships in the segmentation of pathological regions.

In summary, Generative Adversarial Networks (GANs) have brought about a fundamental revolution in the field of medical imaging, offering innovative solutions to longstanding challenges. These breakthroughs hold great promise in addressing the difficulties encountered in MRI image super-resolution, skin lesion classification, brain imaging, and medical image super-resolution. Nonetheless, these advancements are accompanied by obstacles, including the crucial task of accurately representing real-world medical conditions, addressing ethical concerns such as patient privacy and biases in training data, and formulating evaluation metrics that surpass mere visual realism. As we continue to harness the capabilities of GANs in the realm of healthcare, it is of utmost importance that we conscientiously navigate these challenges to ensure the optimization of the technology's benefits while minimizing any potential harm. The future of medical imaging and healthcare is poised to be significantly impacted by the ongoing advancement and ethical application of GANs.

CONCLUSION

In conclusion, the integration of Generative Adversarial Networks (GANs) into the field of medical imaging has brought about a transformative revolution, providing innovative solutions to long-standing challenges. The Fused Attentive Generative Adversarial Network (FA-GAN) holds great potential in enhancing the resolution of MRI images, while GANs utilized for data augmentation and anonymization contribute to strengthening machine learning models. Deep pix2pix GANs have demonstrated effectiveness in generating synthetic medical images, particularly in the domain of brain imaging, and GANs play a pivotal role in the classification of skin lesions, thereby highlighting their versatility [8 - 12]. Nevertheless, these advancements are accompanied by various challenges, including the critical need for accurate representation of real-world medical conditions, ethical concerns about patient privacy and biases in training data, and the necessity for comprehensive evaluation metrics that surpass mere visual realism. As the potential applications of GANs in healthcare continue to unfold, it is of utmost importance to adopt a responsible and ethical approach to optimize the benefits while effectively navigating potential drawbacks.

FUTURE PROSPECTS

In the foreseeable future, the field of medical imaging is expected to undergo a significant transformation due to the emergence of Generative Adversarial Networks (GANs) [3]. GANs have the potential to greatly improve the accuracy of diagnostics, thereby assisting healthcare professionals in providing more precise and timely diagnoses. Furthermore, these networks are positioned to play a crucial role in personalized medicine, as they can customize treatment plans based on the unique characteristics of individual patients [3 - 7]. Additionally, GANs are predicted to expedite the process of drug discovery by generating molecular structures and compounds with specific properties. As a result, the incorporation of these networks is anticipated to streamline the training and evaluation of machine learning models, reducing the need for extensive data annotation [5]. In the realm of telemedicine, GANs are expected to facilitate remote healthcare by enabling real-time assessment of patients' conditions. However, it is essential to acknowledge the ethical considerations surrounding the use of GANs in healthcare systems, including concerns about patient privacy and bias mitigation [8]. Overall, GANs offer significant potential as transformative tools in advancing medical practices.

REFERENCES

[1]   Jiang M, Zhi M, Wei L. FA-GAN: Fused attentive generative adversarial networks for mri image super-resolution science direct. 2021.
[http://dx.doi.org/10.1016/j.compmedimag.2021.101969]

[2]   Bissoto E. GAN-based data augmentation and anonymization for skin-lesion analysis: a critical review arXiv.org, 2021.

[3]   Aljohani A, Alharbe N. Generating synthetic images for healthcare with novel deep Pix2Pix GAN MDPI. 2022.
[http://dx.doi.org/10.3390/electronics11213470]

[4]   Qin Z, Liu Z, Zhu P, Xue Y. A GAN-based image synthesis method for skin lesion classification science direct. 2020.
[http://dx.doi.org/10.1016/j.cmpb.2020.105568]

[5]   Li Y, Sixou B, Peyrin F. A review of the deep learning methods for medical images super resolution problems science direct. 2020.

[6]   Skandarani Y, Jodoin P-M, Lalande A. GANs for medical image synthesis: an empirical study MDPI. 2023.
[http://dx.doi.org/10.3390/jimaging9030069]

[7]   Jiang Y, Chen H, Loew M, Ko H. COVID-19 CT image synthesis with a conditional generative adversarial network arXiv.org. 2020.

[8]   Pollastri F, Bolelli F, Paredes R, Grana C. Augmenting data with GANs to segment melanoma skin lesions ACM Digital Library.

[http://dx.doi.org/10.1007/s11042-019-7717-y]

[9]   Mahapatra D, Bozorgtabar B, Shao L. CVPR 2020 Open Access Repository CVPR. Open Access Repository 2020.

[10]   Levine AB, *et al.* Synthesis of diagnostic quality cancer pathology images. bioRxiv 2020. [http://dx.doi.org/10.1101/2020.02.24.963553]

[11]   Ghorbani A, Natarajan V, Coz D, Liu Y. DermGAN: synthetic generation of clinical skin images with pathology PMLR. 2020.

[12]   Sandfort V, Yan K, Pickhardt P J, Summers R M. Data augmentation using generative adversarial networks (CycleGAN) to improve generalizability in CT segmentation tasks Scientific Reports, Nature. 2019. [http://dx.doi.org/10.1038/s41598-019-52737-x]

[13]   Li X, Jiang Y, Rodriguez-Andina J J, Luo H, Yin S, Kaynak O. When medical images meet generative adversarial network: recent development and research opportunities discover artificial intelligence, springerLink. 2021. [http://dx.doi.org/10.1007/s44163-021-00006-0]

[14]   Kumari S, Singh P. Deep learning for unsupervised domain adaptation in medical imaging: recent advancements and future perspectives arXiv.org. 2023.

[15]   Karim S, Tong G, Li J, Qadir A, Farooq U, Yu Y. Current advances and future perspectives of image fusion: a comprehensive review science direct. 2022.

[16]   Ahmad S, Shakeel I, Mehfuz S, Ahmad J. Deep learning models for cloud, edge, fog, and IoT computing paradigms: Survey, recent advances, and future directions Science direct. 2023. [http://dx.doi.org/10.1016/j.cosrev.2023.100568]

[17]   Lee M. Recent advances in generative adversarial networks for gene expression data: a comprehensive review MDPI. 2023. [http://dx.doi.org/10.3390/math11143055]

[18]   Gao L, Guan L. Interpretability of machine learning: recent advances and future prospects ieee journals & magazine, IEEE Xplore. 2023. [http://dx.doi.org/10.1109/MMUL.2023.3272513]

[19]   Illimoottil M, Ginat D. Recent advances in deep learning and medical imaging for head and neck cancer treatment: MRI, CT, and PET Scans MDPI. 2023. [http://dx.doi.org/10.3390/cancers15133267]

[20]   Paladugu P S, *et al.* Generative adversarial networks in medicine: important considerations for this emerging innovation in artificial intelligence annals of biomedical engineering, springerLink. 2023. [http://dx.doi.org/10.1007/s10439-023-03304-z]

[21]   Wang T, Jiang S, Song P, *et al.* Optical ptychography for biomedical imaging: recent progress and future directions optica publishing group. [http://dx.doi.org/10.1364/BOE.480685]

[22]   Berger L, Haberbusch M, Moscato F. Generative adversarial networks in electrocardiogram synthesis: Recent developments and challenges science direct. 2023. [http://dx.doi.org/10.1016/j.artmed.2023.102632]

[23]   Liu Z, Lv Q, Yang Z, Li Y, Lee C H, Shen L. Recent progress in transformer-based medical image analysis science direct. 2023. [http://dx.doi.org/10.1016/j.compbiomed.2023.107268]

[24]   Lam F, Peng X, Liang ZP. High-dimensional MR spatiospectral imaging by integrating physics-based modeling and data-driven machine learning: current progress and future directions. IEEE Signal Process Mag 2023; 40(2): 101-15. [http://dx.doi.org/10.1109/MSP.2022.3203867] [PMID: 37538148]

[25]   Guo Y, Nie G, Gao W, Liao M. 2D semantic segmentation: recent developments and future directions MDPI. 2023.
       [http://dx.doi.org/10.3390/fi15060205]

[26]   Lin H, Liu Y, Li S, Qu X. How generative adversarial networks promote the development of intelligent transportation systems: a survey ieee/caa journal of automatica sinica. 2023; 10(9): 1781-96.
       [http://dx.doi.org/10.1109/JAS.2023.123744]

[27]   Zhong G, Ding W, Chen L, Wang Y, Yu YF. Multi-scale attention generative adversarial network for medical image enhancement. IEEE Trans Emerg Top Comput Intell 2023; 7(4): 1113-25.
       [http://dx.doi.org/10.1109/TETCI.2023.3243920]

[28]   Zhou T, Li Q, Lu H, Cheng Q, Zhang X. GAN review: Models and medical image fusion applications science direct. 2022.

[29]   Ngo T A, Nguyen T, Thang T C. A survey of recent advances in quantum generative adversarial networks MDPI. 2023.
       [http://dx.doi.org/10.3390/electronics12040856]

[30]   Conze PH, Andrade-Miranda G, Singh VK, Jaouen V, Visvikis D. Current and emerging trends in medical image segmentation with deep learning. IEEE Trans Radiat Plasma Med Sci 2023; 7(6): 545-69.
       [http://dx.doi.org/10.1109/TRPMS.2023.3265863]

CHAPTER 10

AI Revolutionizing Healthcare: Current State and Future Prospects

Poornima N.V.[1] and **Gunavathi C.**[2,*]

[1] *Faculty of Management, Symbiosis Centre for Management Studies, Bengaluru Campus - 560100, Symbiosis International (Deemed University), Pune, India*

[2] *School of Computer Science and Engineering, Vellore Institute of Technology, Vellore - 632014, Tamil Nadu, India*

Abstract: Artificial intelligence (AI) in healthcare is the collection of techniques and resources to improve several facets of the healthcare industry, such as patient care and administrative tasks. Its growing relevance is due to its potential to enhance further healthcare services' efficacy, accuracy, and accessibility. AI can analyze enormous volumes of medical data, assisting healthcare practitioners in making decisions and treatment programs and even forecasting disease outbreaks, which will improve patient outcomes and make healthcare delivery more affordable. As a result of its growing influence, AI is becoming a more significant and transformational force in healthcare.

Keywords: Disease outbreaks, Healthcare practitioners, Healthcare delivery, Personalized treatment programs, Patient outcomes, Transformational force.

INTRODUCTION

Artificial intelligence (AI) refers to a collection of intelligent processes and behaviors produced through computer models, algorithms, or rules that enable machines to replicate human cognitive skills like learning and problem-solving [1, 2]. The term AI refers to computational technologies that replicate the support mechanisms of human intelligence, such as cognition, deep learning, adaptation, engagement, and sensory comprehension [3, 4]. Generally, human interpretation is required for this role, but some devices can perform decision-making [5, 6]. Many fields, including medicine and health, can use these multidisciplinary methods. Since doctors initially attempted to use computer-aided programs to enhance their diagnosis in the 1950s [7, 8], AI has been present in medicine. These developments raise questions about how these talents could assist or per-

---

* **Corresponding author Gunavathi C.:** School of Computer Science and Engineering, Vellore Institute of Technology, Vellore - 632014, Tamil Nadu, India; E-mail: gunavathi.cm@vit.ac.in

Sivakumar Rajagopal, Prakasam P., Konguvel E., Shamala Subramaniam, Ali Safaa Sadiq Al Shakarchi & B. Prabadevi (Eds.)

haps improve human decision-making in health and medical care. The accessibility of pertinent data strongly correlates with AI's potential [9]. The health domain has an abundance of data [10]. Gathering and sharing health data is difficult compared to other forms of data because, on the one hand, it is subject to privacy concerns. For instance, collecting health data in longitudinal research and clinical trials can be costly, necessitating strict security measures once collected.

Additionally, the inability of electronic health record systems to communicate with each other [11], the incompetency to gather relevant social data, and the lack of interoperability among these systems impede the use of even the most basic computational techniques [12]. Meanwhile, as evidenced by the abundance of companies focused on AI in health and health care, the private sector is extremely interested in healthcare data and applications. Customers of CB insights can gather, sort, and map firms based on funding, industry, and a wide range of other factors by using the company's current Expert Collections or creating their custom collections [13]. AI can improve healthcare and empower patients by giving them more control over their health. Recent applications of AI to enhance healthcare delivery include personalized health information, virtual consultations, and remote monitoring [14]. Personalized health information is one of AI's main advantages in healthcare [15].

The anamnesis and lifestyle characteristics are just two examples of the patient data that AI analyzes. AI systems can provide patients with individualized advice on how to stay healthy. With this information, patients can make more informed decisions about their care and gain a better understanding of their health. Remote monitoring is one of the most important AI applications in healthcare. AI-powered remote monitoring devices track and record patients' vital signs, alerting medical personnel to potential problems. As a result, there may be a need for fewer in-person visits to healthcare institutions, earlier intervention, and better patient results. Virtual consultations are another way AI is enhancing healthcare delivery. Patients can obtain medical care without visiting a hospital or facility by offering remote medical care [16]. This can be especially helpful for people with mobility problems or those who live in distant places.

Another area where AI can be useful in empowering individuals is medication administration. AI algorithms can assist healthcare providers in better-managing medicine and reducing the likelihood of adverse drug reactions by analyzing patient data, such as prescription histories and vital signs. This approach can enhance patient safety and improve health outcomes. AI can also promote openness in the healthcare industry by providing consumers with more information about their health and the treatments they are receiving. This can allow patients to take control of their healthcare decisions and foster trust between them and the care provider. The application of AI in medical imaging has the potential to enhance patient outcomes and diagnostic precision. AI can improve medical radiology in many ways because it is critical for diagnosing and treating different medical disorders.

OPPORTUNITIES

AI in healthcare is a rapidly developing topic with various potential prospects. Here are some significant fields where AI is having an impact and opening:

Diagnostic and Medical Imaging

Radiology: AI can help radiologists find anomalies in X-rays, CT scans, and MRIs, possibly increasing accuracy and speeding up diagnosis [17].

Pathology: AI algorithms can detect malignant cells or other quirk in tissue samples and pathology slides.

Ultrasound: AI aids in the understanding of ultrasound pictures for numerous applications, such as obstetrics and cardiology.

Medicines Discovery and Development: By analyzing massive datasets to identify prospective medication candidates and forecast their efficacy and safety, AI speeds up the drug discovery process [18]. Additionally, it can enhance clinical trial designs to make them more effective and economical.

Healthcare Administration and Operations

AI-driven technologies can automate administrative activities such as billing, scheduling, and patient information management, thereby reducing the administrative workload for healthcare practitioners [19]. Predictive analytics may help hospitals and clinics make the best use of staff scheduling and resource allocation.

Specialized Medicine

AI can enhance treatment effectiveness by adjusting treatment regimens and medication dosages based on a patient's genetic and clinical data. It can anticipate how patients react to treatments, minimizing medical trial and error. By predicting patient reactions to therapies, it helps healthcare practitioners avoid costly mistakes.

Disease Forecasting and Preventive Measures

AI may examine patient data to pinpoint people at risk of contracting diseases, enabling early intervention and preventative measures. Machine learning models can monitor disease outbreaks and support resource distribution during epidemics. Models for machine learning can keep track of disease outbreaks and assist with resource distribution during pandemics.

Monitoring from a distance and telemedicine

AI chatbots and virtual health assistants give quick medical advice and support, lowering the pressure on healthcare professionals and enhancing access to healthcare services. AI analytics and remote monitoring tools make it possible to monitor patients' vital signs and conditions [20].

Natural Language Processing (NLP)

NLP algorithms can glean insightful information from unstructured clinical notes, enhancing clinical research and decision-making. Virtual assistants and chatbots can interact with patients, provide information, and make appointments.

Integrity and Adherence

AI can ensure the privacy of patient data and ensure adherence to healthcare regulations, thereby reducing the risk of data breaches.

Psychological Health and Well-Being

AI-powered systems that analyze text or speech to find indications of depression, anxiety, or other mental health disorders can provide mental health assistance. Chatbots and virtual therapists can provide rapid support and resources.

Robotics in Operation

Surgical robots with AI can improve surgical precision and allow for remote operation.

Research and Insights in Healthcare

To give medical experts the most recent knowledge and insights, AI can process enormous amounts of medical literature and research articles.

Education and Training in Healthcare

Virtual reality and AI-based simulations can help medical education by providing a secure and lifelike environment for learning and practicing medical procedures.

Patient Engagement and Behavior Modification

AI-driven wearables and apps can encourage patients to follow their treatment programs and live better by providing individualized recommendations and feedback.

Healthcare Abuse Detection

AI algorithms can examine insurance claims and billing data to find fraudulent activity, saving healthcare providers and insurers billions of dollars.

Pharmaceutical Toxic Event Tracking

AI can monitor patient feedback, social media, and healthcare databases to identify negative drug reactions, expediting drug safety solutions.

Support for the Aging Society

AI-powered robots and gadgets can help the elderly with everyday tasks, keep track of their health, and offer companionship in an aging society.

Medical Data Security

AI can improve cybersecurity by detecting and stopping data breaches and unauthorized access to patient information and medical equipment.

Logistics and Inventory Management

During emergencies like pandemics, AI-driven demand forecasting and inventory management may guarantee the prompt availability of medical supplies and treatments.

The Hospital Room Triage

Depending on the seriousness of a patient's illness, AI can help prioritize patients in emergency rooms, potentially saving lives by cutting wait times for urgent cases.

Electronic Health Records (EHRs) with AI Enhancements

AI can make EHR systems more intuitive making it simpler for medical practitioners to obtain and analyze patient data.

Medical Bots for Preliminary Consultations

By assisting patients in evaluating their symptoms and making initial recommendations for seeking medical attention, chatbots can help individuals avoid unnecessary trips to the hospital.

Assessments of Exotic Disorders

By examining patient data and comparing it to vast databases of medical knowledge, AI can assist in the identification of rare and complex diseases.

Restoration and Physiological Therapies

Virtual reality and AI-powered exoskeletons can support rehabilitation and physical therapy sessions by tailoring exercises to the needs of each patient.

Virtual Medical Assistance

AI provides virtual medical assistance, guiding surgeons through complex procedures, providing real-time feedback, and enhancing surgical accuracy.

Treatment for Behavioral Health Issues and Drug Abuse Disorders

AI can assist in the diagnosis and treatment of behavioral health problems and drug abuse disorders by examining patient behaviors and offering solutions.

Monitoring for Environmental Health

AI can assess environmental elements like water pollution and air quality to forecast and reduce health hazards in certain areas.

Healthcare Equity and Accessibility

By offering accessible and cheap healthcare services to underserved and remote populations, AI can aid in closing healthcare disparities.

Genomic Modeling and Bioinformatics

AI-driven bioinformatics tools can analyze large genomic databases, advancing knowledge of genetic illnesses and tailored therapy. AI presents exciting prospects for enhancing patient care, cutting costs, and expanding medical research. However, as the sector progresses, the issues related to data privacy, ethics, and regulatory compliance should be managed carefully. AI within the medical field is a diverse topic with many potential applications, and it has the power to change the sector. As AI technology advances, it will likely have a significant impact on patient outcomes, healthcare expenditures, and the quality of treatment provided to patients.

RISKS / DISADVANTAGES

This study discovered and explained seven major hazards of AI in medicine and healthcare: 1) patient harm due to AI mistakes; 2) misuse of medical AI tools; 3) bias in AI and the perpetuation of existing injustices; 4) lack of transparency; 5) privacy and security concerns; 6) gaps in accountability; and 7) implementation challenges. As summarized below, each part not only discusses the danger but also suggests viable countermeasures.

Patient Harm Brought on By AI Mistakes

A growing issue in healthcare is patient harm brought on by AI errors. While AI promises significant advantages, it also raises the possibility of mistakes that could endanger patient safety. Underscoring the seriousness of these concerns are AI-driven diagnostic errors, inaccurate therapy suggestions, or surgical complications. Error detection is difficult due to the transparency of AI decision-making processes, also known as the "black box" problem. Additionally, current legislation might not adequately address patient harm brought on by AI, forcing a closer look at the ethical and legal ramifications. Healthcare stakeholders must place a high priority on thorough testing, open algorithms, and interdisciplinary cooperation between regulatory agencies, AI developers, and healthcare practitioners to reduce this risk [21]. The goal is to maximize patient safety in clinical settings while leveraging AI's capabilities to improve patient care.

Medical AI Tools being Misused

Medical AI systems are misused and a growing concern in the healthcare industry. Although these tools have enormous potential to improve patient care, therapy, and diagnostics, their incorrect or ignorant use might have disastrous results. AI is used in place of human expertise and judgment in decision-making is a significant problem. Reliance on AI suggestions that are inappropriate or excessive might result in inaccurate diagnoses, poor treatment decisions, and patient harm. Furthermore, if the potential for bias in AI systems is not addressed, there will be risks in healthcare outcomes. The improper use of medical AI highlights the necessity of thorough training and policies for healthcare practitioners to successfully incorporate AI technologies into their practice while ensuring that human judgment remains a crucial part of healthcare decision-making.

Bias in AI and the Maintenance of Existing Injustice

Biased AI systems have the potential to perpetuate injustices and inequalities, a grave concern. Because historical data frequently contains biases based on race, gender, socioeconomic status, and other factors, AI algorithms often learn from previous data. Training AI systems on such data can inadvertently replicate and even exacerbate these biases, leading to unjust and discriminatory outcomes in fields such as recruiting, lending, and criminal justice. It is essential to establish strict data preparation, openness, and constant monitoring of AI systems to reduce bias in AI and promote justice. To overcome systemic biases and promote more equitable technological solutions, efforts should also concentrate on diversifying AI development teams and incorporating ethical and fairness issues from the outset of AI system design.

Lack of Transparency

In today's technological environment, the lack of transparency in AI systems is a serious concern. It alludes to the uncertainty surrounding the decision-making processes of AI algorithms, which might undermine responsibility and trust. It is difficult to comprehend, validate, or resolve potential biases, errors, or ethical concerns when AI systems behave as "black boxes," without obvious explanations for their actions. Not only is transparency crucial for regulatory compliance, but it also helps users and the general public gain trust in AI technologies. Efforts are underway to develop explainable AI techniques, which aim to ensure AI systems provide clear explanations for their decisions and foster transparency and accountability in their application across various industries, including banking and healthcare.

Privacy and Security Concerns

As more facets of daily life incorporate these technologies, concerns about AI's privacy and security have gained prominence. To operate efficiently, AI systems frequently require enormous volumes of personal data, which raises concerns about data breaches, unauthorized access, and misuse. Strong privacy safeguards and security measures are essential given the risk of intrusive surveillance, profiling, and the mismanagement of sensitive information. It is hard to balance using AI effectively and protecting people's privacy rights. Strict data protection laws, encryption standards, and ethical AI practices are necessary to address this issue and ensure the responsible and secure use of AI technologies.

Gaps in Accountability

Accountability gaps in the AI industry are a major obstacle to ensuring responsible and moral AI adoption. As AI systems become more autonomous and powerful, it becomes difficult to establish clear lines of accountability for their actions. If AI-related mistakes, prejudices, or negative outcomes occur, a lack of accountability may make it challenging to determine who should bear responsibility. The growth of legal frameworks, industry standards, and moral principles that establish accountability mechanisms and place responsibility on developers, operators, and regulatory agencies is necessary to close these gaps. It is crucial to establish a framework that encourages openness and accountability while supporting the responsible application of AI technology to advance society without taking unwarranted risks.

Implementation Challenges

Adopting AI presents a variety of implementation obstacles, including those that are technological, organizational, and ethical. The complexity of AI algorithms and their integration with current systems can pose challenges on the technical front. There are ongoing difficulties in ensuring data quality, removing biases, and reaching scalability. Businesses and organizations may experience organizational challenges in acquiring resources, developing AI knowledge within their personnel, and aligning AI plans with their goals. Clear regulatory frameworks and ethical principles are required because of ethical concerns about accountability, transparency, and justice. To overcome these obstacles and maximize the benefits of AI while minimizing hazards, a comprehensive strategy involving stakeholder cooperation, funding for continuous education and research, and adherence to ethical AI practices is required.

ETHICAL AND SOCIAL CHALLENGES / ISSUES WHEN USING AI

Although there is enormous potential for the application of AI in healthcare, there are also substantial ethical and social concerns. The issue of data security and privacy needs immediate attention. Maintaining the confidentiality and integrity of this data becomes crucial when AI systems process enormous amounts of sensitive patient data. Breach or misuse of medical data can erode patient trust, potentially leading to discrimination or identity theft. Bias and fairness in AI systems present another problem from an ethical standpoint [22]. Developing these systems using biased datasets may sustain and even exacerbate healthcare disparities.

An AI diagnostic tool, for instance, might work well for one demographic but badly for another, unintentionally causing disparities in access to high-quality care. It is crucial to train AI models on the representative data to minimize these biases. In healthcare, where lives are at stake, understanding and trusting AI recommendations is crucial. Achieving transparency enhances accountability and fosters acceptance among healthcare professionals and patients. Transparency and accountability pose additional ethical challenges because many AI algorithms operate as "black boxes," making it difficult to explain the reasoning behind their decisions.

This problem also involves the issue of informed consent. Patients have the right to receive information about the use of AI in their care and voluntarily provide their consent. However, patients may not fully understand how these technologies work because AI algorithms can occasionally make complicated medical decisions. A continuing ethical discussion focuses on how to strike a balance between informed consent and the realities of AI in healthcare. The introduction of AI in healthcare raises social concerns about job displacement. Even though AI can automate some operations, such as administrative tasks, it may cause worries among healthcare personnel about job losses. Managing this transition to allow displaced workers to reskill and uphold patient care standards is a significant societal concern.

Healthcare practitioners, technologists, legislators, and ethicists must work together to overcome these ethical and societal concerns. It is the transformational potential of AI in healthcare, it is crucial to develop strong rules, encourage openness, and give equity and patient welfare priority. To achieve responsible and equitable adoption, various ethical and social issues related to AI in healthcare must be carefully considered.

MORAL AND SOCIAL ISSUES

The following are some of the most important moral and social issues:

- Security of Data and Confidentiality
- Transparency and Prejudice
- Openness and Taking Responsibility
- Informed Consent
- Professional Displacement
- Healthcare Inequality
- Employment Displacement
- Responsible for Errors
- Authorization for Data Sharing and Research
- The Overabundance of AI
- Governmental Obstacles

TRUSTWORTHY

The role of AI will change as its capabilities advance from automating routine, well-defined jobs to assisting humans in making questionable decisions, a task currently performed only by medical professionals. An appropriate trust relationship, calibrated trust, becomes necessary for making good decisions as healthcare providers rely more on AI. AI completely transforms the healthcare industry by enhancing patient care, treatment recommendations, and diagnostics. However, to fully utilize AI in healthcare, reliability is crucial. Regulators, healthcare professionals, and patients all need to have faith that AI systems are trustworthy, private, and safe. The main factors that affect the reliability of AI in healthcare must be examined. To make sure that putting AI into healthcare systems is safe and reliable, this article talks about the main things that make AI trustworthy in healthcare, focusing on how important it is for data privacy, being able to be explained, not being biassed, being strong, working with doctors, and following the rules.

- Data Privacy and Security
- Explainability and Transparency
- Eliminating prejudice: Fairness and Equity
- Robustness and Reliability
- Clinical Collaboration
- Regulatory Compliance

FUTURE RESEARCH DIRECTIONS

AI has made notable advances in the healthcare industry, transforming everything from medication research to diagnostics. However, most of the AI's potential in healthcare remains unrealized. The future of AI in healthcare holds enormous promise, with intriguing research paths on the horizon as technology advances and healthcare demands change.

Precision medicine and personalized healthcare are two areas where artificial intelligence will likely play a significant role in the future of healthcare [23]. Even while AI is helping to customize treatments for specific patients, the next step is to incorporate a wider range of patient data, including genetics, proteomics, and environmental factors. Researchers hope to develop algorithms that can assess complicated datasets and suggest precisely tailored treatment approaches by leveraging the capabilities of machine learning and AI. This study will herald a new era of precision medicine by not only improving patient outcomes but also lowering the likelihood of negative drug reactions and side effects.

Early disease diagnosis and prevention is one of the most important research areas in artificial intelligence for healthcare. Although AI can now spot potential health hazards by looking at patient data, its ultimate potential is to predict illness before any outward symptoms appear. To assess a wide range of health-related data sources, including genetic, lifestyle, and environmental aspects, researchers are developing AI algorithms. This predictive capability has the potential to change healthcare by enabling prompt interventions and proactive preventative actions, saving lives and reducing the burden on healthcare systems.

Explainable AI (XAI) is critical to increasing AI reliability in the healthcare industry. While there are many complex models in use in today's healthcare AI landscape, future research in this field will focus on building more interpretable AI systems. Researchers are working on ways to make AI decision-making transparent and understandable for patients and healthcare practitioners. Making AI models more comprehensible increases confidence, holds clinical judgments to a higher standard of accountability, and enhances the safety and reliability of healthcare procedures. The future research directions are in the following fields:

- Medical Imaging with AI Assistance
- AI in Mental Health
- Ethical AI and Regulatory Frameworks
- Early Disease Detection
- Predictive Analytics
- Personalized Medicine
- AI in Neuroscience [24]
- AI-Enhanced Brain Imaging
- Neurodegenerative Disease Detection
- Neuromorphic Hardware
- Brain-Computer Interfaces (BCIs)
- AI-Driven Neurological Disorder
- AI in Drug Discovery
- Comprehending Neural Networks
- AI and Intellectual Neuroscience
- Personalized Therapy in Psychology
- Ethical Challenges
- AI in Cognition Enhancement
- AI in Surgery [25]
- Robot-Assisted Surgical Procedures
- Scanning and Radiology
- Anesthesia and Medication Dose
- Preventive Analytics and Individual Monitoring

- Telemedicine and Remote Surgery
- Data Security and Privacy
- Regulatory Compliance
- AI in Rehabilitation [26]
- AI in Cardiac Management [27]
- AI in risk Management
- AI in Nursing [28]
- Patients Monitoring
- Organizational Efficiency
- Medicine Management
- Public Interaction
- Education and Training
- The Workflow Optimization for Nurses [29]
- Remote Consultations
- AI in Virtual patient Care [30]
- AI in Precision Care [31, 32]
- AI in Health Monitoring
- AI in Managing Medical Records

CONCLUSION

In conclusion, the application of artificial intelligence (AI) in healthcare is already starting to transform the sector, with its advantages, which improve patient care, expedite procedures, and spur creativity. Promising advancements in AI-powered healthcare are evident in individualized treatment plans, data management, administrative efficiency, and diagnostic instruments. Deep learning and machine learning algorithms are constantly evolving, making it possible for AI systems to evaluate enormous volumes of medical data and provide previously unreachable insights. The potential applications of AI in healthcare appear to be rather promising.

Artificial intelligence (AI) has promise for enhancing clinical decision-making, expediting medication discovery, and augmenting the general efficacy of healthcare systems. AI-powered predictive analytics and preventative care can help with early disease identification and intervention, save lives, and lower healthcare costs. Furthermore, the growing use of AI-powered telemedicine and remote monitoring can improve access to healthcare services, especially in rural or disadvantaged areas.

However, issues such as data protection, ethical concerns, and the need for regulatory frameworks to ensure the secure application of AI in healthcare should be addressed. For AI to fully alter healthcare delivery, it needs to work with researchers, politicians, technology developers, and healthcare practitioners to overcome these obstacles. With everything considered, the AI revolution in healthcare is already well underway and holds the potential to transform the sector into more proactive, efficient, and patient-focused. The combination of human knowledge and artificial intelligence, combined with the rapid advancement of technology, promises to create a healthcare environment that is not only technologically advanced but also essentially centered on enhancing patient outcomes and the general well-being of communities across the globe.

REFERENCES

[1]   Chen M, Decary M. Artificial intelligence in healthcare: An essential guide for health leaders. Healthcare Management Forum. Sage CA: Los Angeles, CA: SAGE Publications 2020; 33: pp. (1)10-8.

[2]   Bajwa J, Munir U, Nori A, Williams B. Artificial intelligence in healthcare: transforming the practice of medicine. Future Healthc J 2021; 8(2): e188-94.
[http://dx.doi.org/10.7861/fhj.2021-0095] [PMID: 34286183]

[3]   Tagliaferri SD, Angelova M, Zhao X, *et al.* Artificial intelligence to improve back pain outcomes and lessons learnt from clinical classification approaches: three systematic reviews. NPJ Digit Med 2020; 3(1): 93.
[http://dx.doi.org/10.1038/s41746-020-0303-x] [PMID: 32665978]

[4]   Tran BX, Vu GT, Ha GH, *et al.* Global evolution of research in artificial intelligence in health and medicine: a bibliometric study. J Clin Med 2019; 8(3): 360.
[http://dx.doi.org/10.3390/jcm8030360] [PMID: 30875745]

[5]   Hamid S. The opportunities and risks of artificial intelligence in medicine and healthcare. 2016.

[6]   Panch T, Szolovits P, Atun R. Artificial intelligence, machine learning and health systems. J Glob Health 2018; 8(2)020303
[http://dx.doi.org/10.7189/jogh.08.020303] [PMID: 30405904]

[7]   Yang X, Wang Y, Byrne R, Schneider G, Yang S. Concepts of artificial intelligence for computer-assisted drug discovery. Chem Rev 2019; 119(18): 10520-94.
[http://dx.doi.org/10.1021/acs.chemrev.8b00728] [PMID: 31294972]

[8]   Burton RJ, Albur M, Eberl M, Cuff SM. Using artificial intelligence to reduce diagnostic workload without compromising detection of urinary tract infections. BMC Med Inform Decis Mak 2019; 19(1): 171.
[http://dx.doi.org/10.1186/s12911-019-0878-9] [PMID: 31443706]

[9]   https://medium.com/machine-intelligence-report/data-not-algorithms-is-key-to-machine-learningsuccess-69c6c4b79f33

[10]   http://www.datasciencecentral.com/profiles/blogs/10-great-healthcare-data-sets

[11]   https://www.healthit.gov/sites/default/files/ptp13-700hhs_white.pdf

[12]   https://irp.fas.org/agency/dod/jason/data-health.pdf

[13]   https://www.cbinsights.com/research/artificial-intelligence-startups-healthcare/

[14]   Parekh ADE, Shaikh OA, Simran , Manan S, Hasibuzzaman MA. Artificial intelligence (AI) in personalized medicine: AI-generated personalized therapy regimens based on genetic and medical history: short communication. Ann Med Surg (Lond) 2023; 85(11): 5831-3.
[http://dx.doi.org/10.1097/MS9.0000000000001320] [PMID: 37915639]

[15]   Al Kuwaiti A, Nazer K, Al-Reedy A, *et al.* A Review of the Role of Artificial Intelligence in

Healthcare. J Pers Med 2023; 13(6): 951.
[http://dx.doi.org/10.3390/jpm13060951] [PMID: 37373940]

[16]   Singh AP, Saxena R, Saxena S, Maurya NK. Artificial Intelligence Revolution in Healthcare: Transforming Diagnosis, Treatment, and Patient Care. Asian Journal of Advances in Research 2024; 7(1): 241-63.

[17]   Alami H, Lehoux P, Denis JL, *et al.* Organizational readiness for artificial intelligence in health care: insights for decision-making and practice. J Health Organ Manag 2020; 35(1): 106-14.
[http://dx.doi.org/10.1108/JHOM-03-2020-0074] [PMID: 33258359]

[18]   Hasselgren C, Oprea TI. Artificial intelligence for drug discovery: Are we there yet? Annu Rev Pharmacol Toxicol 2024; 64(1): 527-50.
[http://dx.doi.org/10.1146/annurev-pharmtox-040323-040828] [PMID: 37738505]

[19]   Hoffman RR, Johnson M, Bradshaw JM, Underbrink A. Trust in Automation. IEEE Intell Syst 2013; 28(1): 84-8.
[http://dx.doi.org/10.1109/MIS.2013.24]

[20]   Zhang X, Ma L, Sun D, Yi M, Wang Z. Artificial Intelligence in Telemedicine: A Global Perspective Visualization Analysis. Telemed J E Health 2024; 30(7): e1909-22.
[http://dx.doi.org/10.1089/tmj.2023.0704] [PMID: 38436235]

[21]   Khan B, Fatima H, Qureshi A, *et al.* Drawbacks of artificial intelligence and their potential solutions in the healthcare sector. Biomedical Materials & Devices 2023; 8: 1-8.
[http://dx.doi.org/10.1007/s44174-023-00063-2]

[22]   Ekampreet Kaur, Akash bans, Uwom Okereke Eze, Jaskaran Singh. Artificial Intelligence in Healthcare: A Prospective Approach.Anveshan: Multidisciplinary Journal of Geeta University. 2023; 1: p. 1.

[23]   Carini C, Seyhan AA. Tribulations and future opportunities for artificial intelligence in precision medicine. J Transl Med 2024; 22(1): 411.
[http://dx.doi.org/10.1186/s12967-024-05067-0] [PMID: 38702711]

[24]   Harris LT. The neuroscience of human and artificial intelligence presence. Annu Rev Psychol 2024; 75(1): 433-66.
[http://dx.doi.org/10.1146/annurev-psych-013123-123421] [PMID: 37906951]

[25]   Guni A, Varma P, Zhang J, Fehervari M, Ashrafian H. Artificial Intelligence in Surgery: The Future is Now. Eur Surg Res 2024; •••: 1.
[http://dx.doi.org/10.1159/000536393] [PMID: 38253041]

[26]   Abedi A, Colella TJF, Pakosh M, Khan SS. Artificial intelligence-driven virtual rehabilitation for people living in the community: A scoping review. NPJ Digit Med 2024; 7(1): 25.
[http://dx.doi.org/10.1038/s41746-024-00998-w] [PMID: 38310158]

[27]   Nazir MB, Hussain I. Cognitive Computing for Cardiac and Neurological Well-being: AI and Deep Learning Perspectives. Rev Esp Doc Cient 2024; 18(02): 180-208.

[28]   Ruksakulpiwat S, Thorngthip S, Niyomyart A, *et al.* A systematic review of the application of artificial intelligence in nursing care: where are we, and what's next? J Multidiscip Healthc 2024; 17: 1603-16.
[http://dx.doi.org/10.2147/JMDH.S459946] [PMID: 38628616]

[29]   Fernandes M, Vieira SM, Leite F, Palos C, Finkelstein S, Sousa JMC. Clinical decision support systems for triage in the emergency department using intelligent systems: a review. Artif Intell Med 2020; 102101762
[http://dx.doi.org/10.1016/j.artmed.2019.101762] [PMID: 31980099]

[30]   Gama F, Tyskbo D, Nygren J, Barlow J, Reed J, Svedberg P. Implementation frameworks for artificial intelligence translation into health care practice: scoping review. J Med Internet Res 2022; 24(1)e32215

[http://dx.doi.org/10.2196/32215] [PMID: 35084349]

[31]   Wolff J, Pauling J, Keck A, Baumbach J. Systematic review of economic impact studies of artificial intelligence in health care. J Med Internet Res 2020; 22(2)e16866
[http://dx.doi.org/10.2196/16866] [PMID: 32130134]

[32]   https://www.simplilearn.com/advantages-and-disadvantages-of-artificial-intelligence-article

CHAPTER 11

Application of Image Processing Methods in the Healthcare Sector

Chilakalapudi Malathi[1] and **Sheela Jayachandran**[1,*]

[1] *School of Computer Science & Engg, VIT-AP University, Vijayawada, Andhra Pradesh, India*

Abstract: The thorough reference book "Image Processing Techniques for Healthcare: Advances and Applications" addresses the fascinating nexus between medical research and cutting-edge image processing techniques. It comprehensively elucidates the concepts, methodologies, and healthcare-related image processing applications. As an essential part of contemporary healthcare, medical imaging supports diagnosing, managing, and overseeing various medical problems. New image processing techniques are required to extract useful information, boost diagnostic precision, and improve patient care from the growing amount and complexity of medical pictures. These critical areas of application and their influence are highlighted in this abstract, which summarises the crucial role that image-processing technologies play in healthcare. It explores how image-processing techniques are essential for enhancing the quality of medical images, supporting clinical decision-making, and advancing medical research in an era where medical imaging is critical for diagnosis, treatment, and patient care. The book covers many subjects, from the core concepts of digital image processing to the most recent developments in medical image analysis. This book covers picture preprocessing, image segmentation, image registration, and three-dimensional reconstruction. A combination of machine learning and artificial intelligence algorithms for automated diagnosis and prognosis is also covered in depth. The main focus is how image processing methods are used in real-world settings in areas of medicine like radiology, cardiology, pathology, and neurology. Disease identification, planning of therapy, and surveillance of patients in healthcare settings. Investigations and actual-life scenarios are used to demonstrate how these strategies are applied.

Keywords: Classification, Feature extraction, Image acquisition, Medical imaging, Registration, Segmentation, Visualization.

---

* **Corresponding author Sheela Jayachandran:** School of Computer Science & Engg, VIT-AP University, Vijayawada, Andhra Pradesh, India; E-mail: sheela.j@vitap.ac.in

Sivakumar Rajagopal, Prakasam P., Konguvel E., Shamala Subramaniam, Ali Safaa Sadiq Al Shakarchi & B. Prabadevi (Eds.)
All rights reserved-© 2025 Bentham Science Publishers

INTRODUCTION

Medical imaging is now essential for diagnosing, planning, and managing many illnesses in contemporary healthcare. The everyday production of enormous numbers of medical images, encompassing X-rays, magnetic resonance imaging, computed tomography scans, and ultrasound images, has increased the need for improved image processing methods [1]. These approaches are essential for raising wellness standards, promoting patient outcomes, and streamlining healthcare provision. This chapter offers an overview of some essential image-processing methods used in healthcare, concentrating on their uses and advantages.

Medical imaging in the realm of healthcare has four main uses. Printing images of the inside parts and body processes is one of these. Beginning with the classification of complex or otherwise undetectable medical illnesses, imaging methods are utilized by healthcare professionals to identify them. Using these photos as a guide facilitates meticulous treatment scheduling and therapy preparation. For example, computed tomography scans assist healthcare professionals in planning chemotherapy for cancer patients by determining the location and size of the tumour [2]. The progression of diseases over time must be monitored *via* medical imaging, which brings us to our third point. Additionally, scientists utilize these pictures to investigate the nuances of human anatomy and physiology and the effects of diseases and treatments on human beings.

Medical imaging (MI) has grown significantly in various biomedical studies and therapeutic applications. Biologists use MI techniques in biology to examine cells and produce detailed 3D confocal imaging information [3]. Researchers in neuroscience use MI techniques such as positron emission tomography, functional Magnetic Resonance Imaging, and nuclear magnetic resonance spectrum imaging to identify specific metabolically engaged areas of the brain. Conversely, Virologists rely on MI to turn micrographs into detailed 3D representations of viruses. Radiologists also use MI to precisely measure and pinpoint tumours in MRI and CT images.

Medical imaging (MI) uses a range of approaches, including X-rays, ultrasounds, and nuclear medicine imaging (NMI). X-ray assessments, which use X-ray radiation, produce images that focus on bone structures [4]. Ultrasound equipment, on the other hand, uses sound waves to visualize the body's inside organs. In nuclear medicine imaging (NMI), a small amount of radioactive material is delivered into the body to provide images of internal organs, blood circulation, and metabolic activities.

On the other hand, medical image analysis continues to be difficult for scientists. Fig. (**1**) depicts seven essential functions commonly included in healthcare imaging processing. The first phase is picture collection, which entails collecting medical pictures from several MI techniques, including X-rays, CT scans, MRI scans, and ultrasound scans. Following that, preprocessing begins, which includes duties like noise reduction, distortion correction, and contrast enhancement, all to improve and perfect medical images. On the contrary, segmentation entails separating areas of interest from the surrounding backdrop and identifying unique structures within the image [5]. Ultimately, the registration procedure comprises synchronizing various medical pictures to form an integrated picture that provides a broader awareness of anatomy and pathology.

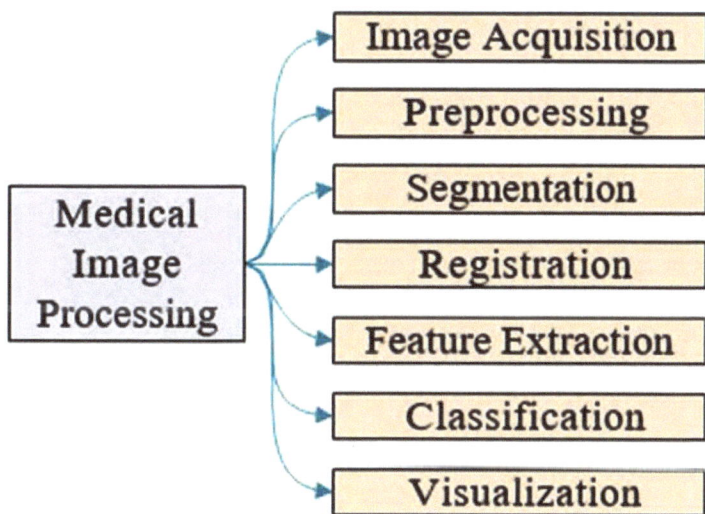

Medical Image Processing
- Image Acquisition
- Preprocessing
- Segmentation
- Registration
- Feature Extraction
- Classification
- Visualization

Fig. (1). Depicts the seven most prevalent operations in medical image processing.

Indeed, feature extraction in medical image processing aims to extract critical features from these pictures, such as the dimension, form, texture, and brightness of various features. On the other hand, identification is in charge of categorizing different structures within medical pictures based on their distinct traits and characteristics. Finally, visualization is critical in producing two-dimensional or multidimensional graphic representations of medical images. These visualizations are essential in assisting research efforts, assisting in the diagnostic process, and simplifying medical therapy preparation [6].

- First and foremost, medical image processing (MIP) is critical in assessing crucial parameters such as tumour or organ dimensions, volume, blood vessels, and blood or other fluid flow properties. Furthermore, using more traditional

methods, can detect slight changes that would otherwise go unnoticed.
- Second, MIP is critical in producing authentic body structure representations and images [7]. These tools are widely used in medical education and training programs.
- Third, MIP provides immediate direction and visualization during operations, considerably improving surgeons' accuracy as they negotiate complicated structural processes.
- Finally, MIP makes major contributions to the creation of personalized prosthetic devices, such as prostheses and orthotics, that are suited to the specific anatomy and needs of particular clients [8].

Despite the introduction of X-rays in 1895, image capture has grown into a standard process in medical diagnosis. This trend has been enhanced by the increased usage of instant digital imaging technologies, showing the growing relevance of digital image processing in the medical field [9]. Furthermore, by integrating digital sensors with technologically native processes such as computed tomography or magnetic resonance imaging, even previously analogue imaging methods such as endoscopy and radiography are being upgraded.

The phrase medical processing of pictures describes the utilization of techniques for processing digital images in healthcare. This field is divided into five fundamental regions, as shown in Fig. (2).

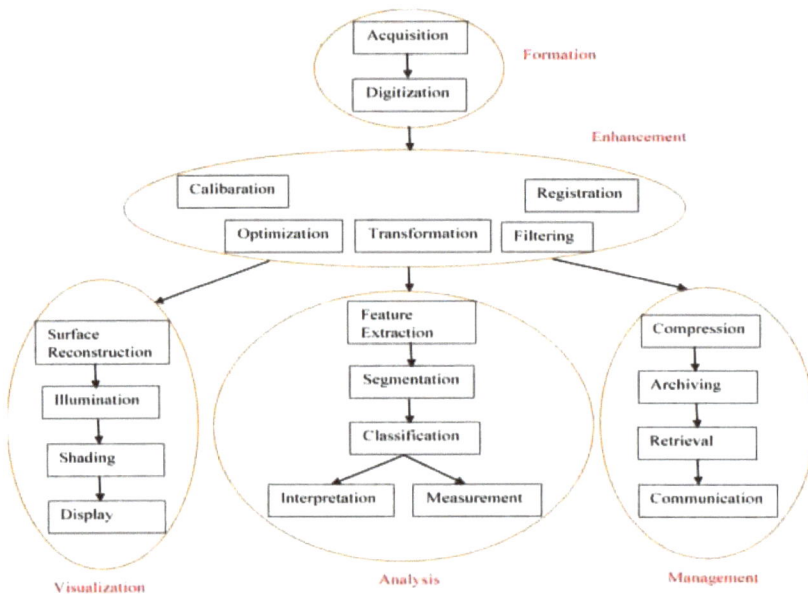

Fig. (2). Methodology for image processing.

Image Formation

This area encompasses the entire image formation process, from picture acquisition to developing a digital image matrix.

Acquisition

In the context of medical treatment, acquisition describes the act of getting health-related data, details, or photographs from numerous sources. This may involve information gathered from patients, medical devices, diagnostic equipment, and healthcare records. Acquisition technology in healthcare comprises a wide range of methods and technologies for gathering, recording, or digitizing health information. Here's how the acquisition technique is applied in medical care:

Medical Imaging

Acquisition technology is critical in healthcare, especially medical imaging. X-rays, Computed Tomography, Magnetic Resonance Imaging, and ultrasound all use specialized equipment to acquire comprehensive pictures of a person's anatomy. These digitized images are critical for identifying and tracking a variety of health disorders.

Electronic Health Records (EHR)

EHR platforms are used by healthcare practitioners to digitally capture, store, and manage health data about patients. This data includes patient demographics, medical history, medications, test findings, and treatment plans. Collecting this information *via* EHRs simplifies record-keeping, allows for data sharing among healthcare providers, and enhances patient care collaboration.

Medical Equipment

Medical equipment such as heart rate monitors, blood pressure cuffs, glucose meters, and wearable health trackers collect real-time patient data [10]. These devices enable continuous monitoring and data collection, providing crucial information to medical professionals for diagnosis and treatment decisions.

Laboratory Information Systems (LIS)

: In clinical testing facilities, LIS software is utilized to collect, manage, and report data from medical tests and analyses. This comprises blood test results, microbiology cultures, pathology reports, and other findings. LIS aids in the automation of data collecting and reporting, which reduces inaccuracies and increases performance.

Telehealth and Remote Monitoring

Telehealth platforms collect patient data remotely. Video consultations, virtual visits, and remote monitoring equipment enable healthcare personnel to collect patients' vital signs, symptoms, and histories from afar, enabling telemedicine consultations and remote care management.

Health Information Exchange (HIE)

HIE technologies allow for the confidential transfer of patient data across healthcare organizations and systems. They collect and send patient data, such as medical records and test results, to healthcare providers.

Data Analytics and Research

Healthcare organizations collect massive amounts of data for research and analysis. The data acquisition system assists in collecting, processing, and analysing clinical and research data to detect trends, perform clinical trials, and improve medical results.

Patient Questionnaires and Surveys

Doctors and nurses frequently use digital platforms to obtain patient feedback and data on their experiences, symptoms, and medical history. This information aids in customizing treatment regimens and enhancing customer satisfaction.

In healthcare, acquisition technology refers to the procedures and technologies used to gather, record, and digitize medical data and information from diverse sources. Technology is critical to modern healthcare since it facilitates diagnosis, treatment, record-keeping, and research while also enhancing patient treatment and results.

Digitization

Digitization transforms analogue data or knowledge into an electronic form, which frequently appears as binary code (0s and 1s). Digitization in healthcare refers to converting various types of medical information, records, photographs, and data from physical or analogue to digital form. This electronic representation facilitates healthcare data storage, transmission, analysis, and sharing [11]. Here are some examples of how digitalization technology is applied in medical care:

Electronic Health Records (EHRs)

The digitization technique is widely utilized to develop and maintain electronic medical records. Patient charts, medical histories, test findings, and therapy plans are digitized and maintained digitally. This enhances data accessibility, removes paper records, decreases errors, and allows healthcare personnel to swiftly get details about patients.

Medical Imaging

Analogue images from traditional film-based modalities, such as X-rays and radiographs, are converted into digital images. Digital medical images (*e.g.*, DICOM format) provide improved clarity, allow for remote access and interpretation, and for computer-aided diagnosis.

Telemedicine and Telehealth

Digitization enables remote consultations, virtual visits, and telehealth services. Patient data, such as vital signs, symptoms, and history of illness, can be communicated electronically between patients and healthcare professionals, making healthcare more accessible to access, particularly in remote or rural areas.

Prescription Management

Computerized pharmaceutical systems use digitization to allow medical professionals to deliver electronic prescriptions straight to pharmacies. This reduces medication mistakes, improves prescription tracking, and improves patient safety.

Medical Devices

Many medical devices, such as ECG machines, pulse oximeters, and infusion pumps, have digital interfaces. These instruments digitally capture and send patient data, allowing real-time monitoring and interaction with EHR systems.

Data Analytics and Research

Digitization makes massive healthcare datasets easier to acquire, store, and analyze for research, clinical trials, and population health management. Massive data machine learning and analytics can spot trends, anticipate disease outbreaks, and improve patient care using digitized medical information.

Health Information Exchange (HIE)

Digitization enables secure patient data sharing across medical organizations and platforms. HIE networks guarantee that authorized healthcare professionals have access to digital medical records and related information, hence improving care coordination.

Patient Portals

Patients can access their health information and engage with healthcare practitioners through secure online portals. They can also become more involved in their healthcare by reviewing test results, scheduling visits, and requesting prescription refills.

Remote Monitoring

Wearable gadgets and remote monitoring tools collect digital health data from patients. This data can be communicated to healthcare providers, allowing for continuous monitoring and early intervention in chronic illnesses.

Digitization is critical in contemporary healthcare because it improves data accuracy, accessibility, and interoperability, resulting in better patient care, faster administrative operations, and enhanced healthcare practices.

Image Visualization

Image Visualization entails making various changes to this matrix to improve the image's appearance. The visualization and interpretation of healthcare data and pictures to improve diagnosis, treatment planning, and communication among healthcare professionals is called image visualization in healthcare [12]. Here is a quick rundown of how image visualization is applied in healthcare:

Diagnostic Imaging

Radiologists and other healthcare workers use image visualization to interpret medical pictures such as X-rays, Computed Tomography scans, Magnetic Resonance Imaging, and ultrasounds. Advanced visualization techniques extensively investigate anatomical features and anomalies, assisting in precise diagnosis.

Surgical Planning

Surgeons utilize 3D image visualization to plan complex surgical procedures such as organ transplants or tumour excision. These visualizations provide insights into the anatomy of the patient, directing surgical procedures and reducing risks.

Treatment Monitoring

Medical pictures can be examined over time, utilizing picture visualization to track illness development or treatment effectiveness. This is especially significant in oncology, where tumour size and response to treatment may be tracked.

Education and Training

Medical educators use image visualization to teach students and train healthcare professionals. Interactive 3D models and simulations aid in communicating complex medical concepts and procedures.

Videoconferencing

In telemedicine, healthcare providers use image visualization to exchange medical pictures with remote colleagues for consultations. This allows professionals from all over the world to collaborate on diagnoses and treatment recommendations.

Patient Education

Image visualization tools enable healthcare providers to better clarify health issues and treatment alternatives to patients. Individuals can improve their understanding of their medical issues and make more educated choices regarding health care.

Innovation and Research

Image visualization aids in the study of experimental data and clinical trials in medical research. It is employed in the visualization of cellular architecture, the tracking of the progression of disease, and the evaluation of therapy effects.

Less-invasive Operations

During less-invasive operations (*e.g.*, laparoscopy or endoscopy), real-time image visualization helps guide instruments and visualize inside structures, minimizing the invasiveness of surgery and improving patient outcomes.

Radiotherapy Treatment

Image visualization is critical in radiation therapy planning. It allows for precision tumour targeting while minimizing injury to adjacent healthy tissue.

Dental and Orthopaedic Functions

Dentists and orthopaedic surgeons employ image visualization for arranging treatment, placing implants, and orthodontic examinations, assuring correct and successful therapies.

Image visualization methods and technologies are crucial in improving diagnostic accuracy, enhancing the handling of patients, and promoting medical study and instruction in the medical profession.

Image Analysis

This category includes all the analysis procedures followed for quantitative evaluations and abstract assessments of medical pictures. These processes rely on a prior understanding of the properties and content of the image, which must be integrated into advanced abstraction algorithms. As a result, image analysis is a highly specialized topic with few algorithms that are immediately transferrable to other fields of application [13]. In healthcare, image analysis entails the computerized or manual study and comprehension of medical pictures to extract relevant information for diagnosis, treatment, and research. Here is a quick rundown of how image analysis is utilized in medicine:

Disease Treatment

Image analysis assists physicians in detecting and diagnosing diseases or abnormalities in medical pictures such as X-rays, Computed Tomography, Magnetic Resonance Imaging, and histopathological slides. It aids in diagnosing illnesses like cancer, fractures, and neurological disorders.

Tumour Identification and Quantification

In various medical pictures, image analysis algorithms can detect, quantify, and analyze tumour growth. This information is critical for cancer therapy planning and patient monitoring.

Risk Assessment

Image analysis is utilized to analyze risk variables such as the severity of calcium deposits in coronary arteries (coronary calcium score), which can assist in forecasting the likelihood of heart disease.

Treatment Planning

It aids in planning surgical treatments, radiation therapy, and interventions by providing comprehensive anatomical data and finding appropriate courses or targets within the body.

Illness Progress Monitoring

Image analysis allows medical professionals to track illness progression or therapy effectiveness over time, allowing them to change treatment strategies accordingly.

Pattern Recognition of Achievement

It detects patterns and anomalies in medical images, assisting in diagnosing unusual diseases or ailments with distinct visual features.

Quality Control

Image analysis helps improve the reliability and precision of equipment used for medical imaging by finding artefacts or anomalies in images.

Image Registration

Registration techniques match various pictures from distinct modalities or periods, allowing for a more accurate evaluation and therapy schedule.

Drug Production

Image analysis is utilized in preclinical studies in drug development to examine the effects of novel medications or therapies on disease models.

Quantitative Measurements

Image analysis gives exact quantitative assessments of anatomical structures or biomarkers, which can be critical in research investigations and therapeutic trials.

Personalized Medicine

It promotes personalized medicine by customizing treatment strategies based on a person's specific health imaging data.

Machine Learning and AI

Modern machine learning and artificial intelligence (AI) techniques improve forecasting and diagnosis precision and effectiveness.

Image analysis in healthcare improves diagnosis accuracy, treatment planning, and research capacities by collecting valuable data from medical photographs [14]. It is critical to enhancing patient care, illness management, and medical progress.

Image management

Image management includes strategies for the optimal preservation, interaction, exchange, archiving, and retrieval of image data. A typical grayscale radiograph, for example, may require several megabytes of storage capacity in its native form, necessitating the adoption of compression methods. Image management also incorporates telehealth practices. The effective storing, recovery, and distribution of clinical pictures such as X-rays, Computed Tomography, Magnetic Resonance Imaging, and ultrasounds is the goal of image management technology in healthcare. Here is how it works in a nutshell:

Image Capture

Medical imaging equipment such as X-ray machines, CT scanners, and MRIs take images of a patient's anatomy.

Digitization

These photographs are digitized, which means they are converted from analogue to digital data for simple preservation and analysis.

Storage

Digital photographs are saved in a Picture Archiving and Communication System (PACS) or a Vendor-Neutral Archive (VNA). PACS is a specialized system for storing and retrieving medical images, whereas VNA provides more flexible storage across systems and vendors.

Metadata and Indexing

Every picture has been linked to patient data, research information, and metadata such as date, duration, and picture type. This metadata is critical for retrieval efficiency.

Access Control

To maintain the confidentiality of patients and adherence to healthcare standards (*e.g.*, HIPAA in the United States), entry to these photos is strictly limited.

Image Retrieval

Physicians and medical professionals can view these photos *via* secure, authorized systems, typically accessible *via* a web-based interface or connected with their Electronic Health Record (EHR) system.

Viewing and Analysis

Healthcare workers can view and analyze images using specialized medical picture-viewing software. They can modify the photos to zoom in, increase contrast, and use various diagnostic tools.

Sharing

Health photos can be simply transferred with other medical centres or professionals for discussions or additional perspectives if required. This can be done safely *via* the network.

Backup and Redundancy

Recovery and duplication are essential for ensuring that medical pictures are not destroyed by technology failures or calamities.

Long-Term Success Preservation

Some photos must be kept for a more extended period, and technologies such as digital photography and communications in Medicine (DICOM) aid in preserving medical pictures.

Connectivity with EHR

Integrating with the health information system guarantees that medical pictures are easily accessible to healthcare practitioners as part of a patient's medical record.

AI and Automation

Contemporary image management platforms may contain artificial intelligence (AI) for image evaluation and digitization, supporting doctors' interpretations or detecting irregularities.

Safety and Conformity

Strict safety precautions and adherence to healthcare standards are required to secure patient data and preserve trust in the system.

Ultimately, image management software in healthcare plays a critical role in enhancing patient care by offering healthcare workers instant access to reliable medical images, assisting in treatment planning and patient monitoring [15], and improving the general effectiveness of medical operations.

Image Enhancement

In contrast to high-level image processing, which includes picture analysis, low-level image processing includes human or automated approaches that can be used without prior knowledge of the image's specific content. The algorithms above provide consistent effects independent of the image's portrayed content. Image enhancement technological innovation in healthcare refers to procedures and instruments that boost the appearance and visibility of medical pictures to facilitate accurate diagnosis and evaluation. Here is a quick rundown of how it works:

Image Production

Cinematography devices capture medical pictures such as X-rays, Computed Tomography scans, Magnetic Resonance Imaging, and ultrasound scans. Because of factors such as equipment constraints and patient circumstances, these pictures may contain intrinsic sound, artefacts, or inadequate contrast.

Preprocessing

Preprocessing processes are frequently performed on acquired pictures before augmentation [16]. Noise diminution, geometric modification, and validation to rectify deformities and flaws may be included.

Techniques for Image Enhancement:

Luminance Enhancement

Algorithms can adjust an image's sharpness to make fine details more noticeable. This entails adjusting the pixel's brightness levels to emphasize object distinctions.

Luminance Adjustment

An image's luminosity can be altered to improve the visibility of specific features or tissues.

Probabilistic Equalization

This approach shifts pixel intensity values to obtain a more consistent equitable dispersion, hence improving image contrast.

Filtering

Spatial filters, such as Gaussian or median filters, are used to eliminate noise and smooth the picture.

Deconvolution

Deconvolution algorithms are implemented to reverse the blurring effects induced by imaging techniques, resulting in sharper images.

Multi-Modality Fusion

When many different types of imaging are employed, fusion methods can merge data from various sources to generate a single image.

Multi-Modality Fusion

When many different types of imaging are utilized, fusion methods may merge data from various sources to generate a more useful hybrid picture.

Deep Learning-Based Enhancement

Deep neural networks, such as convolutional neural networks, can be trained to enhance medical pictures by learning patterns and features from vast datasets. They are capable of handling complex picture-enhancing tasks such as denoising and super-resolution.

Actual-Time Image Augmentation

In some medical contexts, immediate-time image augmentation is critical. Imaging systems, for instance, may use enhancement methods during operations or interventional surgeries to provide instant input to the medical group.

Interactive Tools

Radiologists and medical staff may utilise interactive instruments that allow them to fine-tune picture-enhancing settings based on their needs and preferences. These programs offer a dynamic technique for image quality optimization.

Quality Control

Methods for quality assurance are frequently used to guarantee that the improvement procedure does not generate artefacts or errors that might confuse medical professionals.

Clinical Translation

Imaging specialists and doctors use augmented pictures for evaluation, therapy planning, and patient tracking. The enhanced quality of photographs may result in a more precise and secure diagnosis.

Documentation

Augmented pictures are often stored in a person's electronic health record (EHR) for further review and reference.

Image enhancement technology in healthcare is critical to improving the accuracy and effectiveness of medical imaging, resulting in better treatment for patients and results [17]. It allows healthcare practitioners to visualize and comprehend medical pictures more clearly, assisting in early identification, diagnosis, and treatment planning.

Benefits of Medical Image Processing in Healthcare

Image processing in healthcare refers to the use and study of three-dimensional information produced from magnetic resonance imaging or computed tomography of the human body. These datasets are used for various objectives, including detecting medical disorders, facilitating medical operations such as surgery planning, and assisting in research. Medical image processing is used by radiation therapists, engineers, and doctors to better understand the anatomical features found in individual patients and larger patient cohorts.

Methods for image processing have several advantages in the medical field, enhancing patient care, diagnosis, treatment, and research. Among the main advantages are:

The capacity to undertake detailed, non-invasive studies of internal anatomy is an essential benefit of medical image processing. Healthcare practitioners can enhance patient outcomes, develop improved medical equipment and drug administration strategies, and obtain more exact diagnoses by generating and assessing 3D models of relevant anatomies. It has grown into one of the most important tools for expanding the science of medicine over the last few decades.

Ongoing picture quality enhancement and cutting-edge software tools allow perfect electronic reconstruction of anatomical components at various scales and with diverse characteristics spanning from bone to soft tissues. This development allows a better understanding of patient anatomy and medical technology interplay. This accomplishment results from measurement, statistical analysis, and the development of simulation models that include authentic anatomical geometry.

Improved Diagnostic Accuracy

Image processing techniques enhance the quality of medical images, making it simpler for medical personnel to accurately identify and classify illnesses or irregularities. This results in improved patient outcomes and earlier intervention.

Objective Measurements

Measurements that can be quantified objectively include the size of tumours, the number of tissues, and circulation. This unbiased data helps track the illness's development and the therapy's efficacy.

Treatment Planning

The planning of treatments is aided by medical images that have undergone enhanced processing. By using 3D reconstructions to visualize anatomical components, surgeons can perform accurate surgical procedures with fewer risks.

Reduced Irradiation Lighting

Image processing enables the use of reduced radiation doses in imaging modalities like CT scans while maintaining the appearance of the image, reducing the risk of patient **damage**.

Telemedicine and Remote Consultation

Remote visits and ophthalmology are made possible by the simple transmission of generated medical pictures over encrypted networks. Patients in remote locations or in times of emergency can significantly benefit from this.

Process Productivity

Image processing improves workflow efficiency in the healthcare industry by streamlining procedures like picture registration, segmentation, and analysis. This enables prompt patient treatment while saving healthcare personnel time.

Disease Monitoring

Healthcare professionals can monitor a disease's course or treatment response by continuously observing medical images. This is especially crucial for cancer patients and people with chronic diseases.

Personalized Medicine

Image processing helps create individualized treatment plans by considering each person's particular anatomical and physiological traits, which results in more successful therapy.

Research Developments

Sophisticated processing of medical imaging data enables medical research. It allows for the identification of fresh biomarkers, the creation of cutting-edge treatments, and a better comprehension of illness mechanisms.

Data Administration

Image processing tools make it easier to store, retrieve, and manage massive amounts of data related to medicine. This guarantees both usability and information security.

Expense-Effective Care

Image analysis can lessen the need for invasive operations, hospital stays, and pointless tests, thus cutting down on healthcare expenses. This is because it helps with precise treatment and diagnosis planning.

Patient Education

Altered medical photos are useful resources for teaching patients. They allow medical providers to provide patients with more thorough explanations of diseases and available treatments.

Image processing methods have revolutionized healthcare by delivering instruments and insights that improve patient care, support medical research, and boost diagnostic capacities. They have also promoted the general efficacy and productivity of healthcare organizations.

The Fundamental Components of Medical Image Processing

Image Acquisition

The procedure starts with the gathering of health-related pictures utilizing X-rays, Computed Tomography, ultrasound, endoscopy, and Magnetic Resonance Imaging. The majority of these photographs are in electronic form.

Image Preprocessing

The raw healthcare images are frequently noisy and may contain artefacts. Image preprocessing methods, including noise elimination and contrast augmentation, are used to boost image quality.

Image Enhancement

Methods are employed to improve the images' perceived value and make structures and anomalies more obvious.

Image Segmentation

Image separation: Techniques for segmenting photographs locate and outline particular objects of interest—such as organs, tumours, or blood vessels—or places of interest within the pictures.

Feature Extraction

The segmented sections are then processed to extract pertinent properties, including size, shape, texture, and intensity. These qualities are crucial for a later examination.

Image Registration

Aligning and registering several images from various modalities or periods can sometimes create a unified view. Picture registration prevents incorrect positioning and picture fusion.

Image Evaluation

This crucial stage involves extracting the photos' quantitative measurements and diagnostic data. Several strategies, including machine learning and deep learning, are employed for tasks including disease diagnosis, categorization, and monitoring.

3D Image Reconstruction

Restoration of 3D photographs from 2D slices is used in imaging modalities like CT and MRI to provide a more thorough view of the anatomical characteristics.

Clinical Decision Support

Clinical decision-making assistance uses analyzed pictures and information to help medical personnel make defensible clinical judgements regarding handling patients, medical planning, and surgical direction.

Information Preservation and Retrieval

Picture Archiving and Communication Systems (PACS) are frequently used for saving and retrieving processed healthcare images digitally. Patient records and photographs can now be easily retrieved and shared.

Telehealth

Analyzed imaging files can be safely transferred *via* networks, allowing for remote consultations and telemedicine, especially in distant or underdeveloped locations.

Research and Education

Medicine relies heavily on analyzed medical photographs to better understand illnesses, create new therapies, and increase medical understanding. They also provide instructional functions, assisting in the training of medical specialists.

Moral and Security Issues

It is crucial to protect patients' confidentiality and use imaging services ethically. Medical imaging processing in healthcare must comply with ethical standards and have strong security measures.

Interdisciplinary Collaboration

To efficiently develop and deploy cutting-edge approaches, medical imaging analysis frequently entails working together among medical practitioners, radiologists, computer scientists, and engineers.

Continuous Advancements

Investigations and technology breakthroughs are underway in the dynamic field of processing medical photographs. For the finest patient treatment, staying current with the newest methods and tools is crucial.

Healthcare image processing methods are a comprehensive strategy that utilizes technology, computational algorithms, and domain expertise to enhance the medical quality of images, support evaluation and therapy, and contribute to medical study and instruction while maintaining the ethical, safe management of patient information.

LIVER DISEASE PREDICTION

Liver illness forecasting involves combining clinical data, such as histories of patients, laboratory results, imaging, and sometimes, genetic data, in conjunction with modern computer approaches, such as machine learning, to determine the chance of an individual acquiring liver-related health concerns. The goal is to detect at-risk patients early on, allowing for prompt treatment and successful management of liver illnesses. This prediction technique improves patient results and lowers the burden of liver disorders on medical facilities. Table **1** shows the literature review of liver disease detection.

Table 1. Literature review.

S.No	Author	Method	Pros	Cons
1	Sanjay Kumar *et al.* (2018) [26]	Naive-Bayes, Random forest, K-means, C5.0 and K-Nearest Neighbors(KNN).	Five classification algorithms were assessed for precision, recall, and accuracy, showing promising results for early diagnosis.	Notable are the absence of precise information, the acceptance of constraints, and the numerical outcomes.
2	R. Kalaviselvi *et al.* (2019) [27]	Decision tree, SVM, RF	A comprehensive investigation and evaluation of multiple prediction algorithms was conducted to determine the most precise and effective techniques.	Large amounts of data from clinical testing are difficult for prediction systems to handle, resulting in inefficiencies.
3	L. Alice Auxilia (2018) [28]	Decision Tree; Naive Bayes; Support Vector Machine; Random Forest; Artificial Neural Network.	The superiority of grouping systems over individual classifiers and the practical value of using actual data from Indian liver disease patients for accurate liver illness predictions.	Prediction algorithms experience inefficiencies when dealing with enormous amounts of data from hospital tests.
4	B. Venkata Ramana *et al.* (2011) [29]	SVM, KNN, Naive Bayes Backpropagation algorthim	To reduce the burden on healthcare professionals by assessing accuracy, precision, sensitivity, and specificity.	These methods successfully identified diseases using outdoor images but struggled with similar diseases.

(Table 1) cont.....

S.No	Author	Method	Pros	Cons
5	Veena *et al.* (2018) [30]	C5.0, KNN, Random Forest, K means, Naive Bayes	It points out potential directions for early diagnosis and better patient care.	Big data sets from hospital testing present inefficiencies for prediction algorithms to process.
6	Anju Gulia *et al.* (2014) [31]	J-48, Multi-Layer Perceptron, Support Vector Machine, Random Forest, Bayesian Network,	It implements a hybrid model construction approach and adopts a comprehensive approach to address the problem.	The substantial accuracy gain from feature selection and the dataset's restricted context raises the possibility of overfitting.

For this project, we employed the dataset referred to as the "Indian Liver Patient Records" from Kaggle. Our major goal is to create a predictive model that can determine whether a patient has liver disease based on particular variables presented in the dataset.

We investigated the possible links between liver illness and numerous parameters such as total proteins, albumin, *etc.* By analyzing these parameters, we hope to uncover patterns and associations that might be used for forecasting reasons for the existence or absence of liver disease in patients.

We developed a predictive model to aid healthcare providers in making more informed decisions about liver disease diagnosis and treatment using data analysis, statistical approaches, and potentially machine learning algorithms [18]. This initiative has the prospect of enhancing early identification and patient care for liver-related illnesses. The flowchart of the proposed methodology is shown in Fig. (3).

Reading the Dataset

The collection contains 583 patient records from Andhra Pradesh's northeast region. 416 of these records are for patients with liver disease (liver patients), whereas the remaining 167 are for people who do not have liver disease (non-liver patients). The information is further divided by gender, with 441 records representing male patients and 142 records representing female patients. For consistency and privacy, any patient over 89 has been documented as "90." Fig. (2) shows the patient's information.

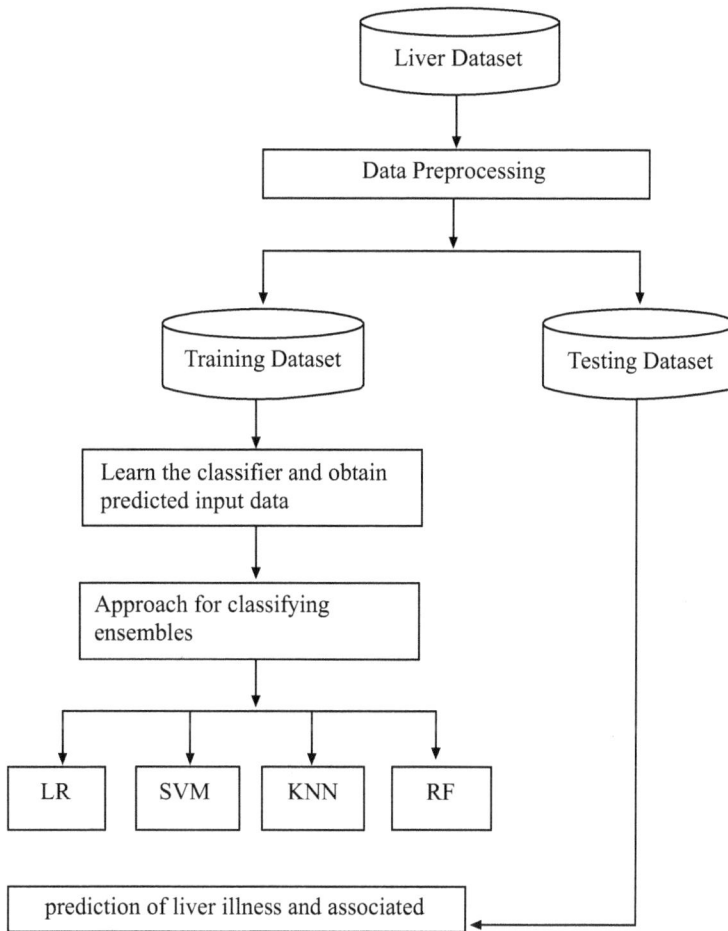

Fig. (3). The Flowchart of the proposed methodology.

The dataset has 583 rows and 11 columns, resulting in 583 data points, each containing information on 11 traits or features in shows in Table **2**.

Table 2. Patient Information.

S. No.	Age	Gender	Total Bilirubin	Direct Bilirubin	Alkaline Phosphatase	Alanine Aminotransferase	Aspartate Aminotransferase	Total Proteins	Albumin	Albumin and Globulin Ratio	Data Set
0	63	Female	0.6	0.2	177	14	17	6.9	3.2	0.89	1
1	61	Male	9.9	4.5	599	62	99	7.4	3.1	0.73	1
2	61	Male	6.9	3.9	390	58	67	7.3	3.1	0.90	1
3	59	Male	1.3	0.5	172	12	19	6.9	3.3	1.02	1
4	71	Male	3.5	2.5	185	26	58	7.2	2.5	0.39	1

Data Exploration

Based on the graph above, Figs. (**4** and **5**) show that the dataset has more males than females.

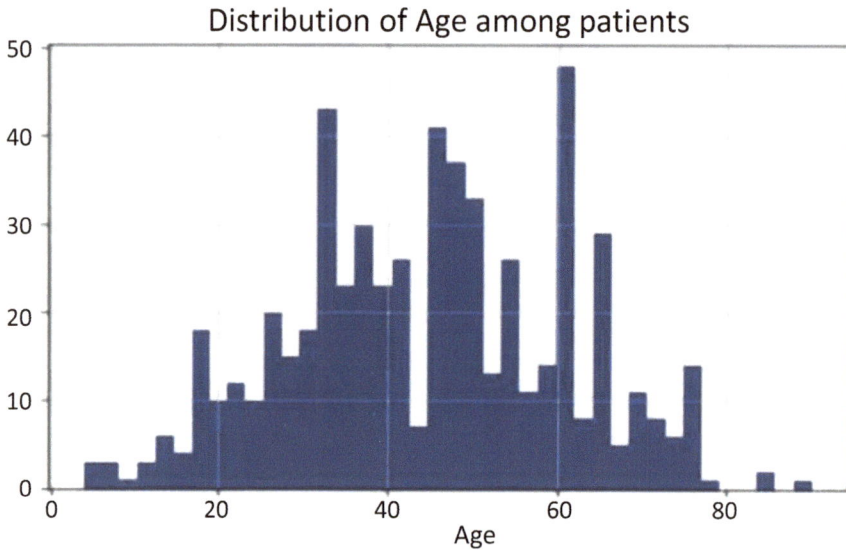

Fig. (4). The distribution of patients' ages.

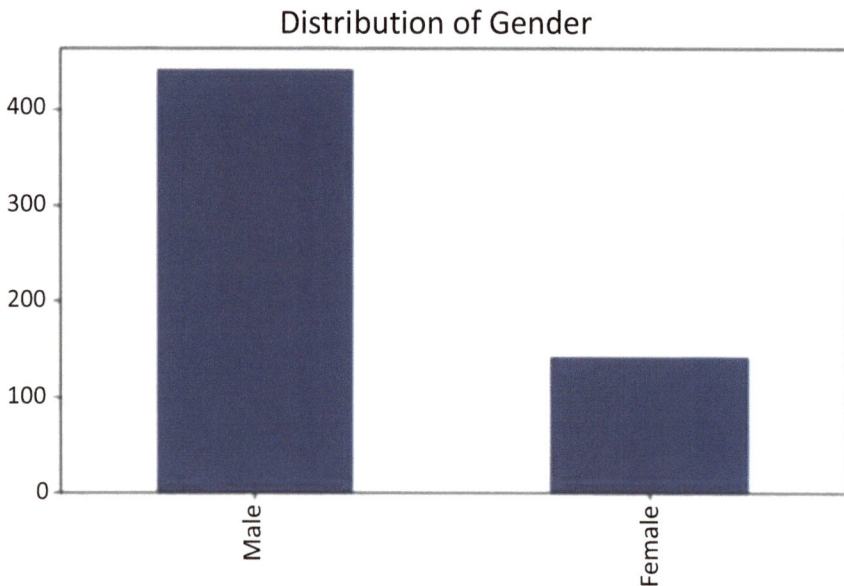

Fig. (5). The gender distribution among the patients in the dataset.

In our dataset, Fig. (**6**) shows that a more significant number of patients have been diagnosed with liver disease compared to those who do not have the disease.

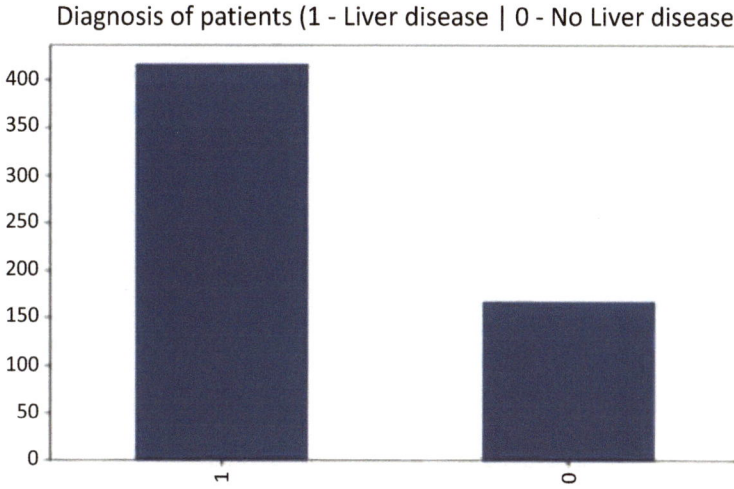

Fig. (6). The determination of the patient's medical condition or diagnosis.

Fig. (**7**) represents the patient data pair plot depending on male and female.

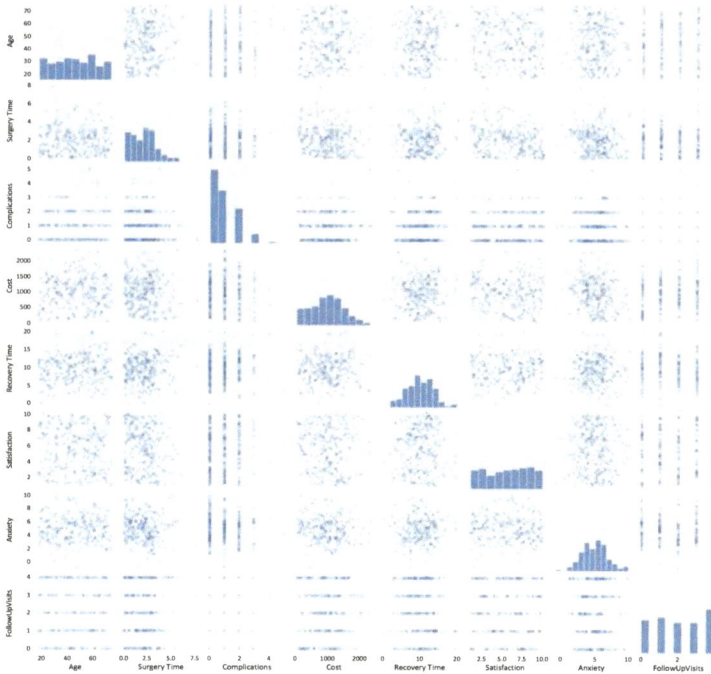

Fig. (7). The patient data pair plot depending on gender.

The following are the protein intake levels for both men and women. Fig. (8) shows that males consume far more protein than females. The albumin levels for both genders. Fig. (9) represents that males have higher albumin levels than females.

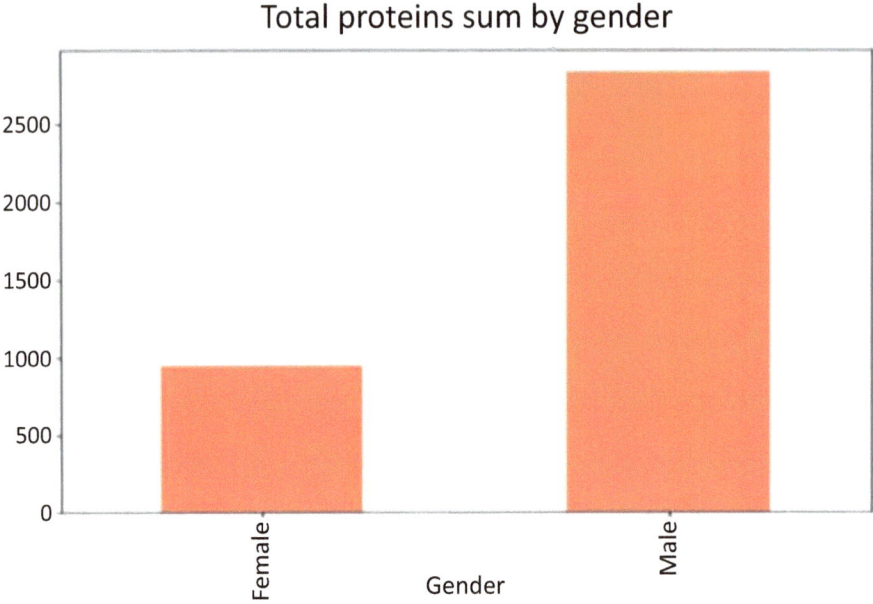

Fig. (8). A gender-based analysis comparing various metrics.

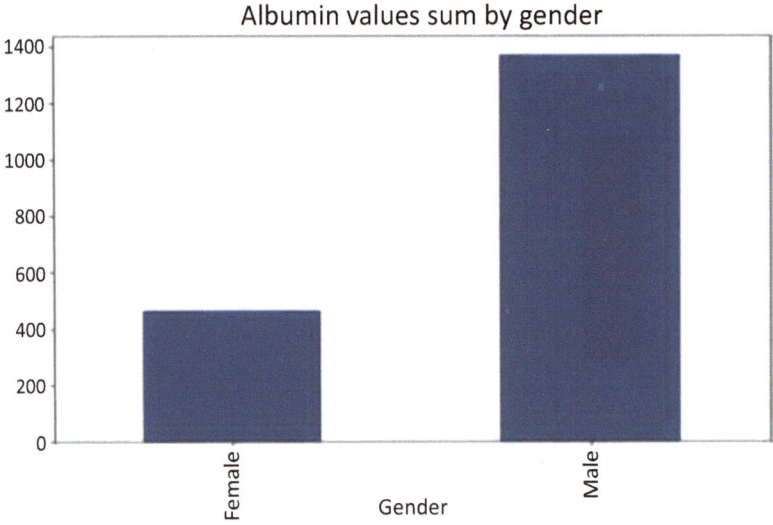

Fig. (9). The albumin levels for both genders.

Fig. (**10**) shows the bilirubin levels in both males and females, with males having higher levels. Table **3** converts the categorical column into binary numerical data. Fig. (**11**) shows the correlation between the features using a heat map.

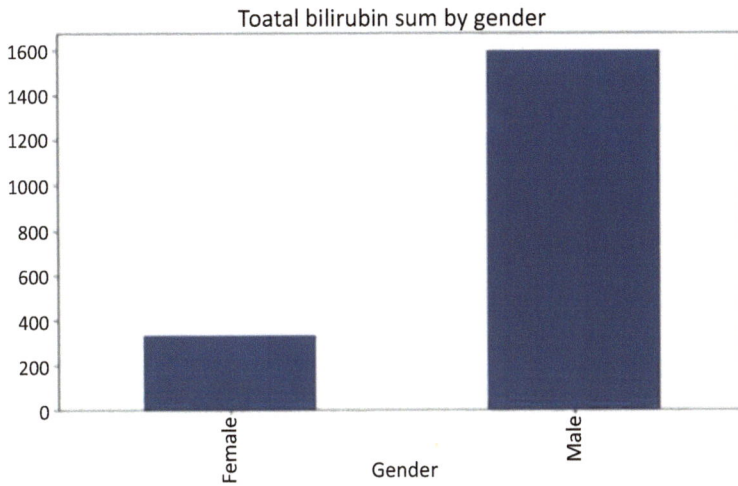

Fig. (10). The bilirubin levels in both males and females.

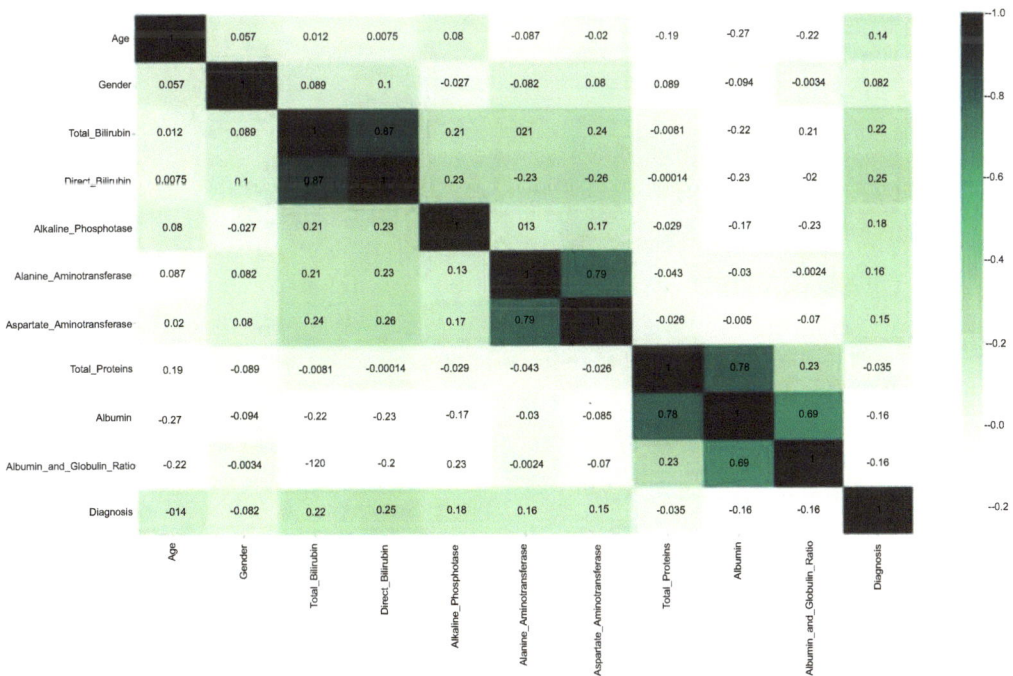

Fig. (11). The correlation between the features using a heat map.

Table 3. Converting categorical column to binary numerical.

	Age	Gender	Total Bilirubin	Direct Bilirubin	Alkaline Phosphatase	Alanine_ Aminotransferase	Aspartate_ Aminotransferase	Total_ Proteins	Albumin	Albumin_and_ Globulin_Ratio	Diagnosis
0	63	Female	0.6	0.2	177	14	17	6.9	3.2	0.89	1
1	61	Male	9.9	4.5	599	62	99	7.4	3.1	0.73	1
2	61	Male	6.9	3.9	390	58	67	7.3	3.1	0.90	1
3	59	Male	1.3	0.5	172	12	19	6.9	3.3	1.02	1
4	71	Male	3.5	2.5	185	26	58	7.2	2.5	0.39	1

Several machine learning algorithms can be utilized to develop predictive models for liver disease. Here's a quick rundown of some popular algorithms:

Logistic Regression

Logistic regression is a binary classification system that predicts liver disease [19]. It estimates the probability of belonging to one of the two groups by modelling the link between a binary outcome (presence or absence of liver disease) and a set of predictor variables (*e.g.*, biomarkers, demographic data). It is convenient when working with linearly separable data.

$$y = \frac{1}{1 + e1^{-(b_0 + b_1 x_1 + \cdots \ldots\ldots b_n x_n)}} \ldots \ldots \ldots \ldots \ldots \ldots \qquad (1)$$

Support Vector Machine (SVM)

Support Vector Machine is a sophisticated technique for binary and multiclass classification. It determines the ideal hyperplane that best divides data points of distinct classes. In liver disease prediction, a Support Vector Machine can establish a decision boundary that successfully distinguishes between patients with and without liver disease based on data such as lab findings and patient characteristics.

$$\frac{1}{2} X^T X + c \sum_{i=1}^{N} \varsigma_i \ldots \ldots \ldots \ldots \ldots \ldots \ldots . \qquad (2)$$

$$\frac{1}{2} X^T X - vp + \frac{1}{N} \sum_{i=1}^{N} \varsigma_i \ldots \ldots \ldots \ldots \ldots \ldots . . \qquad (3)$$

The Random Forest

Random Forest is an ensemble learning method that uses numerous decision trees to produce predictions. It's handy for dealing with large, multidimensional datasets [20]. A Random Forest model can be trained to assess many features and their significance in predicting the presence or absence of liver disease.

K-Nearest Neighbours (KNN) Classifier

K-Nearest Neighbours is an essential and intuitive classification technique that classifies data points based on the majority class among their nearest neighbours in a feature space. In the context of liver disease prediction, K-Nearest Neighbours can classify patients by comparing their feature values (*e.g.*, lab results) to those of their nearest neighbours, assisting in identifying trends among similar patients.

These algorithms can be used to analyze the Indian Liver Patient Records dataset or comparable datasets to construct predictive models for liver disease diagnosis. The algorithm chosen is determined by criteria such as the size, complexity, and specific needs of the prediction task. It is frequently advantageous to evaluate several algorithms' performance to see which offers the best results.

CHALLENGES FACED BY IMAGE PROCESSING TECHNOLOGY IN HEALTHCARE

Data Volume and Collection

- *Big data:* The volume of medical images produced is enormous, posing storage and data management issues. Terabytes or even petabytes of data demand a robust infrastructure and efficient data preservation solutions.
- *Extended-term preservation:* Medical pictures must be kept for long periods for patient records and reference in the future. Maintaining the authenticity and usability of preserved material over time can be difficult.

Data Privacy and Security

- *HIPAA compliance (or equivalent regulations):* Biomedical data, especially clinical photographs, is subject to stringent confidentiality laws. It is difficult to ensure compliance while granting authorized workers access.
- *Data breach hazards:* Due to the high value of medical data, it is a prime target for hackers. To safeguard data about patients, strong safeguards are required.

Compatibility

- *Compatibility with EHRs:* Imaging technology must be seamlessly integrated with electronic health records (EHRs) to offer an in-depth view of an individual's medical condition. However, compatibility between different systems and providers might be difficult to achieve.
- *Standardization:* Differences in picture formats (DICOM versus non-DICOM) and procedures might impede data sharing and interoperability between medical centres.

Processing Speed

- *Actual-time analysis:* In some medical applications, such as surgery or radiation therapy procedures, image analysis must be performed in the present moment [21]. Ensuring speedy and efficient processing is vital for patient security and procedure success.

Quality Assurance

- *Artefact amendment:* Artefacts in medical photographs can occur due to various causes, such as equipment limits, movement, or patient circumstances. Creating algorithms that will fix or minimize existing artefacts without producing new faults is difficult.
- *Validation:* It is critical to ensure that image processing methods do not lead to incorrect interpretations or diagnostic errors. Verification and certification will be comprehensive.

Clinical Adoption and Training

- *Doctor recognition:* Medical professionals may be hesitant about embracing new technologies or lack trust in the outcomes of computerized image processing. Clinicians must be educated and trained to comprehend and trust the technology.
- *Accessibility:* To promote uptake, software for processing images should be easy to use and integrated into the workflow for clinical imaging.

Ethical and Legal Concerns

- *Moral utilisation of AI:* AI-based processing of pictures generates moral issues regarding how judgements are reached and their influence on patient treatment.
- *Culpability:* Establishing culpability when image processing methods are used in the diagnostic procedure can be technically complicated.

Resource Constraints

- *Cost:* Setting up and operating image processing structures, particularly those that utilize advanced AI, can be costly.
- *Technology:* Many medical centres, particularly those in underprivileged areas, may lack the technology and assets required for high-end image processing.

Government Conformity

- *FDA permission:* Getting government permission, such as clearance from the FDA, for healthcare imaging processing software and algorithms based on AI, may be time-consuming and costly.

Overcoming these difficulties would necessitate coordination among technology designers, medical professionals, lawmakers, and governing bodies. Furthermore, continued advancements in research are required to advance methodologies for image processing and surmount these barriers for the advancement of the treatment of patients.

APPLICATIONS OF IMAGE PROCESSING TECHNIQUES IN HEALTH CARE

Strategies for image processing are essential in many aspects of healthcare, including medical diagnosis, therapy, and investigation [22]. Here are some significant computational imaging applications in health care:

Medical Imaging and Diagnosis

- *X-ray and CT scan analysis:* Image analysis improves and analyzes X-ray and CT scanner pictures to discover irregularities such as broken bones, tumours, and injuries inside the body.
- *MRI image processing:* Image processing aids in the diagnosis of neurological conditions, tumours, and muscular problems by increasing the quality of MRI images, decreasing noise, and supporting the detection of orthopaedic concerns.
- *Image processing methods can improve sonar image quality*: Image processing techniques can improve the clarity of ultrasound pictures, making it more straightforward to recognize and evaluate disorders in pregnancy, cardiology, and other fields.
- *Mammography:* This technique aids in the early identification of breast cancer by analyzing photographs of the breast to recognize problematic areas.

Disease Detection and Diagnosis

- *Histopathology:* Image processing helps pathologists analyze and measure tissue specimens, which aids in identifying diseases such as cancer.
- *Dermatology* aids physicians in examining skin lesions and moles for evidence of malignancy or additional skin conditions.
- *Ophthalmology*: Through retinal examination, image processing is utilized to identify and track eye illnesses such as diabetes-related retinopathy and hypertension.
- *Surgical navigation*: Image processing assists doctors during operations by offering immediate instruction and 3D visualization of inside structures, allowing for more accurate and less invasive surgeries.
- *Radiotherapy therapeutic:* Image processing aids in chemotherapy schedules, enabling precise tumour targeting while minimizing harm to normal tissue.
- *Telehealth and remote tracking:* Healthcare professionals in remote or impoverished locations can employ image processing to send medical images to experts for remote evaluation and evaluation.
- *Health monitoring devices:* Wearable gadgets such as smartwatches and cameras employ image processing to track blood pressure, detect deviations, and provide advance notifications about health issues.
- *Drug discovery and development* involves analyzing tiny pictures of tissues and cells to assess the impact of proposed medications and explore disease causes.
- *Clinical trials:* Image processing aids in quantifying and analysing medical images used in clinical studies, allowing investigators to determine the efficacy of freshly developed therapies [23].
- *Genomic imaging*: Image processing is used in genomics to analyze photographs of DNA microarrays and find mutations tied to illnesses.
- *Cerebral spectroscopy:* fMRI (functional Magnetic Resonance Imaging) methods are used to investigate brain function, and picture processing assists in analysing complex neural activity sequences [24].
- *Personalized medicine*: Image processing can be utilized to alter suggestions for therapy based on unique patient information and imaging results in customized medicine.
- *Public health surveillance* helps track the spread of illness by following the development of plagues using information from medical imaging.
- *Rehabilitation*: Physical rehabilitation professionals use computational imaging to measure patient improvement and customize treatment plans.

Image processing strategies enhance the visualization and interpretation of medical pictures in all of these areas and contribute to more precise medical diagnoses, better treatment organization, and superior patient outcomes [25].

CONCLUSION

The healthcare industry is undergoing a dynamic transformation in image processing, with an ongoing introduction of fresh techniques and technology. As imaging for medical purposes becomes more important in diagnosis and treatment planning, there is a rising need for innovative image-processing approaches. The combination of artificial intelligence, deep learning, and big data analytics has an opportunity to significantly improve the accuracy and efficiency of medical picture analysis, resulting in considerable benefits for patients and medical professionals. This chapter has provided an overview of the wide range of methods for image processing now used in medical care. However, it is crucial to remember that the vista of opportunities remains limitless as technology evolves at a breakneck pace.

REFERENCES

[1]   Jin T, Song H, Mon-Nzongo DL, Ipoum-Ngome PG, Liao H, Zhu M. Virtual three-level model predictive flux control with reduced computational burden and switching frequency for induction motors. IEEE Trans Power Electron 2023; 38(2): 1571-82.
[http://dx.doi.org/10.1109/TPEL.2022.3210388]

[2]   Chen Y, Xu C, Ding W, Sun S, Yue X, Fujita H. Target-aware U-Net with fuzzy skip connections for refined pancreas segmentation. Appl Soft Comput 2022; 131: 109818.
[http://dx.doi.org/10.1016/j.asoc.2022.109818]

[3]   Xu Y, Hu S, Du Y. Research on optimization scheme for blocking artefacts after patch-based medical image reconstruction. Comput Math Methods Med 2022; 2022.

[4]   Altaf F, Islam SMS, Akhtar N, Janjua NK. Going deep in medical image analysis: concepts, methods, challenges, and future directions. IEEE Access 2019; 7: 99540-72.
[http://dx.doi.org/10.1109/ACCESS.2019.2929365]

[5]   Ciuntu BM, Georgescu SO, Toma S, Zabara M, TROFIN AM, Vintila D, Vasilescu A, LOZNEANU L, Lupascu CD. Role of the prosthetic medical devices in the management of abdominal parietal defects. Medical-Surgical Journal 2022; 126(4): 528-42.

[6]   Wilson J, Agha O, Wiggins AJ, *et al.* Gender and racial diversity among women's professional sports leagues' head medical and athletic training staff. Orthop J Sports Med 2023; 11(2): 23259671221150447.
[http://dx.doi.org/10.1177/23259671221150447] [PMID: 36846816]

[7]   Al-Griffi TA, Al-Saif AS. Analytical investigations for the joint impacts of electro-osmotic and some relevant parameters to blood flow in mildly stenosis artery. J Appl Comput Mech 2023; 9(1): 274-93.

[8]   Laso S, Flores-Martin D, Herrera JL, Galán-Jiménez J, Berrocal J. Identification and visualization of a patient's medical record *via* mobile devices without an internet connection. Electronics (Basel) 2022; 12(1): 75.
[http://dx.doi.org/10.3390/electronics12010075]

[9]   Zhang Y, Dong Z. Medical imaging and image processing. Technologies (Basel) 2023; 11(2): 54.
[http://dx.doi.org/10.3390/technologies11020054]

[10]   Heo GS, Diekmann J, Thackeray JT, Liu Y. Nuclear methods for immune cell imaging: Bridging molecular imaging and individualized medicine. Circ Cardiovasc Imaging 2023; 16(1): e014067.
[http://dx.doi.org/10.1161/CIRCIMAGING.122.014067] [PMID: 36649445]

[11]   Yan Y, Yao XJ, Wang SH, Zhang YD. A survey of computer-aided tumour diagnosis based on convolutional neural network. Biology (Basel) 2021; 10(11): 1084.
[http://dx.doi.org/10.3390/biology10111084] [PMID: 34827077]

[12]   Kakkar C, Gupta S, Kakkar S, Gupta K, Saggar K. Spectrum of magnetic resonance abnormalities in leigh syndrome with emphasis on correlation of diffusion-weighted imaging findings with clinical presentation. Ann Afr Med 2022; 21(4): 426-31.
[http://dx.doi.org/10.4103/aam.aam_160_21] [PMID: 36412346]

[13]   Tozer DJ, Brown RB, Walsh J, *et al.* Do regions of increased inflammation progress to new white matter hyperintensities?: A longitudinal positron emission tomography-magnetic resonance imaging study. Stroke 2023; 54(2): 549-57.
[http://dx.doi.org/10.1161/STROKEAHA.122.039517] [PMID: 36621823]

[14]   Nzekwe S, Morakinyo A, Ntwasa M, Oguntibeju O, Oyedapo O, Ayeleso A. Influence of flavonoid-rich fraction of *Monodora tenuifolia* seed extract on blood biochemical parameters in streptozotocin-induced diabetes mellitus in male Wistar rats. Metabolites 2023; 13(2): 292.
[http://dx.doi.org/10.3390/metabo13020292] [PMID: 36837910]

[15]   Cheslerean-Boghiu T, Hofmann FC, Schulthei M, Pfeiffer F, Pfeiffer D, Lasser T. Wnet: A data-driven dual-domain denoising model for sparse-view computed tomography with a trainable reconstruction layer. IEEE Trans Comput Imaging 2023; 9: 120-32.
[http://dx.doi.org/10.1109/TCI.2023.3240078]

[16]   López-Jaime FJ, Benitez O, Caballero N, *et al.* Esplorhem: evaluation of spanish experience of using florio® haemo digital medical device for treatment monitoring in hemophilia patients. a preliminary report. Blood 2022; 140 (Suppl. 1): 8460-1.
[http://dx.doi.org/10.1182/blood-2022-170821]

[17]   Hussain Ali Y, Chinnaperumal S, Marappan R, *et al.* Multi-layered non-local Bayes model for lung cancer early diagnosis prediction with the internet of medical things. Bioengineering (Basel) 2023; 10(2): 138.
[http://dx.doi.org/10.3390/bioengineering10020138] [PMID: 36829633]

[18]   Abreu de Souza M, Alka Cordeiro DC, Oliveira J, Oliveira MFA, Bonafini BL. 3d multi-modality medical imaging: Combining anatomical and infrared thermal images for 3d reconstruction. Sensors (Basel) 2023; 23(3): 1610.
[http://dx.doi.org/10.3390/s23031610] [PMID: 36772650]

[19]   Miotto R, Wang F, Wang S, Jiang X, Dudley JT. Deep learning for healthcare: review, opportunities and challenges. Brief Bioinform 2018; 19(6): 1236-46.
[http://dx.doi.org/10.1093/bib/bbx044] [PMID: 28481991]

[20]   Esteva A, Robicquet A, Ramsundar B, *et al.* A guide to deep learning in healthcare. Nat Med 2019; 25(1): 24-9.
[http://dx.doi.org/10.1038/s41591-018-0316-z] [PMID: 30617335]

[21]   Ravì D, Wong C, Deligianni F, *et al.* Deep learning for health informatics. IEEE J Biomed Health Inform 2017; 21(1): 4-21.
[http://dx.doi.org/10.1109/JBHI.2016.2636665] [PMID: 28055930]

[22]   Chen YW, Jain LC. Deep learning in healthcare Paradigms and applications. Heidelberg: Springer 2020.
[http://dx.doi.org/10.1007/978-3-030-32606-7]

[23]   Bordoloi D, Singh V, Sanober S, Buhari SM, Ujjan JA, Boddu R. Deep learning in the healthcare system for the quality of service. J Healthc Eng 2022; 1-11.
[http://dx.doi.org/10.1155/2022/8169203] [PMID: 35281541]

[24]   Xiao C, Sun J. Introduction to deep learning for healthcare 2021; 11.
[http://dx.doi.org/10.1007/978-3-030-82184-5]

[25]   Rana M, Bhushan M. Advancements in healthcare services using deep learning techniques. In2022 International mobile and embedded technology conference (MECON) 2022; 157-61.
[http://dx.doi.org/10.1109/MECON53876.2022.9752020]

[26]   Kumar S, Katyal S. Effective analysis and diagnosis of liver disorder by data mining. In 2018 international conference on inventive research in computing applications (ICIRCA). 2018; 1047-51.
[http://dx.doi.org/10.1109/ICIRCA.2018.8596817]

[27]   Kalaviselvi R, Santhoshni G. A comparative study on predicting the probability of liver disease. International Journal of Engineering Research and Technology 2019; 8(10): 560-4.

[28]   Auxilia LA. Accuracy prediction using machine learning techniques for indian patient liver disease. In 2018 2nd International Conference on Trends in Electronics and Informatics (ICOEI) 2018; 45-50.
[http://dx.doi.org/10.1109/ICOEI.2018.8553682]

[29]   Venkata Ramana B, Babu MSP, Venkateswarlu NB. A critical study of selected classification algorithms for liver disease diagnosis. International Journal of Database Management Systems 2011; 3(2): 101-14.
[http://dx.doi.org/10.5121/ijdms.2011.3207]

[30]   Veena GS, Sneha D, Basavaraju D, Tanvi T. Effective analysis and diagnosis of liver disorder. In 2018 International Conference on Communication and Signal Processing (ICCSP) 2018; 0086-90.
[http://dx.doi.org/10.1109/ICCSP.2018.8524347]

[31]   Gulia A, Vohra R, Rani P. Liver patient classification using intelligent techniques. Int J Comput Sci Inf Technol 2014; 5(4): 5110-5.

CHAPTER 12

Augmented Reality (AR) and Virtual Reality (VR): A Study on Exploring the Emerging Applications and Future Directions in Healthcare

Anitej Chander Sood[1], Nishant Kumar Singh[1], Dhruv Jain[1], Mayank Kumar Dubey[1] and Iyappan Perumal[1,*]

[1] *School of Computer Science & Engineering, Vellore Institute of Technology University, Vellore, Tamil Nadu, India*

Abstract: Recent advances in science and technology investigate the emerging operations and untapped potential of Augmented Reality (AR) and Virtual Reality (VR) in the healthcare industry. AR and VR technologies have the potential to revolutionize healthcare by improving medical education and training, refining patient treatment and recovery, and optimizing surgical techniques. Based on an overview of the description and introduction concepts of AR and VR, it emphasizes their features and operations in the healthcare industry while also comparing those technologies, highlighting their respective strengths and limits. The study focuses on two key areas: medical education training, patient care and recovery. In medical education training, AR and VR technologies provide immersive technology for simulating medical procedures, improving anatomical understanding, and creating interactive literate environments. Case studies and exemplifications indicate successful medical education implementation. AR and VR technologies provide evidence-based therapies, pain relief improved treatment technology in patient care and recovery. It also demonstrates the implicit benefits through academic exemplifications and case studies as well as investigates the emerging applications of AR and VR in surgical planning and visualization. It examines how these technologies can aid with preoperative planning, surgical process simulation and training, and real-time surgical guiding. The influence of AR and VR on surgical concerns is demonstrated through case studies and examples.

Keywords: Augmented Reality(AR), Healthcare, Patient care, Surgical planning, Virtual Reality(VR).

---

* **Corresponding author Iyappan Perumal:** School of Computer Science & Engineering, Vellore Institute of Technology University, Vellore, Tamil Nadu, India; E-mail: iyappan.perumal@vit.ac.in

Sivakumar Rajagopal, Prakasam P., Konguvel E., Shamala Subramaniam, Ali Safaa Sadiq Al Shakarchi & B. Prabadevi (Eds.)

INTRODUCTION

Augmented Reality (AR) and Virtual Reality (VR) technologies have gained significant attention in recent times due to their eventuality to revise healthcare assistance. It offers immersive and interactive technology that can transfigure medical education, enhance patient care, and facilitate surgical procedures. The integration of these technologies in healthcare has the implicit ability to improve individual delicacy, ameliorate treatment issues, and give innovative results for patient engagement and recovery. The capability of augmented reality (AR) and virtual reality (VR) to ground the gap between the real world and the virtual world, giving healthcare professionals cutting-edge tools and chops to improve patient care, is the main significance of AR and VR in healthcare [1]. These technologies give fresh approaches to imaging medical information, exercising medical ways, and delivering supported curatives. AR and VR have the eventuality to ameliorate medical training, boost individual delicacy, and transfigure patient gestures by furnishing the healthcare labor force with better visualization, real-time guidance, and interactive knowledge exploits. The main objective of this work is to explore the rising operations and unborn directions of AR and VR in healthcare, dissect their impact on medical opinion, patient treatment, and surgical procedures, and show the implicit benefits and challenges associated with their perpetration. It also aims to give perceptivity into the transformative eventuality of AR and VR technologies and their applicability in the healthcare industry. It also gives a comprehensive overview and covers the principles of AR and VR, their comparison, and the applicability and implicit impact of these technologies in the healthcare sector. By this, the study aims to contribute to the understanding of the transformative eventuality of AR and VR in healthcare and give precious perceptivity to healthcare professionals, experimenters, and policymakers.

Section 2 presents a detailed review done over the recent works on classifying and examining the use of AR and VR in healthcare assistance. Section 3 explains about emerging applications of AR and VR in various healthcare applications. Section 4 demonstrates the future directions and ethical considerations to be followed. Section 5 concludes with the use of AR and VR in various healthcare operations and provides the way for healthcare assistance.

RELATED WORKS

The use of Augmented Reality (AR) and Virtual Reality (VR) in healthcare has garnered significant interest in recent times, leading to extensive exploration of their potential applications and benefits [1]. A substantial body of work has emerged, examining the effectiveness of AR and VR in various healthcare disciplines, including medical education, patient treatment and rehabilitation, and surgical procedures [2, 3]. In medical education and training, researchers have focused on using AR and VR technologies to enhance the learning experience for medical students and professionals [3, 4]. Numerous studies have explored the impact of immersive simulations and virtual environments on improving anatomical understanding, procedural skills, and diagnostic abilities [1, 4]. These studies have shown that AR and VR can create realistic and engaging educational environments, enabling interactive exploration of anatomical structures, virtual patient interactions, and realistic surgical simulations [1, 4]. The use of AR and VR in medical education has been found to enhance knowledge retention, spatial understanding, and decision-making skills, ultimately contributing to the development of competent healthcare professionals [4].

In the field of patient treatment and recovery, researchers have explored the potential benefits of AR and VR technologies [5]. Studies have investigated their effectiveness in managing pain, reducing anxiety, and facilitating the recovery process [5]. VR has been used to create immersive experiences that distract patients from painful procedures, such as wound care or dental treatments, resulting in reduced pain perception and increased patient comfort [5]. AR has been employed to provide personalized interventions and remote monitoring, enabling healthcare professionals to deliver targeted treatments and support patients in their recovery journeys [5]. The integration of AR and VR into therapy sessions has shown promise in improving patient engagement and motivation, as well as facilitating motor recovery and cognitive training [5]. Research in this area has demonstrated the potential of AR and VR to enhance patient outcomes, increase treatment adherence, and improve overall patient satisfaction [5].

In the field of surgical procedures, researchers have focused on employing the capabilities of AR and VR for surgical planning, simulation, and real-time guidance [6]. Studies have explored the use of AR to overlay patient-specific information, such as preoperative imaging data or vital signs, onto the surgical field, providing surgeons with valuable visual guidance during procedures [6]. VR simulations have been used to practice and refine surgical techniques, allowing surgeons to gain experience in a risk-free environment and enhancing their skills and confidence [6]. The integration of AR and VR in surgical workflows has the potential to improve surgical precision, reduce operative time, and minimize complications [6]. Research in this area has highlighted the benefits of AR and VR technologies in enhancing surgical outcomes, optimizing resource utilization, and supporting surgical training and skill acquisition [6].

While the body of research has shed light on the potential applications and benefits of AR and VR in healthcare, several challenges and limitations need to be addressed [7]. These include issues related to privacy and security, technological feasibility, cost-effectiveness, user acceptance, and standardization [7]. Further research is needed to explore the long-term impact, scalability, and integration challenges associated with the widespread adoption of AR and VR technologies in healthcare settings [7]. Future studies should also focus on developing guidelines, best practices, and evidence-based frameworks to ensure the ethical and responsible use of AR and VR in healthcare [7].

In summary, the expansive study conducted on AR and VR in healthcare has handed precious perceptivity into their implicit operations and benefits [1 - 3]. The findings indicate that these technologies can transfigure medical education, enhance patient treatment and recuperation, and ameliorate surgical procedures [1 - 3]. Still, continued exploration is demanded to address the challenges and limitations, and to foster the effective and ethical integration of AR and VR into routine healthcare practices [7].

EMERGING APPLICATIONS OF AR AND VR IN HEALTHCARE

Medical Education and Training

AR and VR technologies have the potential to revolutionize medical education and training by providing immersive and interactive experiences for healthcare professionals [1]. These technologies can simulate medical procedures and surgeries, allowing students and practitioners to gain hands-on experience in a controlled and risk-free environment [1]. Through AR and VR simulations, medical professionals can practice complex procedures, refine their skills, and improve their decision-making abilities [1]. This enhances their competence and confidence when performing procedures in real-life clinical settings. Additionally, AR and VR offer unique opportunities for enhancing anatomical understanding and visualization [1]. Medical students can use these technologies to explore the human body in three-dimensional (3D) space, visualizing complex anatomical structures with precision [1]. This interactive learning experience allows students to grasp spatial relationships and anatomical variations more effectively, leading to a deeper understanding of the human body [1]. AR and VR also enable students to engage with realistic patient scenarios, fostering critical thinking and problem-solving skills [1].

Patient Treatment and Rehabilitation

AR and VR technologies have promising applications in patient treatment and recovery, offering personalized interventions and improved outcomes [5]. These technologies can create immersive and interactive experiences tailored to individual patient needs [5]. AR and VR interventions can be designed to provide personalized therapy sessions, assist in pain management, and support patient recovery [5]. By incorporating real-time data and feedback, healthcare professionals can monitor patients and adjust treatment plans accordingly [5]. Immersive VR technology has been shown to effectively reduce pain and anxiety in various medical settings [5]. By transporting patients to virtual environments that engage their senses, VR can distract them from physical discomfort or anxiety-inducing procedures [5]. For example, VR has been used to create calming virtual landscapes for patients undergoing painful procedures or treatments [5]. By immersing patients in these environments, VR can help alleviate pain and anxiety, reducing the need for traditional pain medications and improving the overall patient experience [5].

AR and VR technologies also find applications in rehabilitation and therapy [5]. These technologies can provide interactive exercises and simulations to aid in physical and cognitive recovery [5]. Fig. (**1**) shows the illustration of AR and VR technology in real time user interaction. For example, AR can overlay virtual objects onto the real-world environment to help patients recover motor control or improve balance [5]. VR can simulate real-life scenarios, allowing patients to practice everyday activities and develop skills in a safe and controlled environment [5]. These immersive technologies promote engagement, motivation, and adherence to rehabilitation programs, ultimately enhancing the recovery process [5].

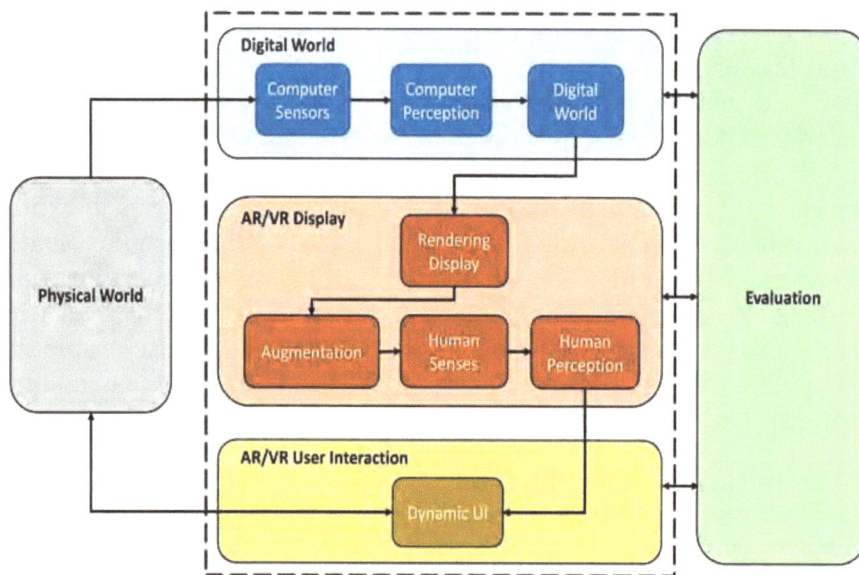

Fig. (1). Illustration of AR and VR technology in real-time user interaction

Several case studies and academic examples have demonstrated the benefits of AR and VR inpatient treatment and recovery [5]. For instance, AR-based applications have been used to provide visual cues and guidance during physical therapy sessions for patients recovering from musculoskeletal injuries [5]. VR-based programs have shown promise in rehabilitating stroke patients by offering interactive exercises that facilitate motor recovery [5]. These examples highlight the potential of AR and VR technologies to improve patient outcomes, increase engagement in treatment, and enhance the overall quality of care [5].

Surgical Planning and Visualization

AR and VR technologies offer significant advancements in surgical planning and visualization, providing healthcare professionals with valuable tools to improve surgical procedures [6]. These technologies can assist in preoperative planning, enhance visualization of patient anatomy, and provide real-time guidance during surgery [6]. AR can overlay digital information onto the surgical field, offering surgeons vital information and visual cues during procedures [6]. VR, on the other hand, allows surgeons to immerse themselves in a virtual environment that replicates patient anatomy, enabling them to practice complex procedures and improve surgical precision [6]. AR and VR are particularly valuable in surgical planning, as they allow surgeons to visualize patient-specific anatomical structures and plan surgical approaches [6]. With AR, surgeons can view patient images, such as medical scans or X-rays, overlaid onto the patient's body, aiding in the identification of critical structures and precise incision placement [6]. VR complements this process by providing a virtual representation of patient anatomy, allowing surgeons to explore and simulate surgical scenarios, plan the sequence of steps, and assess potential risks or complications [6].

In addition to planning, AR and VR enable the simulation and training of surgical procedures in a virtual environment [6]. Surgeons can use VR to repeatedly practice complex surgeries, refining their techniques and improving their skills without putting actual patients at risk [6]. These simulations provide an opportunity for surgeons to familiarize themselves with new procedures, evaluate different approaches, and gain confidence before performing surgeries in the operating room [6]. By incorporating realistic haptic feedback and accurate anatomical models, VR simulations offer an invaluable platform for surgical training and skill development [6]. During surgery, AR and VR can provide real-time assistance through overlays and visualizations [6]. AR overlays can project relevant information, such as patient vitals, instrument tracking, or anatomical structures, directly onto the surgeon's field of view, enhancing situational awareness and precision [6]. VR visualization, combined with advanced imaging techniques, allows surgeons to navigate complex anatomical structures and perform procedures with greater accuracy [6]. These technologies have the potential to reduce surgical errors, enhance patient safety, and improve surgical outcomes [6].

FUTURE DIRECTIONS AND CHALLENGES

Integration with Emerging Technologies

The integration of Augmented Reality (AR) and Virtual Reality (VR) with emerging technologies opens up innovative possibilities for the future of healthcare [1]. One of the most promising areas of integration is with artificial intelligence (AI) and machine learning (ML) [2]. AI and ML algorithms can enhance the capabilities of AR and VR by providing real-time data analysis, personalized recommendations, and predictive analytics [2]. For example, AI algorithms can analyze patient data in real time during augmented reality-supported surgeries, providing surgeons with valuable insights and decision support [2]. ML algorithms can also improve the accuracy of VR simulations by adapting to individual patient characteristics and adjusting the virtual environment accordingly. The combination of AR, VR, and AI/ML also has the potential to transform healthcare delivery, enabling evidence-based and data-driven interventions [2]. Another area of application is robotics and telemedicine [3]. By integrating AR and VR with robotic systems, surgeons can perform minimally invasive procedures with enhanced precision and control [3]. AR overlays can provide surgeons with real-time guidance and feedback, while robotic systems carry out the surgical tasks [3]. This synergy between AR, VR, and robotics can lead to safer, more efficient surgeries and improved patient outcomes [3]. Similarly, the integration of AR and VR with telemedicine can expand access to healthcare services by enabling remote consultations, diagnostics, and interventions [10].

Patients in remote areas or with limited mobility can benefit from virtual healthcare technologies that bring medical expertise directly to their homes [10]. Fig. (2) demonstrates the interaction between doctors and patient through AR and VR technologies. The combination of AR, VR, robotics, and telemedicine holds great promise for transforming healthcare delivery, bridging gaps in access, and improving patient outcomes [3, 10, 11]. The integration of AR and VR with emerging technologies has implications for enhanced healthcare delivery and patient care [1, 4]. By integrating these technologies, healthcare professionals can access real-time information, improve decision-making, and provide personalized care [1, 4]. AR and VR enable immersive and interactive experiences, promoting patient engagement, education, and adherence to treatment plans [4]. Integration with AI/ML and robotics can enhance procedural precision, reduce errors, and optimize surgical outcomes [2, 4]. Telemedicine integration expands access to healthcare services, particularly in underserved areas, and facilitates remote monitoring and interventions [3, 10]. Overall, the convergence of AR, VR, and emerging technologies has the potential to revolutionize healthcare by delivering more effective, personalized, and accessible care [1 - 4, 10, 12].

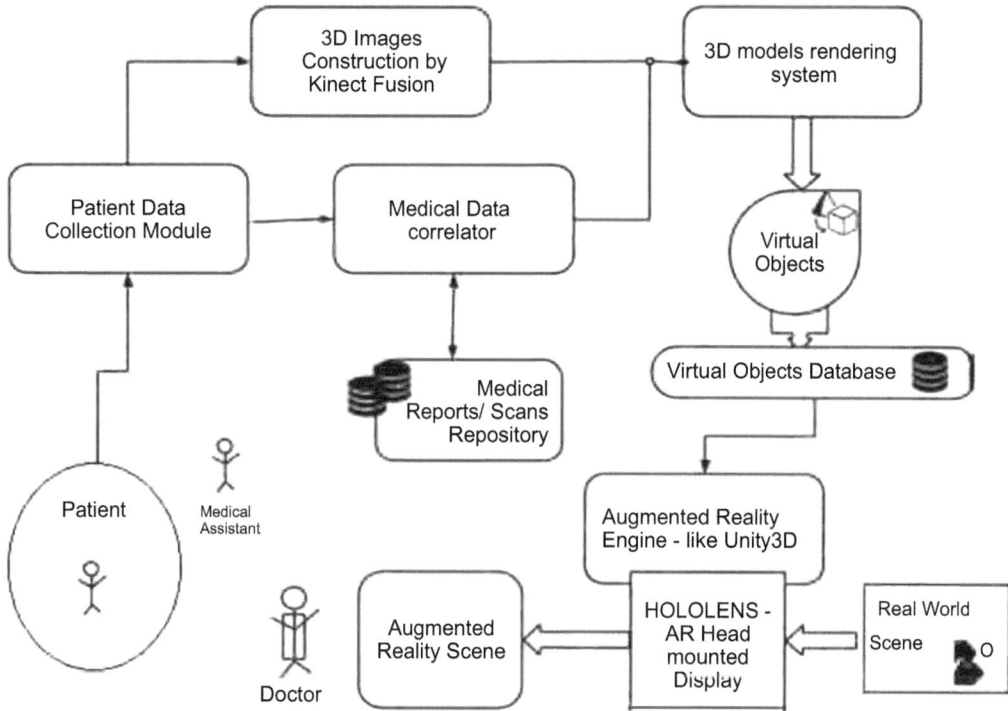

Fig. (2). Demonstration of interaction between doctor and patient through AR and VR technologies.

Ethical and Privacy Considerations

While the integration of Augmented Reality (AR) and Virtual Reality (VR) in healthcare holds immense potential, it also brings forward critical ethical and privacy considerations [5, 7]. The collection and utilization of patient data in AR and VR applications necessitate careful handling to ensure patient privacy and data security [5]. These technologies often involve the capture and analysis of sensitive medical information, such as patient images, scans, and health records [5]. Healthcare organizations must strictly adhere to privacy protocols and comply with data protection regulations to safeguard patient confidentiality [5]. Transparency in data collection, storage practices, and the purposes of data use is essential to build trust among patients, healthcare providers, and technology developers [7 - 9]. Moreover, robust security measures should be implemented to protect patient data from unauthorized access or breaches [5]. Encryption, secure network protocols, and strong authentication mechanisms are crucial to mitigate potential risks associated with data handling in AR and VR environments [5]. Clear policies for data retention and disposal should be established to prevent unnecessary data retention and minimize the risk of data misuse [5].

Beyond privacy and security, the adoption of immersive technologies in healthcare raises ethical concerns [4, 8, 9]. It is imperative to ensure that the implementation of AR and VR adheres to ethical guidelines and respects patient autonomy [8 - 10]. Informed consent from patients should be obtained before they participate in AR and VR experiences, ensuring they have comprehensive information about the technology, its potential benefits, and associated risks [8]. Healthcare professionals must prioritize patient welfare and ensure that the use of AR and VR does not compromise patient safety or the quality of care delivered [4]. Regular monitoring and evaluation of these technologies are essential to assess their effectiveness, address potential biases, and mitigate unintended consequences [8, 15]. In conclusion, while the integration of AR and VR in healthcare offers transformative possibilities, it necessitates careful attention to ethical and privacy considerations [5, 7, 15]. By prioritizing patient privacy, implementing robust security measures, and adhering to ethical guidelines, healthcare organizations can harness the potential of AR and VR while safeguarding patient trust and rights [5, 7 - 10, 13, 14].

CONCLUSION

In conclusion, this study explored the emerging applications and future directions of Augmented Reality (AR) and Virtual Reality (VR) in the healthcare industry. The findings underscore the transformative potential of AR and VR technologies across diverse healthcare disciplines. In medical education and training, AR and VR offer immersive and interactive technologies that enhance learning, simulate medical procedures, and improve anatomical understanding. Patient treatment and recovery benefit from personalized interventions, pain management, and advanced rehabilitation techniques facilitated by AR and VR. In surgical planning and visualization, these technologies enable precise preoperative planning, realistic simulations, and real-time guidance during surgeries. The integration of AR and VR in healthcare has the potential to revolutionize the industry by improving patient outcomes, enhancing healthcare delivery, and increasing accessibility. By providing realistic and immersive experiences, AR and VR technologies can enhance the skills of healthcare professionals and improve patient safety. Similarly, these technologies empower patients by offering personalized interventions, reducing pain and anxiety, and facilitating recovery. Moreover, the integration of AR and VR with emerging technologies such as artificial intelligence and robotics further expands the possibilities for personalized care, remote monitoring, and enhanced surgical precision. However, despite these advancements and potential benefits, there are significant challenges that need to be addressed. Ethical and privacy considerations related to patient data collection and protection require careful attention. Implementing robust security measures and complying with data protection regulations are essential to ensure patient privacy and confidentiality. Clear ethical guidelines should also be established to uphold patient autonomy, informed consent, and patient welfare in the use of AR and VR technologies. To fully harness the transformative potential of AR and VR in healthcare, further research and exploration are needed. Continued studies and evaluations are necessary to assess the long-term effectiveness, safety, and cost-effectiveness of these technologies. Interdisciplinary collaborations between healthcare professionals, technologists, and policymakers can drive innovation and address the challenges associated with the integration of AR and VR in healthcare. In conclusion, AR and VR technologies have the potential to transform healthcare by enhancing education and training, improving patient treatment and recovery, and optimizing surgical procedures. Through ongoing research, innovation, and ethical considerations, AR and VR can contribute to advancing healthcare practices, ultimately improving patient outcomes and transforming the delivery of healthcare services.

REFERENCES

[1] Dorri M. Healthcare research: VR and AR. Br Dent J 2017; 222(4): 224-5.
[http://dx.doi.org/10.1038/sj.bdj.2017.145] [PMID: 28232703]

[2] Kitaria D, Mwadulo M. Adoption of Augmented Reality (AR) and Virtual Reality (VR) in healthcare systems. African Journal of Science, Technology and Social Sciences. 2022 Sep 26;1(1).

[3] Xiong J, Hsiang EL, He Z, Zhan T, Wu ST. Augmented reality and virtual reality displays: emerging technologies and future perspectives. Light Sci Appl 2021; 10(1): 216.
[http://dx.doi.org/10.1038/s41377-021-00658-8] [PMID: 34697292]

[4] Mekacher L. Augmented Reality (AR) and Virtual Reality (VR): The future of interactive vocational education and training for people with handicaps. International Journal of Teaching. Education and Learning 2019; 3(1): 118-29.

[5] Grieco P. Virtual and augmented reality applications in medicinal chemistry. Future Med Chem 2022; 14(20): 1417-9.
[http://dx.doi.org/10.4155/fmc-2022-0213] [PMID: 36196876]

[6] Jo YJ, Choi JS, Kim J, Kim HJ, Moon SY. Virtual reality (VR) simulation and augmented reality (AR) navigation in orthognathic surgery: a case report. Appl Sci (Basel) 2021; 11(12): 5673.
[http://dx.doi.org/10.3390/app11125673]

[7] Huang TK, Yang CH, Hsieh YH, Wang JC, Hung CC. Augmented reality (AR) and virtual reality (VR) applied in dentistry. Kaohsiung J Med Sci 2018; 34(4): 243-8.
[http://dx.doi.org/10.1016/j.kjms.2018.01.009] [PMID: 29655414]

[8] Sugimoto M. Augmented holographic HPB surgical navigation using extended reality: XR (VR/AR/MR). HPB (Oxford) 2021; 23: S393.
[http://dx.doi.org/10.1016/j.hpb.2020.11.1011]

[9] Scavarelli A, Arya A, Teather RJ. Virtual reality and augmented reality in social learning spaces: a

literature review. Virtual Real (Walth Cross) 2021; 25(1): 257-77.
[http://dx.doi.org/10.1007/s10055-020-00444-8]

[10]   Bhugaonkar K, Bhugaonkar R, Masne N. The trend of metaverse and augmented & virtual reality extends to the healthcare system. Cureus 2022; 14(9): e29071.
[PMID: 36258985]

[11]   Mehta K, Aayushi C. Revolutionizing healthcare by accessing the opportunities for virtual and augmented reality 2023 7th international conference on intelligent computing and control systems (ICICCS), Madurai, India, 2023, pp. 836-841.

[12]   Ullah H, Manickam S, Obaidat M, Laghari SUA, Uddin M. Exploring the potential of metaverse technology in healthcare: Applications, challenges, and future directions. IEEE Access 2023; 11: 69686-707.
[http://dx.doi.org/10.1109/ACCESS.2023.3286696]

[13]   Saudagar AKJ, Kumar A, Khan MB. Mediverse beyond boundaries: A comprehensive analysis of AR and VR integration in medical education for diverse abilities. Journal of Disability Research 2024; 3(1).
[http://dx.doi.org/10.57197/JDR-2023-0066]

[14]   Raja M, Priya GGL. The role of augmented reality and virtual reality in smart health education: State of the art and perspectives. Artificial intelligence for smart healthcare EAI/springer innovations in communication and computing. Cham: Springer 2023.
[http://dx.doi.org/10.1007/978-3-031-23602-0_18]

[15]   King PH. Extended reality for health care systems: Recent advances in contemporary research. IEEE Pulse 2023; 14(2): 39-40.
[http://dx.doi.org/10.1109/MPULS.2023.3269762]

CHAPTER 13

Chem-bioinformatics: Computational Alternatives to Clinical Diagnosis, Treatment and Preventative Measures

Samiha Nuzhat[1]**, Mahtabin Rodela Rozbu**[1]**, Christine Thevamirtha**[1]**, Maryam Wardeh**[1]**, Fatema -Tuz-Zohora**[1]**, AKM Moniruzzaman Mollah**[1]**, Sweety Angela Kuldeep**[1] **and Mosae Selvakumar Paulraj**[1,*]

[1] *Environmental Sciences Program, Asian University for Women, Chittagong-4000, Bangladesh*

Abstract: Nowadays, chem-bioinformatics tools are widely used for genomic and proteomic data analysis, gene prediction, genome annotation, expression profiling, biological network building, and many more purposes. Clinical applications of such computational approaches are also needed to ensure real-life implementation of findings from the fields of cheminformatics and bioinformatics. Despite being a new field of science, studies found huge significance and interconnectivity of cheminformatics and bioinformatics approaches in modern medical science. Identification of cancer biomarkers, for instance, has been possible *via* bioinformatics tools mediated in-depth genome analysis, resulting in cancer susceptibility being easily calculated nowadays using a bioinformatics approach. In addition, bioinformatics tools are helping docking studies in the prediction of anticancer drug structures as well. Also, genome analysis of patients using bioinformatics techniques is the initial requirement for personalized therapeutics designing in cancer treatment. Additionally, in recent times, computer-aided drug designing has benefited since bioinformatics tools offer easier determination of effective active sites and potential side-effects of the predicted drug on system biology and genetics constitution. Besides, diagnosis and treatment of infectious diseases often require a suitable bioinformatics approach to study host-pathogen interaction. Moreover, treatment of metabolic disorders, complex genetic disorders, point of care diagnostics, observation of drug efficacy, *etc.* are controlled, monitored, and modified using multiple bioinformatics tools by manipulating the biological data sets. Such various applications can benefit the medical sector in multiple clinical processes. Realizing these, this book chapter aims to explore some of such major applications of chem-bioinformatics studies in the medical sector; mostly in terms of diagnosis, treatment, and prevention of diseases.

Keywords: Bioinformatics, Cheminformatics, Diagnosis and treatment, Genomic data, Proteomic data, Medical science.

* **Corresponding author Mosae Selvakumar Paulraj:** Environmental Sciences Program, Asian University for Women, Chittagong-4000, Bangladesh; E-mail: p.selvakumar@auw.edu.bd

Sivakumar Rajagopal, Prakasam P., Konguvel E., Shamala Subramaniam, Ali Safaa Sadiq Al Shakarchi & B. Prabadevi (Eds.)

BACKGROUND

Today's world highly relies on data science and not a single sector is refrained from using mass-scale data to advance its traditional workflows. The biomedical and overall health sector is, thus, no different from being highly data-driven. The study of biological and biochemical data has drastically gained much popularity over the past few decades [1]. Being an interdisciplinary field of science, cheminformatics, and bioinformatics explore and analyze biochemical data using software and analyzing tools. Cheminformatics, an area of information technology closely associated with chemical information, is also essential to collect, retain, examine, and organize chemical aspects of biological data [2]. Cheminformatics has digitized and reshaped traditional clinical processes drastically with the help of bioinformatics applications. Bioinformatics tools are widely used for genomic and proteomic data analysis, gene prediction, genome annotation, expression profiling, biological network building, and many more. Researchers are constantly exploring further opportunities to be created by chem-bioinformatics tools and bridging the gaps, particularly to save resources and time needed for lengthy and uncertain on-hand experiments [3]. The current scope of chem-bioinformatic research is also abundant as evidenced by the wide number of tools available for various biological and biochemical analyses.

Apart from its huge research prospects, clinical applications of bioinformatics are one of the most crucial sectors that require real-life implementation of findings from the field of bioinformatics. Due to the rapid accumulation of large quantities of biological data sets, computer-based approaches associated with machine learning and other tools offer huge opportunities for manipulation to extract and analyze useful information and newer data dimensions [4]. Though this arena of study experience lacks expert manpower and study materials, it is yet an emerging field with huge prospects not only to lessen the workload of existing biological research-based implications but also to open wide opportunities for biological research at a minimal cost and timespan [5]. This is why, despite the challenges and ethical concerns associated with bioinformatic-based approaches, their popularity is still expanding.

Considering the wide scope of bioinformatic applications, the huge utility of bioinformatics-related discoveries has been noticed in multiple stages of medical science as well. Disease diagnosis, treatment, and prevention-all these stages have experienced multiple applications of bioinformatics. The implications of such activities are too vast to wrap up in a few hundred pages. Also, the variety of opportunities offered by medical science-related approaches to bioinformatics is a matter of attraction for many researchers. For instance, cancer biomarkers have been identified through in-depth analysis of the genome using bioinformatics tools. As a result, the susceptibility of cancer can be easily calculated nowadays using a bioinformatics approach [6]. At the same time, personalized therapeutics designing, a potential means of cancer treatment, requires genome analysis of patients using bioinformatics techniques [7]. Such dependency on bioinformatics is also noticed in the case of computer-aided drug designing for other diseases, as bioinformatics tools help to design more effective active sites and determine potential side-effects of the designed drug on system biology and genetics constitution [8]. Besides, diagnosis and treatment of microbial diseases may require a suitable bioinformatics approach so that host-pathogen interaction is studied properly before deciding on corresponding treatment and preventative strategies [9]. Gene therapy, controlled manipulation over genes' constitution and expression to design treatment of several diseases is another emerging field of bioinformatics, as it requires in-depth genetic analysis using bioinformatics tools [10]. Moreover, the treatment of metabolic disorders, complex genetic disorders, point-of-care diagnostics, observation of drug efficacy, *etc.* is controlled, monitored, and modified using multiple bioinformatics tools by manipulating the data sets as per the professionals' interests.

Considering all such different applications and connective associations between bioinformatics and medical science, this book chapter aims to explore some of the major applications of bioinformatics studies in medical science; especially in terms of diagnosis, treatment, and prevention of diseases.

DISEASE DIAGNOSIS

Disease diagnosis has multiple stages that are traditionally conducted using biochemical or microbiological approaches. Adopting a chem-bioinformatic approach to the diagnosis of diseases is becoming popular because of the reduced cost and time associated with this process [11]. The process of disease diagnosis solely depends on the type of disease to be diagnosed; however, bioinformatics-based approaches have often made diagnosis-related tasks easier, cheaper, and handier. Major contributions of bioinformatics for disease diagnosis have been briefly discussed in the following subsections.

Pathogen Identification

Next-Generation Sequencing (NGS) is the rapid and sophisticated mechanism for sequencing which is getting popular for research and clinical activities that puts huge public health and clinical importance on the process. Metagenomic NGS helps to detect pathogens present in the infected body fluids. This uses cell-free DNA of the body fluids to perform 16s rRNA PCR [12]. In this regard, NGS works under certain stages *i.e.* PCR amplification, sequence assembly, and finally pathogen identification. Then for clinical diagnosis using other microbiological and biochemical approaches, other required tests are conducted [13]. However, sometimes through such analysis, some unexpected strains of pathogens can also be detected for which genomic information, evolutionary tracing, strain comparison, and other suitable activities are performed using biological data and bioinformatics tools, which often help to detect new pathogens and predict their characteristics and interactions with the body [14]. This process often involves the utilization of multiple bioinformatics tools. For instance, PATRIC (Pathosystems Resource Integration Center) is an infectious-disease-related database that provides genomic and other related data comprehensively [15]. Its application in clinical diagnosis is yet not common, still, its utilization is hoped to assist clinical diagnosis drastically. PAIPLine is another bioinformatics tool designed to sort out difficulties in detecting pathogens from diagnostic samples, assuring precise, robust, and adaptable findings for downstream analyses using NGS samples [16]. Many such tools are often used for pathogen identification in the clinical sector.

Gene Selection

High-throughput data are commonly generated in recent times for multiple bioinformatics analyses and the use of conventional technologies often imposes many challenges regarding the analysis process of these data. Data mining is the key approach through which the task of gene selection is conducted where genetic data is analyzed one by one to identify the targeted genetic component [17]. One of the most effective methods to process such high throughput data is to use the univariate/multivariate analysis methods. Univariate/multivariate analyses are usually used for gene selection to discover the relationship between traits and genes. Univariate analyses study the information of each gene individually based on a single criterion to find its relation to the desired traits. There are many methods used to perform the univariate analysis like the Signal-to-Noise Ratio (SNR) [18], the threshold number of Misclassification (TNOM) score [19], and many other methods. The multivariate analysis defines the search and analysis path based on the information of a group of genes in contact with the univariate analysis whose search path is based on individual genes. There are many methods for multivariate analysis listed in the literature like greedy forward search, which is considered as one of the easiest methods to conduct multivariate analysis [20]. Floating searches and genetics algorithms are considered to be more complex methods [21]. Another method, Recursive Feature Elimination (RFE), used for Alzheimer's, Parkinson's, and other genetic disease detection, uses a machine vector classifier to assign a weight to each gene in the set, and then the genes with the smallest weight are considered the least informative genes and they are eliminated out of the set, the process keeps repeating until there are no genes left

[22]. Univariate and multivariate analyses thus assist significantly in gene detection.

Further studies on such analyses discovered many more dimensions of bioinformatics-approach-based gene selection. Lai and colleagues compared the results of the multivariate and univariate analysis in gene selection for cancer classification [23]. It has been found that univariate analysis produces better results than multivariate analysis, which leads to more stable performance in many types of cancer gene sets. Whereas the multivariate approach did not perform as better than the univariate approach and its performance was not as stable as other methods [23]. As the process of univariate analysis is considered to be simpler than multivariate analysis, it is widely used when looking for disease-related genes. The complexity of the gene expression process may lead to many false-negative results when using the univariate approach, meaning that many genes will be shown to be directly affecting the occurrence of disease while it is related to the disease outcome. When the rate of false discovery is high, the produced results become less accurate and not replicable [24]. While performing the gene-selection experiment, researchers try to minimize the false discovery rate using a process called Knockoff. The knockoff procedure framework can be simple or complicated depending on the used method like Gaussian distribution and lasso coefficient difference statistics [25]. This process is quite popular in developed countries for cancer biomarker identification and corresponding gene selection.

Pathogenicity of Microbes

Lately, Bioinformatics tools have been developed to predict the pathogenicity of newly discovered microbes. In some cases, detecting a new bacteria in the human body could cause unknown diseases and symptoms leading to a negative impact on the body and humanity overall. Metagenomic data is used to detect pathogenic characteristics of microbes instead of using culture-based or PCR-based molecular experiments [26]. The decreased cost of producing high throughput data and whole-genome sequencing is making the process of classifying and detecting gene families easier.

New software was developed to predict the pathogenicity of bacteria with 88.6% accuracy, PathogenFinder [27]. PathogenFinder is a web server that allows the fast identification of human bacterial pathogenicity using user-uploaded proteomes. This website contains a protein family database, which contains groups of protein families (PF) associated with pathogenic and non-pathogenic organisms from 885 bacterial complete genomes. Each of these PFs is tagged as pathogenic or non-pathogenic with a calculated weight (z-score). This software contains 10 different models available to use depending on the type of data uploaded by the users. Users can upload either raw reads, complete or a draft genome of the organism to the software and choose the appropriate model to predict its pathogenicity. When analyzing the uploaded data, the software will calculate the weight (z-score) of the PF present in the organism and according to the z-score, it will predict the pathogenicity. When the z-score is greater than 0, the organism is considered to be a human pathogen and when it is below 0 then it is a human non-pathogenic bacteria [28].

Pathogenicity Prediction for Bacterial Genome (PaPrBaG) is a machine learning-based approach that helps to predict the pathogenic phenotype of unknown bacteria, available for users as an R-package. This approach is different from the previous existing tools as these methods are highly dependent on genome assembly and annotation, which are time-consuming and hold a high risk of error (especially in low-coverage reads) [29]. The PaPrBaG combines a set of genome data and metadata into a system of rules that identify the photogenic and non-pathogenic bacteria [30]. It uses the random forest classifier, which gives more accurate results, a shorter prediction period, and can deal with noisy data. The accuracy of this approach comes from the prediction process, which includes a cross-validation strategy followed by a read simulation; the final results are compared with other methods for further accuracy. The evaluation metric of this program is measured using the Majority prediction rule and Minimum detection threshold. The majority prediction rule calculated the overall read-based prediction probability based on the single reads of the genome, if the value is greater than 0.5, then the sample is considered to be pathogenic [31]. The Minimum detection threshold measurement takes into consideration the uncertainty associated with the majority prediction rule and allows the user to set the minimum reads fraction that should be considered for predicting a phenotype. Thus, using this approach, pathogenicity can be predicted.

DISEASE TREATMENT

Bioinformatics tools have wide applications in disease treatment that can be incorporated with clinical sectors for drastic improvement and modernization of the co-current processes. Suitable treatment strategy designing using bioinformatics tools highly depends on the type and status of the disease. The utilization of such tools lies in drug assigning to personalized treatment. Even if needed, bioinformatics contributes to the design of new drugs, and vaccines, and determines therapeutic applications as well. For instance, tools of bioinformatics have been used to predict the structure of suitable vaccines for the SARS-CoV-2 virus caused by the COVID-19 pandemic. Even researchers compared the efficacy, effectiveness, pros and cons of the proposed vaccines using bioinformatics tools [32, 33]. Multi-dimensional research has also been conducted to propose suitable drugs to improve the physical conditions of the infected patients, which also required the utilization of bioinformatics tools [34]. Disease treatment approaches might benefit from bioinformatics in multiple ways; some of such ways have been illustrated in the following sub-sections.

Drug Designing and Significance of Cheminformatics

Along with bioinformatic tools and approaches, the emergence of cheminformatics can significantly help the treatment of diseases by predicting suitable drugs or biomolecule–drug interactions based on bioinformatic or cheminformatic modeling [35]. Nowadays, pharmaceutical industries largely depend on chemical data to assess the efficacy of predicted drugs. This is commonly known as a process called computer-aided drug designing (CADD) [36]. For instance, drugs can be flexibly designed using scaffold-related information when poly-pharmacological designing is needed [37]. Also, bioinformatics tools contain high-throughput data that contribute to drug discovery organizing those data based on the researchers' interests. Some of such commonly used types of data for drug discovery include epigenetic effects, genome architecture, cistromes, ribosome profiling, DNA/RNA/Protein structures, exome sequencing, and so on [8]. Based on these data, researchers often predict suitable drugs that can act at this molecular level. Their suitable docking and interaction trends are also predicted using bioinformatics tools. For instance, ICM, AutoDock, QXP, *etc.* are the commonly used software to predict molecular docking. HADDOCK, another popular docking prediction software can model docking characteristics of huge protein-protein complexes under the effect of suitable conditions [38]. Argus lab, Macro Model, *etc.* are software that has drug modeling options to predict its interactions with biomolecules and to assume its potential effects on system biology [39]. Also, bioinformatics tools help to predict the destination of such drugs and therapeutics and re-define the proposed structures if any unexpected interactions are noted from bioinformatics-mediated system biology and pathway analysis. Aside from such modern approaches to drug designing, historically used herbal medications can also benefit from bioinformatics tools.

Herbal medication can be termed an eternal treatment that most often harbors zero side effects, deeming it to be a good choice for disease treatment and prevention as well. The intersection of biochemistry, bioinformatics, and cheminformatics allows us to understand the various chemicals in herbal leaves, garlic, turmeric, *etc.* along with their interaction in the biological system [35, 40, 41]. Studying and discovering antibiotics, antiseptics, and anti-cancer agents present in herbs and foods, learning and understanding their effects on gene expression and their ability to boost immunity, would significantly be a guiding knowledge to develop drugs and unveil knowledge on having the expression of oncogenes, vascular diseases controlled [35, 41, 42]. Also, disease symptom-mediated treatments for less-known health issues can be assured using bioinformatics simulation. For instance, a study predicted the drug or corresponding derivative associated with the agonist molecules of degrading taste receptor sensing during COVID-19 infection [43]. Such symptom treatment approaches can also be designed using bioinformatics simulation for such health issues whose specific treatment has not been found yet. Therefore, to ensure multidimensional treatment, studies and further research on genomics, food chemicals interactions with various metabolites and molecules, the discovery of more herbal leaves and their impact on our bodies, learning about ancient medication, together with bioinformatics, cheminformatics, and biochemical approaches to create and manage meaningful data, we would have the knowledge and understanding on how our body works and can be influenced, thus being an essential insight to prevent and treat diseases, extending life-span.

Personalized Treatment

Personalized treatment is an emerging field of clinical studies where treatments (medicines, therapies, diet plans, *etc.*) are suggested based on very specific conditions of every individual. Growing applications of bioinformatics made personalized treatment more available and easily adaptable. Personalized treatments are suggested with high specificity and sensitivity. DNA microarray is often used in this regard since the microscopic DNA sports are organized for every individual in this chip. Based on the expression trend of the genes in the microarray, pharmacologists often determine potential benefits or side effects that certain medications may cause [44]. Especially for tumor and cancer treatment, such microarray-based personalized treatment assists therapeutics identification very promptly [45]. Such treatments require pathway analysis and expression trend analysis using multiple bioinformatics tools that reduce risks, identify suitable target genes, and predict all the potential effects that assure highly specific treatment for the patients and it has been found safer than other traditionally used treatment strategies [46]. Bioinformatics tools also help in personalized treatment design in many other ways.

Also, with the constant innovation and research on next-generation sequencing, one's genome can easily be sequenced at a relatively cheaper rate . The sequenced genome is then processed by various computational tools and software that utilizes various gene and proteomic networks, to predict the susceptibility of disease for that person. The prediction or diagnosis of any disease *via* such a bioinformatics approach caters to personalized medical steps to be adopted by the patient, to have the disease suppressed and inactivated.

DISEASE PREVENTION

Early diagnosis of diseases, and facilitating prevention, is one of the most prominent applications of bioinformatics tools. Prevention is better than cure and when it comes to health, it is an instinctive priority to secure good health and take proper action ahead of time, before any illness cripples the body. Usually, symptoms signal the onset of disease, and in case of serious diseases such as cancer, HIV, dementia, or neurodegenerative diseases, especially those that are genetically susceptible, early detection can significantly increase survival chances and recovery rates. However, for such chronic diseases and illnesses, it can be difficult and expensive to detect the formation and development of the diseases early on, before it is too late. The ability to screen an embryo or fetus for possibilities of developing cancer or being HIV-prone would allow the medicine to take necessary measures early on, preventing far worse-case scenarios, stress, expense, *etc.* after the child is born. Taking vital steps to suppress and cure a disease when at a manageable stage significantly increases the efficiency of the treatment and chances of survival, lowering the sufferance of the patient. Analyzing protein-protein interactions, gene networks, and systems, comparing known and unknown disease-associated proteins and genes, and unmasking new candidate genes, bioinformatics is used to perform a systematic investigation to identify disease-related complexes, to make predictions of disease-causing genes, prominently for proper prevention measures [47, 48]. The following subsections discuss some of the major disease prevention aspects offered by bioinformatics approaches.

Bioinformatics Tools for Disease Prevention

Gene Ontology (GO) is a system that aids in classifying genes, genomic data, associated functions, and annotations that help to understand traits, used to identify hub genes for diseases [49]. Another resourceful service for bioinformatics, is the Kyoto Encyclopedia of Genes and Genomes (KEGG), a database, that helps with complex genomic information, providing associated chemical and biological pathways [49, 50]. Additionally, other large communal data repositories such as Array Express [51] and Gene Expression Omnibus (GEO) [52] also facilitate gene expression annotations [53]. Furthermore, to identify functions, cellular components, and pathways of differentially expressed genes responsible for specific diseases, the tool Database for Annotation, Visualization and Integrated Discovery (DAVID) can be used, an online service that facilitates visual representations [54, 55]. However, usually, these bioinformatic tools often require R-programming or other languages for exclusive annotative operations. Such advancement weighs the promise of increased lifespan, opening a new era for the health and medical industry. In short, humanity's propagation understands the correlation of genomics, pathogens, and immunology, enabling the development of remarkable tools that demonstrate their usefulness, providing potential applications, particularly in the prediction of diseases and screening, essentially leads towards on-set early detection and treatments, allows necessary cheaper preventative measures to be taken.

Disease Trend Analysis to Design Preventative Measures

Recent advancements and studies enhancing the development of various bioinformatics tools and technologies accelerated the understanding of disease evolution and associated interactions with host immune systems that help take preventative measures beforehand [56]. For instance, bioinformatic technologies assisted in the understanding of the underlying progression and evasion mechanism of human papillomavirus and other oncogenic viruses on the host immune system [56]. Furthermore, using innovative bioinformatics tools, the ability of next-generation sequencing and analyzing 'Big Data', detection and hence prevention of disease as early as an embryo, in the 21st century and for the future years to come is a forerunning achievement. The 'Big Data' approach in bioinformatics consequently enables researchers to gain novel insights into pathology biology, understanding the type of environment pathogens thrive in, related weaknesses, *etc.* For example, influenza researchers were able to utilize unstructured social media data, taken from Google searches and Tweets, to predict influenza patterns ahead of the Centers for Disease Control and Prevention (CDC) [57, 58]. Such availability of data allows the development of new tools, methods, and Application software to address pandemics and epidemics such as COVID-19 and HIV, respectively [57 - 61]. The ability to use data and information to formulate structured actions, strategies, and prevention policies, vastly helps to solidify public health measures and associated preparedness to tackle an outbreak of disease [58, 59, 62 - 65]. For the COVID-19 pandemic, governments and NGOs too relied on data, submitted by various health institutions and citizens themselves to keep track of patients, infected regions, and recovery rates [60, 66]. Proper utilization and monitoring of patient health data can help prevent the spread of diseases and ensure control, fostering better public health in a community (COVID-19 and digital health). Such approaches are often hoped to benefit stakeholders based on the bioinformatics tools' predicted results.

Additionally, the use of Artificial Intelligence (AI), Machine Learning (ML), and the incorporation of neural networks have significantly guided the precision of disease diagnosis and target specific gene/protein detections within samples [67,

68]. Artificial Neural Networking (ANN) detects common patterns, discrepancies, and relationships in raw data and is trained to provide precise results [68, 69]. Hence, these AI, ML, and ANN incorporated algorithms are developed to diagnose and predict diseases, like the thought process of a human brain, with the hope of avoiding man-made errors, while making predictions and diagnoses of diseases upon observing the health data of a patient [68]. Computational biology and bioinformatics have revolutionized the healthcare industry, enabling molecularization of the body, studying biology at the DNA level, and understanding the complex interaction of proteomics, genetics, pathogens, and immunity.

Genetic Screening for Disease Prediction

Whole Exome Sequencing (WES) and Whole Genome Sequencing (WGS) tools, together with GO analysis can be used for genetic screening of a fetus that can grow being highly susceptible to cancer or other diseases after birth. WES and WGS promise prominent discoveries on Mendelian disease genes, which enable the diagnosis of underlying genetic disorders, causes of birth defects, and many more [70, 71]. The use of such technologies is promising for obstetrics, gynecology, prenatal genetic diagnosis, and various other medical fields, although there lies a particular challenge on informed consent and ethical issues [72 - 76]. Early detection allows clinicians and doctors to have a head start on preparing methods and develop techniques to suppress the activation of the oncogenes, for preventing cancer development in the child after birth. Moreover, having the genome decoded for the baby, possible biomarkers and drugs can be modified and prepared that efficiently tackle the activation of cancer, again ensuring prevention. For example, bio computational tools such as Correct, omniBioMarker, GoMiner, *etc.*, significantly facilitate such diagnosis and preparation of personalized therapies and medication as per the child's molecular profile [54]. In another example, a genome sequence analysis of a woman's genome may reveal the presence of an oncogene that harbors the code for cervical cancer activation.

Detecting the presence of such genes, *via* various bioinformatics tools and technologies allows the woman to take necessary precautions throughout her life to ensure that the activation of the malignant gene does not take place, saving her money, stress, and trauma, even before the appearance of symptoms. Furthermore, it enables the woman to plan her life accordingly, not jumping into unexpected shock or diagnostic reports of cancer at a later age. Similarly, for various diseases such as breast cancer, dementia, lung cancer, HIV, Taysachs, *etc.* with the development of bioinformatics tools, the possibility of detecting and predicting the presence of disease and pathogen-inducing genes can be diagnosed and treated [29, 48, 49, 53, 58, 70, 77 - 79]. As a result, bioinformatics-based disease prediction warns the potential patients ahead of time so that advanced treatments can minimize the risks, and even can prevent the corresponding disease beforehand.

Predicting Cancer Using Bioinformatics Approach

According to WHO, cancer is one of the leading causes of premature death. Lung cancer, breast cancer, prostate cancer, cervical cancer, leukemia, *etc.* can be prevented and treated, only if diagnosed early. As Ismaeel *et al.*, agrees, early prediction of cancer cases, depending on accurate and early diagnosis, provides insights about the human genome and allows early prevention and treatment procedures to be adopted [68]. For example, SELDI-TOF-MS Protein Chip enables the detection of unique serum proteins present in breast cancer patients and can be further categorized into malignant and benign cases using associated bioinformatics tools and software [78]. SELDI-TOF-MS protein chip diagnostic model was developed using ANNs, together with discriminant analysis and significantly four candidate biomarkers of breast cancer could be detected [68, 78]. Search Tool for the Retrieval of Interacting Genes (STRING), an online tool, evaluates complex protein-protein interactions and provides information [49]. The protein chip takes in the sequenced sample of the patient reads the data, performs computation and comparison analysis, and retrieves data from its database servers on pre-stored genetic codes that mark breast cancer, finally providing an output [76]. The method is sensitive and specific [68], hence demonstrating the potential for early detection of breast cancer and its subsequent prevention.

For the detection of lung cancer, a protein-protein interaction network was studied [77]. The system retrieved interacting genes and proteins from an online database and also utilized search tools and Cytoscape software, visualizing the protein-protein interactions [77]. The Cytotype MCODE analysis helped provide prominent screened hub genes [49]. For instance, the system provided 7 hub genes based on connectivity degree, whose presence was further validated in lung adenocarcinoma (LUAD) and lung squamous cell carcinoma (LUSC), using Gene Expression Profiling Interactive Analysis (GEPIA), an online service [77]. GEPIA helped to verify the expression differences of the screened hub genes between cancerous and normal tissues [49, 77]. The predicted development of the disease and its overall survival were detected using the Kaplan-Meier plotter, which shortlisted the potential malignant genes to five from seven [77]. However, provided that such bio computational approaches are in fact predictions, validation of the predicted hub genes is always necessary and for that reverse, a transcription-quantitative polymerase chain reaction was performed [77, 80]. Such testing of the bioinformatics prediction approach of detecting lung cancer genes provides insights towards therapeutic development for metastasis cells, including potential biomarkers for non-small cell lung cancer tissues. Thus, bioinformatics approaches contribute to cancer studies.

Table 1. Glossary of recently popular computational tools and approaches for disease diagnosis, treatment, and prevention.

Tools and Approaches	Specific Applications in Medicine
Argus lab	Molecular-level modeling and drug-designing platform.
Array Express	Microarray gene expression database by EBI.
Artificial Neural Networking (ANN)	Computing biological neural networks associated with animal brains.
AutoDock	Predicts the interactions among substrates, drugs, *etc.* in a 3D model.
caCorrect	Quality control methods for gene expression-related processes.
COPASI	Modelling of the epidemic dynamics and predicting intervention outcomes.
CORDITE	Meta-analysis tool for drug predictions, commonly used for clinical trials.
CoVex	Existing-approved drug analysis to treat Covid-19 infection.
CoV-GLUE	Change detectors in the SARS-CoV-2 genome.
Covidex	Identifies subtypes of the SARS-CoV-2 genome.
COVIDSIM	Epidemiological tools to analyze Covid infection protection measures for policymaking and political decision-making.
CoVPipe	High throughput analysis of NGS data of SARS-CoV-2.
Cytoscape	Open-source software to visualize complex biological networking data.
Cytotype Molecular Complex Detection (MCODE) Analysis	Tool to detect interconnected regions in biological networks.
Gene Expression Omnibus (GEO)	Database for high throughput gene expression data for hybridization arrays, microarrays, and DNA chips.
Gene Expression Profiling Interactive Analysis (GEPIA)	Used for applying statistical techniques to detect differentially expressed genes and associated chromosome numbers dynamically.
Gene Ontology (GO)	Systemic and quality annotations to organize huge biological data.
GoMiner	Identifying biological components, functions, and processes for molecular biology.
HADDOCK	Biomolecular complex modeling platform.
Haploflow	Detection and reconstruction of multi-strain infections.
ICM	Shows the molecular environment of small molecules with or without corresponding textures.
Kaplan-Meier plotter	The process to detect genetic effects on cancer prognosis.
Knockoff procedure framework	False discovery rate minimization process.

(Table 1) cont.....

Tools and Approaches	Specific Applications in Medicine
Kyoto Encyclopedia of Genes and Genomes (KEGG)	Analysis tool for high-level functions and utilities of biological system.
Macro Model	Organic compounds and biopolymer modeling platform.
Next Generation Sequencing (NGS)	Organizes genomic data and creates scope for manipulation for the targeted regions of DNA/RNA.
omniBioMarker	Web tool for NCI cancer gene index analysis.
PAIPLine	Automatic pathogen identification tool.
Pangolin	Comparison tool for query genome.
PathogenFinder	Detects human-bacterial pathogenicity based on proteomic data.
Pathogenicity Prediction for Bacterial Genome (PaPrBaG)	Prediction of pathogenic phenotypes for non-identified or unknown bacteria.
Pathosystems Resource Integration Center (PATRIC)	Genetic data analysis tool, especially for pathogens and infectious disease research.
Pfam	Detection of certain proteins to track outbreaks, evolution, and mutation.
P-HIPSTer	Bayesian framework-based modeling to calculate the likelihood of protein-protein interaction.
poreCov	Reduces bioinformatic bottlenecks when sequencing runs.
PoSeiDon	Detection and analysis of protein-coding genes.
PriSeT	Calculating SARS-CoV-2 related primers derived from RT-PCR tests.
QXP	Data manipulation as per docking algorithms.
Rfam COVID-19	Annotates structured RNAs from the coronavirus sequence database and predicts mutated/secondary structures.
Search Tool for the Retrieval of Interacting Genes (STRING)	Biological database for protein-protein interaction analysis.
SELDI-TOF-MS	Novel mass spectrometric technique for biomarkers discovery connecting chromatography and mass spectrometric analysis.
UniProt	Updated and reliable proteomic data source for pathogens.
VADR	Validates and annotates of SARS-CoV-2 sequences.
VBRC genome analysis tools	Determines sequence difference of coronavirus under multiple resolutions.
VirHostNet	Analyzes molecular mechanisms for virus replication and pathogenesis.
VIRify	Virus identification from clinical samples.
VIRULIGN	Prompt and accurate sequence alignment and annotations for virus genome.
Visualization and Integrated Discovery (DAVID)	Visualization tools for pathway mapping, functional classification, and other types of analyses.
V-Pipe	Reproduces NGS-based based on end-to-end analysis of genomic diversity for intra-host viruses.

(Table 1) cont.....

Tools and Approaches	Specific Applications in Medicine
Whole Exome Sequencing (WES)	Sequencing technique for the protein-coding regions of DNA.
Whole Genome Sequencing (WGS)	Determination of the order of bases in the genome of an organism.

ISSUES AND ETHICAL DILEMMAS ASSOCIATED WITH MEDICAL APPLICATIONS OF BIOINFORMATICS

Although bioinformaticians work to increase the accuracy of their diagnosis and prediction of disease, wet lab experiments must always be accompanied to increase the confidence of the predicted results obtained. Secondly, the possibility of false positives is always a concern when it comes to predictions, using a computational approach. For instance, understating the interactions of a plethora of human papilloma-virus types and subsequent neoantigens and making certain predictions of the protein functionalities and disease whereabouts, using bioinformatics tools, it should be noted that the most predicted neoantigens are determined based on the data and resources on known existing neoantigen presentation pathways [56]. The tool NetMHC only predicts neoantigens depending on the binding affinity with MHC class I proteins, excluding T cell receptors and a few others [56]. Hence, the health data generated by the software and tools demands proper analysis, validation, and verification, given that depending on the results further diagnoses and treatments are to be made and suggested [80]. Therefore, the modern healthcare industry requires more medical health informaticians, neuroinformatics, and bioinformaticians to decipher the health data provided by the software. However, given that people fear being diagnosed positively early on, there lies a responsibility of health informaticians, clinicians, and doctors to encourage early diagnosis, to level up their chances of recovery. It is also important that people can trust such diagnostic and predicting tools to ensure the prevention of diseases, by increasing confidence in the accuracy of bioinformatics tools, teaching and educating the technology's associated functions and reliability to the common masses.

Vitally enough, another significant concern surrounding the prevention of diseases using early screening, prediction, and diagnosis of diseases *via* computation means can be an ethical concern. Bioethics is a rising concern in the biomedical arena, particularly gene editing with CRISPR-Cas9 technology, which demands huge attention. With the technological ability to screen disease-inducing genes and to aimfully remove those genes from the genome, editing in germline cells, the question of consent, the possibility of unprecedented damage of cells in a fetus and germlines leaves a huge ethical dilemma that needs to be addressed

[71]. The birth of Lu Lu and Na Na, the world's first CRISPR-Cas9 babies, who were hoped to have their HIV-prone genes removed, is a significant example that demonstrates the necessity of bioethics concerning disease prevention, medications, and treatments in the modern medical industry. With CRISPR-Cas9 technology implemented to edit the germlines, Lu Lu and Na Na are at risk for developing unwanted mutations, which the research community is yet unaware of [81]. Hence, more studies, research, and a collaborative merge of both scientists and ethical communities are needed to draw clear boundaries in the utilization of various bioinformatics and gene editing tools for ensuring the prevention of diseases. Although the development of such tools and technologies signifies humanity's advancement in medical research, far more studies are essential to ensure complete safety and efficient use of such technologies and tools, for diagnosis, treatment, and ultimately prevention of diseases [82].

Another ethical concern is the privacy of health data. The COVID-19 pandemic demonstrated the necessity of people sharing their health data [60, 65, 66]. Software and Apps that required people to submit their health data, such as temperature, and COVID-19 tested positive or negative reports, it was observed that in some cases people were reluctant to share truthful information about their data with organizations and governments even when it came to personal health safety [60, 61]. With growing data, organizations, medical institutions, and governments must ensure the protection of citizen's health data, when the citizens trust these institutions to provide them guidance in maintaining their health. Leakage or misuse of data without permission would be a significant ethical question and would require specific policies on handling such data, provided that proper ethical guidelines when handling people's data would build trust and allow people to take in such prediction tests and preventive measures for better health care [66, 81 - 83].

Precise and accurate prediction of diseases, and enhancing the most efficient early personalized treatments, therapies, medications, and advisories are the future of the medical, biotech, and pharmaceutical industries. With computational innovations revolutionizing the globe, as humanity strides ahead, understanding more about their biological data, systems' networks, and their sophisticated interactions, more bioinformatic tools facilitating drug therapies and innovative diagnostic and prediction technologies and software are imminent promises. Briefly put, the inventions and innovations to scan the biological body and swiftly decode a person's genome within seconds, then tailoring a personalized therapy or medicine is undoubtedly the future of medicine and disease diagnosis.

CONCLUSION

Bioinformatics approaches, while connected with different stages of clinical studies, assist in more sophisticated and prompt clinical diagnosis, treatment strategies designing, and preventative measures. Huge biological data manipulation properties of bioinformatics tools are opening new aspects to utilize these data based on the users' interests. However, despite having huge potential, bioinformatics approaches are not frequently incorporated in clinical practice, mostly due to the lack of experienced manpower and ethical concerns associated with bioinformatics approaches. If enough regulations and confidentiality can be assured, clinical studies can lend bioinformatics to update their processes and accelerate innovations. From disease diagnosis to treatment and prevention, bioinformatics tools bring about newer dimensions of findings using their data manipulation properties. Therefore, it is hoped that adopting bioinformatics tools in medical applications will expand the implications of such modern tools for medicine-related innovations.

REFERENCES

[1]   Iqbal N, Kumar P. From Data Science to Bioscience: Emerging era of bioinformatics applications, tools and challenges. Procedia Comput Sci 2023; 218: 1516-28.
[http://dx.doi.org/10.1016/j.procs.2023.01.130]

[2]   Raslan MA, Raslan SA, Shehata EM, Mahmoud AS, Sabri NA. Advances in the applications of bioinformatics and chemoinformatics. Pharmaceuticals (Basel) 2023; 16(7): 1050.
[http://dx.doi.org/10.3390/ph16071050] [PMID: 37513961]

[3]   Kaushik AC, Mehmood A, Wei DQ, Nawab S, Sahi S, Kumar A. Cheminformatics and bioinformatics at the interface with systems biology: bridging chemistry and medicine. Royal Society of Chemistry 2023.
[http://dx.doi.org/10.1039/9781839166037]

[4]   Liu L, Tang L, Dong W, Yao S, Zhou W. An overview of topic modeling and its current applications in bioinformatics. Springerplus 2016; 5(1): 1608.
[http://dx.doi.org/10.1186/s40064-016-3252-8] [PMID: 27652181]

[5]   Merelli I, Pérez-Sánchez H, Gesing S, D'Agostino D. Managing, analyzing, and integrating big data in medical bioinformatics: open problems and future perspectives. BioMed Res Int 2014; 2014: 1-13.
[http://dx.doi.org/10.1155/2014/134023]

[6]   Hsin KY, Ghosh S, Kitano H. Combining machine learning systems and multiple docking simulation packages to improve docking prediction reliability for network pharmacology. PLoS One 2013; 8(12): e83922.
[http://dx.doi.org/10.1371/journal.pone.0083922] [PMID: 24391846]

[7]   Yan Q. Translational bioinformatics and systems biology approaches for personalized medicine.Systems Biology in Drug Discovery, and Development. Totowa: Humana Press 2010; pp. 167-78.
[http://dx.doi.org/10.1007/978-1-60761-800-3_8]

[8]   Xia X. Bioinformatics and drug discovery. Curr Top Med Chem 2017; 17(15): 1709-26.
[http://dx.doi.org/10.2174/1568026617666161116143440] [PMID: 27848897]

[9]   Chen H, Guo W, Shen J, *et al.* Structural principles analysis of host-pathogen protein-protein interactions: A structural bioinformatics survey. IEEE Access 2018; 6: 11760-71.

[http://dx.doi.org/10.1109/ACCESS.2018.2807881]

[10]   Fan Y, Zhang L, Sun Y, *et al.* Expression profile and bioinformatics analysis of COMMD10 in BALB/C mice and human. Cancer Gene Ther 2020; 27(3-4): 216-25.
[http://dx.doi.org/10.1038/s41417-019-0087-9] [PMID: 30787448]

[11]   Sreeraman S, Kannan MP, Singh Kushwah RB, *et al.* Drug design and disease diagnosis: the potential of deep learning models in biology. Curr Bioinform 2023; 18(3): 208-20.
[http://dx.doi.org/10.2174/1574893618666230227105703]

[12]   Gu W, Deng X, Lee M, *et al.* Rapid pathogen detection by metagenomic next-generation sequencing of infected body fluids. Nat Med 2021; 27(1): 115-24.
[http://dx.doi.org/10.1038/s41591-020-1105-z] [PMID: 33169017]

[13]   Gwinn M, MacCannell D, Armstrong GL. Next-generation sequencing of infectious pathogens. JAMA 2019; 321(9): 893-4.
[http://dx.doi.org/10.1001/jama.2018.21669] [PMID: 30763433]

[14]   Gu W, Miller S, Chiu CY. Clinical metagenomic next-generation sequencing for pathogen detection. Annu Rev Pathol 2019; 14(1): 319-38.
[http://dx.doi.org/10.1146/annurev-pathmechdis-012418-012751] [PMID: 30355154]

[15]   Gillespie JJ, Wattam AR, Cammer SA, *et al.* PATRIC: the comprehensive bacterial bioinformatics resource with a focus on human pathogenic species. Infect Immun 2011; 79(11): 4286-98.
[http://dx.doi.org/10.1128/IAI.00207-11] [PMID: 21896772]

[16]   Andrusch A, Dabrowski PW, Klenner J, *et al.* PAIPline: pathogen identification in metagenomic and clinical next generation sequencing samples. Bioinformatics 2018; 34(17): i715-21.
[http://dx.doi.org/10.1093/bioinformatics/bty595] [PMID: 30423069]

[17]   Singh P, Singh N. Role of data mining techniques in bioinformatics.In Research Anthology on Bioinformatics, Genomics, and Computational Biology. IGI Global 2024.

[18]   Golub TR, Slonim DK, Tamayo P, *et al.* Molecular classification of cancer: class discovery and class prediction by gene expression monitoring. Science 1999; 286(5439): 531-7.
[http://dx.doi.org/10.1126/science.286.5439.531] [PMID: 10521349]

[19]   Ben-Dor A, Bruhn L, Friedman N, *et al.* Tissue classification with gene expression profiles. Proceedings of the Fourth Annual International Conference on Computational Molecular Biology. 54-64.
[http://dx.doi.org/10.1145/332306.332328]

[20]   Blanco R, Larrañaga P, Inza I, Sierra B. Gene selection for cancer classification using wrapper approaches. Int J Pattern Recognit Artif Intell 2004; 18(8): 1373-90.
[http://dx.doi.org/10.1142/S0218001404003800]

[21]   Chow ML, Moler EJ, Mian IS. Identifying marker genes in transcription profiling data using a mixture of feature relevance experts. Physiol Genomics 2001; 5(2): 99-111.
[http://dx.doi.org/10.1152/physiolgenomics.2001.5.2.99] [PMID: 11242594]

[22]   Richhariya B, Tanveer M, Rashid AH. Diagnosis of Alzheimer's disease using universum support vector machine based recursive feature elimination (USVM-RFE). Biomed Signal Process Control 2020; 59: 101903.
[http://dx.doi.org/10.1016/j.bspc.2020.101903]

[23]   Lai C, Reinders MJT, van't Veer LJ, Wessels LFA. A comparison of univariate and multivariate gene selection techniques for classification of cancer datasets. BMC Bioinformatics 2006; 7(1): 235.
[http://dx.doi.org/10.1186/1471-2105-7-235] [PMID: 16670007]

[24]   Shen A, Fu H, He K, Jiang H. False discovery rate control in cancer biomarker selection using knockoffs. Cancers (Basel) 2019; 11(6): 744.
[http://dx.doi.org/10.3390/cancers11060744] [PMID: 31146393]

[25]   Sechidis K, Kormaksson M, Ohlssen D. Using knockoffs for controlled predictive biomarker identification. Stat Med 2021; 40(25): 5453-73.
[http://dx.doi.org/10.1002/sim.9134] [PMID: 34328655]

[26]   Lindner BG, Gerhardt K, Feistel DJ, *et al.* A user's guide to the bioinformatic analysis of shotgun metagenomic sequence data for bacterial pathogen detection. International Journal of Food Microbiology 2023; p. 110488.

[27]   Cosentino S, Voldby Larsen M, Møller Aarestrup F, Lund O. PathogenFinder-distinguishing friend from foe using bacterial whole genome sequence data. PLoS One 2013; 8(10): e77302.
[http://dx.doi.org/10.1371/journal.pone.0077302] [PMID: 24204795]

[28]   Ekwanzala MD, Dewar JB. Genome sequence of carbapenem-resistant Citrobacter. 2019 [29] Saeb AT. Current Bioinformatics resources in combating infectious diseases. Bioinformation 2018; 14(1): 31. 2019.

[29]   Saeb ATM. Current Bioinformatics resources in combating infectious diseases. Bioinformation 2018; 14(1): 031-5.
[http://dx.doi.org/10.6026/97320630014031] [PMID: 29497257]

[30]   Deneke C, Rentzsch R, Renard BY. PaPrBaG: A machine learning approach for the detection of novel pathogens from NGS data. Sci Rep 2017; 7(1): 39194.
[http://dx.doi.org/10.1038/srep39194] [PMID: 28051068]

[31]   Deneke C, Rentzsch R, Renard BY. PaPrBaG: A random forest approach for the detection of novel pathogens from NGS data. PeerJ Preprints 2016.

[32]   Robson B. Computers and viral diseases. Preliminary bioinformatics studies on the design of a synthetic vaccine and a preventative peptidomimetic antagonist against the SARS-CoV-2 (2019-nCoV, COVID-19) coronavirus. Comput Biol Med 2020; 119: 103670.
[http://dx.doi.org/10.1016/j.compbiomed.2020.103670] [PMID: 32209231]

[33]   Sadat SM, Aghadadeghi MR, Yousefi M, Khodaei A, Sadat Larijani M, Bahramali G. Bioinformatics analysis of SARS-CoV-2 to approach an effective vaccine candidate against COVID-19. Mol Biotechnol 2021; 63(5): 389-409.
[http://dx.doi.org/10.1007/s12033-021-00303-0] [PMID: 33625681]

[34]   Chukwudozie OS, Duru VC, Ndiribe CC, Aborode AT, Oyebanji VO, Emikpe BO. The relevance of bioinformatics applications in the discovery of vaccine candidates and potential drugs for COVID-19 treatment. Bioinform Biol Insights 2021; 15
[http://dx.doi.org/10.1177/11779322211002168] [PMID: 33795932]

[35]   Avram S, Puia A, Udrea AM, *et al.* Natural compounds therapeutic features in brain disorders by experimental, bioinformatics, and cheminformatics methods. Curr Med Chem 2020; 27(1): 78-98.
[http://dx.doi.org/10.2174/0929867325666181031123127] [PMID: 30378477]

[36]   Parikh PK, Savjani JK, Gajjar AK, *et al.* Bioinformatics and cheminformatics tools in early drug discovery. Bioinformatics tools for pharmaceutical drug product development 2023: 147-81.
[http://dx.doi.org/10.1002/9781119865728.ch8]

[37]   Chen Z, Yu J, Wang H, *et al.* Flexible scaffold-based cheminformatics approach for polypharmacological drug design. Cell 2024; 187(9): 2194-2208.e22.
[http://dx.doi.org/10.1016/j.cell.2024.02.034] [PMID: 38552625]

[38]   Pagadala NS, Syed K, Tuszynski J. Software for molecular docking: a review. Biophys Rev 2017; 9(2): 91-102.
[http://dx.doi.org/10.1007/s12551-016-0247-1] [PMID: 28510083]

[39]   Bisht N, Sah AN, Bisht S, Joshi H. Emerging need of today: Significant utilization of various databases and softwares in drug design and development. Mini Rev Med Chem 2021; 21(8): 1025-32.
[http://dx.doi.org/10.2174/1389557520666201214101329] [PMID: 33319657]

[40]   Aalikhani Pour M, Sardari S, Eslamifar A, Rezvani M, Azhar A, Nazari M. Evaluating the anticoagulant effect of medicinal plants *in vitro* by cheminformatics methods. J Herb Med 2016; 6(3): 128-36.
[http://dx.doi.org/10.1016/j.hermed.2016.05.002]

[41]   Rahgozar N, Bakhshi Khaniki G, Sardari S. Evaluation of antimycobacterial and synergistic activity of plants selected based on cheminformatic parameters. Iran Biomed J 2018; 22(6): 401-7.
[http://dx.doi.org/10.29252/.22.6.401] [PMID: 29510602]

[42]   Bhowmick S, AlFaris NA, ALTamimi JZ, *et al.* Screening and analysis of bioactive food compounds for modulating the CDK2 protein for cell cycle arrest: Multi-cheminformatics approaches for anticancer therapeutics. J Mol Struct 2020; 1216: 128316.
[http://dx.doi.org/10.1016/j.molstruc.2020.128316]

[43]   Dhanaraj P, Muthiah I, Rozbu MR, Nuzhat S, Paulraj MS. Computational studies on T2Rs Agonist based anti-Covid-19 drug design. Front Mol Biosci 2021; 8: 637124.
[http://dx.doi.org/10.3389/fmolb.2021.637124] [PMID: 34485378]

[44]   Pavlovic S, Kotur N, Stankovic B, Zukic B, Gasic V, Dokmanovic L. Pharmacogenomic and pharmaco-transcriptomic profiling of childhood acute lymphoblastic leukemia: paving the way to personalized treatment. Genes (Basel) 2019; 10(3): 191.
[http://dx.doi.org/10.3390/genes10030191] [PMID: 30832275]

[45]   Ye Q, Raese RA, Luo D, *et al.* MicroRNA-based discovery of biomarkers, therapeutic targets, and repositioning drugs for breast cancer. Cells 2023; 12(14): 1917.
[http://dx.doi.org/10.3390/cells12141917] [PMID: 37508580]

[46]   Emde M, Beck S, Benes V, Moreaux J, Seckinger A, Hose D. RNA-sequencing based assessment of targets, risk and long term survival for personalized treatment of multiple myeloma. Blood 2019; 134 (Suppl. 1): 1801.
[http://dx.doi.org/10.1182/blood-2019-131159]

[47]   Lage K, Karlberg EO, Størling ZM, *et al.* A human phenome-interactome network of protein complexes implicated in genetic disorders. Nat Biotechnol 2007; 25(3): 309-16.
[http://dx.doi.org/10.1038/nbt1295] [PMID: 17344885]

[48]   Banasik K, Justesen JM, Hornbak M, *et al.* Bioinformatics-driven identification and examination of candidate genes for non-alcoholic fatty liver disease. PLoS One 2011; 6(1): e16542.
[http://dx.doi.org/10.1371/journal.pone.0016542] [PMID: 21339799]

[49]   Liu J, Ma L, Chen Z, *et al.* Identification of critical genes in gastric cancer to predict prognosis using bioinformatics analysis methods. Ann Transl Med 2020; 8(14): 884.
[http://dx.doi.org/10.21037/atm-20-4427] [PMID: 32793728]

[50]   Chen L, Zhang YH, Wang S, Zhang Y, Huang T, Cai YD. Prediction and analysis of essential genes using the enrichments of gene ontology and KEGG pathways. PLoS One 2017; 12(9): e0184129.
[http://dx.doi.org/10.1371/journal.pone.0184129] [PMID: 28873455]

[51]   Parkinson H, Kapushesky M, Shojatalab M, *et al.* ArrayExpress-a public database of microarray experiments and gene expression profiles. Nucleic Acids Res 2007; 35(Database): D747-50.
[http://dx.doi.org/10.1093/nar/gkl995] [PMID: 17132828]

[52]   Edgar R, Domrachev M, Lash AE. Gene Expression Omnibus: NCBI gene expression and hybridization array data repository. Nucleic Acids Res 2002; 30(1): 207-10.
[http://dx.doi.org/10.1093/nar/30.1.207] [PMID: 11752295]

[53]   Phan JH, Moffitt RA, Stokes TH, *et al.* Convergence of biomarkers, bioinformatics and nanotechnology for individualized cancer treatment. Trends Biotechnol 2009; 27(6): 350-8.
[http://dx.doi.org/10.1016/j.tibtech.2009.02.010] [PMID: 19409634]

[54]   Ma L, Zou D, Liu L, *et al.* Database commons: a catalog of worldwide biological databases. Genomics Proteomics Bioinformatics 2023; 21(5): 1054-8.

[http://dx.doi.org/10.1016/j.gpb.2022.12.004] [PMID: 36572336]

[55]   Aryal S, Anand D, Huang H, *et al.* Proteomic profiling of retina and retinal pigment epithelium combined embryonic tissue to facilitate ocular disease gene discovery. Hum Genet 2023; 142(7): 927-47.
[http://dx.doi.org/10.1007/s00439-023-02570-0] [PMID: 37191732]

[56]   Gensterblum-Miller E, Brenner JC. Protecting tumors by preventing human papillomavirus antigen presentation: insights from emerging bioinformatics algorithms. Cancers (Basel) 2019; 11(10): 1543.
[http://dx.doi.org/10.3390/cancers11101543] [PMID: 31614809]

[57]   McIntyre RS, Cha DS, Jerrell JM, *et al.* Advancing biomarker research: utilizing 'Big Data' approaches for the characterization and prevention of bipolar disorder. Bipolar Disord 2014; 16(5): 531-47.
[http://dx.doi.org/10.1111/bdi.12162] [PMID: 24330342]

[58]   Young SD. A "big data" approach to HIV epidemiology and prevention. Prev Med 2015; 70: 17-8.
[http://dx.doi.org/10.1016/j.ypmed.2014.11.002] [PMID: 25449693]

[59]   Vaishya R, Javaid M, Khan IH, Haleem A. Artificial Intelligence (AI) applications for COVID-19 pandemic. Diabetes Metab Syndr 2020; 14(4): 337-9.
[http://dx.doi.org/10.1016/j.dsx.2020.04.012] [PMID: 32305024]

[60]   Amit M, Kimhi H, Bader T, Chen J, Glassberg E, Benov A. Mass-surveillance technologies to fight coronavirus spread: the case of Israel. Nat Med 2020; 26(8): 1167-9.
[http://dx.doi.org/10.1038/s41591-020-0927-z] [PMID: 32457444]

[61]   Fahey RA, Hino A. COVID-19, digital privacy, and the social limits on data-focused public health responses. Int J Inf Manage 2020; 55: 102181.
[http://dx.doi.org/10.1016/j.ijinfomgt.2020.102181] [PMID: 32836638]

[62]   Broniatowski DA, Paul MJ, Dredze M. National and local influenza surveillance through Twitter: an analysis of the 2012-2013 influenza epidemic. PLoS One 2013; 8(12): e83672.
[http://dx.doi.org/10.1371/journal.pone.0083672] [PMID: 24349542]

[63]   Ginsberg J, Mohebbi MH, Patel RS, Brammer L, Smolinski MS, Brilliant L. Detecting influenza epidemics using search engine query data. Nature 2009; 457(7232): 1012-4.
[http://dx.doi.org/10.1038/nature07634] [PMID: 19020500]

[64]   Polgreen PM, Chen Y, Pennock DM, Nelson FD. Using internet searches for influenza surveillance. Clin Infect Dis 2008; 47(11): 1443-8.
[http://dx.doi.org/10.1086/593098] [PMID: 18954267]

[65]   COVID-19 and digital health: What can digital health offer for COVID-19? World Health Organization. who. int. Available from: https://www.who.int/china/news/feature-stories/detail/covid-19-and-digital-health-what-can-digital-health-offer-for-covid-19

[66]   Segal E, Zhang F, Lin X, *et al.* Building an international consortium for tracking coronavirus health status. Nat Med 2020; 26(8): 1161-5.
[http://dx.doi.org/10.1038/s41591-020-0929-x]

[67]   Lisboa PJ, Taktak AFG. The use of artificial neural networks in decision support in cancer: A systematic review. Neural Netw 2006; 19(4): 408-15.
[http://dx.doi.org/10.1016/j.neunet.2005.10.007] [PMID: 16483741]

[68]   Ismaeel AG, Ablahad AA. Novel method for mutational disease prediction using bioinformatics techniques and backpropagation algorithm. arXiv preprint 2013.

[69]   Azmi MSBM, Cob ZC. Breast cancer prediction based on backpropagation algorithm. IEEE Student Conference on Research and Development. 164-68.

[70]   Altarescu G. Prevention is the Best Therapy: The Geneticist's Approach. Pediatr Endocrinol Rev 2016; 13 (Suppl. 1): 649-54.

[PMID: 27491212]

[71]   Van den Veyver IB, Eng CM. Genome-wide sequencing for prenatal detection of fetal single-gene disorders. Cold Spring Harb Perspect Med 2015; 5(10): a023077.
[http://dx.doi.org/10.1101/cshperspect.a023077] [PMID: 26253094]

[72]   McGuire AL, Lupski JR. Personal genome research : what should the participant be told? Trends Genet 2010; 26(5): 199-201.
[http://dx.doi.org/10.1016/j.tig.2009.12.007] [PMID: 20381895]

[73]   McGuire AL, McCullough LB, Evans JP. The indispensable role of professional judgment in genomic medicine. JAMA 2013; 309(14): 1465-6.
[http://dx.doi.org/10.1001/jama.2013.1438] [PMID: 23571582]

[74]   Appelbaum PS, Waldman CR, Fyer A, *et al.* Informed consent for return of incidental findings in genomic research. Genet Med 2014; 16(5): 367-73.
[http://dx.doi.org/10.1038/gim.2013.145] [PMID: 24158054]

[75]   Holm IA, Savage SK, Green RC, *et al.* Guidelines for return of research results from pediatric genomic studies: deliberations of the Boston Children's Hospital Gene Partnership Informed Cohort Oversight Board. Genet Med 2014; 16(7): 547-52.
[http://dx.doi.org/10.1038/gim.2013.190] [PMID: 24406460]

[76]   Scollon S, Bergstrom K, Kerstein RA, *et al.* Obtaining informed consent for clinical tumor and germline exome sequencing of newly diagnosed childhood cancer patients. Genome Med 2014; 6(9): 69.
[http://dx.doi.org/10.1186/s13073-014-0069-3] [PMID: 25317207]

[77]   Sun R, Meng X, Wang W, *et al.* Five genes may predict metastasis in non-small cell lung cancer using bioinformatics analysis. Oncol Lett 2019; 18(2): 1723-32.
[http://dx.doi.org/10.3892/ol.2019.10498] [PMID: 31423239]

[78]   Hu Y, Zhang S, Yu J, Liu J, Zheng S. SELDI-TOF-MS: the proteomics and bioinformatics approaches in the diagnosis of breast cancer. Breast 2005; 14(4): 250-5.
[http://dx.doi.org/10.1016/j.breast.2005.01.008] [PMID: 16085230]

[79]   Cavallo F, Astolfi A, Iezzi M, *et al.* An integrated approach of immunogenomics and bioinformatics to identify new Tumor Associated Antigens (TAA) for mammary cancer immunological prevention. BMC Bioinformatics 2005; 6(S4) (Suppl. 4): S7.
[http://dx.doi.org/10.1186/1471-2105-6-S4-S7] [PMID: 16351756]

[80]   Chen YP, Wang YQ, Lv JW, *et al.* Identification and validation of novel microenvironment-based immune molecular subgroups of head and neck squamous cell carcinoma: implications for immunotherapy. Ann Oncol 2019; 30(1): 68-75.
[http://dx.doi.org/10.1093/annonc/mdy470] [PMID: 30407504]

[81]   Greely HT. CRISPR'd babies: human germline genome editing in the 'He Jiankui affair'. J Law Biosci 2019; 6(1): 111-83.
[http://dx.doi.org/10.1093/jlb/lsz010] [PMID: 31666967]

[82]   Greely HT. CRISPR babies: human germline genome editing in the 'He Jiankuivaffair'. Journal of Law and the Biosciences 2019; 6(1): 111–83.

[83]   Dash S, Shakyawar SK, Sharma M, Kaushik S. Big data in healthcare: management, analysis and future prospects. J Big Data 2019; 6(1): 54.
[http://dx.doi.org/10.1186/s40537-019-0217-0]

CHAPTER 14

Computer-aided Diagnosis Model for White Blood Cell Leukemia and Myeloma Classification using Deep Convolutional Neural Network

K. P. Sujith[1], P. Vetrivelan[1], P. Prakasam[2,*] and **T. R. Sureshkumar[2]**

[1] *School of Electronics Engineering, Vellore Institute of Technology, Chennai, India*

[2] *School of Electronics Engineering, Vellore Institute of Technology, Vellore, India*

Abstract: Diagnosing white blood cell (leukocyte) diseases (Leukemia and Myeloma) is a thought-provoking task in the body. The abnormal growth of the leukocytes leads to an unbalanced immune system. Therefore, the automatic detection and classification of leukocytes will be the best aiding tool for the physician. This research work proposes a Computer-aided Diagnosis (CAD) model using the Deep Convolutional Neural Network (DCNN) to classify the white blood cell Acute Myeloid Leukemia (AML), Acute lymphoblastic leukemia (ALL), Myeloma, and its sub-types. The Gaussian distribution and k-means clustering segment the input image for future extraction. We utilized the Gray Level Covariance Matrix method to attain the texture features required to train the proposed DCNN model. The DCNN classifier is trained and tested with the mined features, and it detects the early stage of leukocyte cancer and achieves a classification accuracy of 97.8%. The precision, recall, and F1 score are achieved as 0.977, 80.955, and 0.966, respectively. We compared the performance of the proposed CAD model with the existing deep-learning classifier models. The analysis reveals that the proposed CAD model outperforms the existing methods.

Keywords: Computer-aided diagnosis, Convolutional neural networks, Gray level covariance matrix, Leukemia, Myeloma, White blood cells.

INTRODUCTION

The blood is vital for many organs to function correctly. We assessed the healthiness of the blood by using blood cells. Out of three blood cells, white cells (leukocytes) play a significant role in the organs' proper function [1]. The leukocytes are generated in the bone marrow, responsible for the body's immune system. The statistical report released by the World Health Organization (WHO) stated that the most deaths occur due to cancer, and led to 10 million deaths in the

---

[*] **Corresponding author P. Prakasam** School of Electronics Engineering, Vellore Institute of Technology, Vellore, India; E-mail: prakasamp@gmail.com

Sivakumar Rajagopal, Prakasam P., Konguvel E., Shamala Subramaniam, Ali Safaa Sadiq Al Shakarchi & B. Prabadevi (Eds.)

year 2020. If cancer is diagnosed and treated at an early stage, then it can be cured easily, but if not diagnosed early, it can be life-threatening.

The symptoms of white blood cell cancer diseases are not easy to detect. These diseases have many subcategories, each having almost similar symptoms. So, even if the disease is detected at an early stage, chances for misdiagnosis by the doctor are very high. A detailed survey of feature extractions using traditional image processing techniques to classify white blood cells was presented for better understanding [2]. As we know, the first step to solving the problem is to identify the problem accurately. Nowadays, we rely on automated systems to diagnose such diseases so they can be cured.

A digital image is the processing of a 2-D picture by a digital computing system. It is represented using the finite value of bits, which may be natural or complex numbers. We can process this digitized Image for further directions. Image processing is a technique used to extract the required features. White blood cell cancer is also one of the major diseases that can affect many people in the world. Leukemia and Myeloma are the major diseases in blood cells [3].

The bone marrow causes Leukemia, which generates abnormal leukocytes. Therefore, it does not function normally, leading to chronic or acute disease. Acute Myeloid Leukemia (AML) is further divided into various stages as M0, M1, M2, M3, M4, M5, M6, and M7 levels, whereas Acute lymphoblastic Leukemia (ALL) is divided into various stages as L1, L2, and L3 levels. Myeloma disease is a cancer formed in the plasma cells in the bone marrow.

The identification and classification of leukocytes have been exciting in the recent past. Due to technological advancements, many automated leukocyte disease prediction methods have been proposed by many researchers [4 - 10]. The autonomous detection of leukocytes is generally achieved using advanced image preprocessing techniques such as image enhancement, segmentation, feature extraction, and classification. Typically, the automatic leukocyte classification method's performance depends on suitable segmentation techniques. The researchers suggested many segmentation techniques, such as segmentation by clustering using the standard features [4], segmentation by various thresholding mechanisms [5, 6], morphological-based segmentation [7], feature extraction using edge detection, and region-growing-based segmentation [8]. The various reported techniques have their own merits and demerits.

Many researchers have concentrated on fuzzy-based and machine learning (ML) based white blood cell cancer detection for efficient implementation. An image processing-based detection method was proposed for early-stage detection [9 - 11]. The authors suggested using the k-means nearest neighbor (k-NN) clustering technique for segmentation and a support vector machine to classify the various stages of blood leukemia. Mohapatra *et al*. [10] demonstrated a Fuzzy logic-based segmentation to detect and classify blood Leukemia. The classification of AML M1 and M2 using ML was proposed and the various morphological, radiomics, and clinical features were extracted to train and validate the classifier [11]. The random forest classifier was proposed to classify the ALL and its subcategory from microscopic images [12].

Due to the era of artificial intelligence, researchers have deployed many autonomous systems to solve classification problems. Researchers developed and deployed many machines and deep learning-based models to analyze and predict medical-related issues [13 - 15]. Due to their intelligent nature, the Deep Neural Networks can predict the region of interest and mine the essential features for the provided dataset [16, 17]. The neural network-based classification [18] and Generative Adversarial Network [19] based classification have been deployed. Yao *et al*. [20] proposed a white blood cell classification technique using the two-module weighted CNN model. The convolutional neural networks have been demonstrated to effectively classify multiclass and binary blood cell samples fully automated [21, 22].

Various optimization techniques, such as particle swarm optimization, genetic algorithm, grey wolf optimization algorithms, and blood cell morphology, were used to optimize the CNN model to better classify white blood cells and predict ALL and AML subcategories [23, 24]. Khandekar *et al*. trained and implemented YOLO V4 using the Object Detection algorithm to detect and classify ALL blood cells from microscopic images [25]. The Chronological Sine Cosine method-based DCNN model from single blood cell smear images was proposed to detect and classify the ALL and its subtypes [26]. They have used mutual information-based hybrid methods to segment the images. The authors used the robust global and local extracted features to train and test the Al-Net model, which was suggested for detecting ALL cells [27]. The advantages and disadvantages of the existing state-of-the-art (SOTA) methods and algorithms are summarized in Table **1**.

Research Contributions

Despite notable technological advancements, the existing methods still have a few limitations/drawbacks. Most reported techniques detected and classified white blood cancer cells as AML or ALL. However, those methods failed to identify the subcategories of each type. Therefore, effectively detecting and classifying Myeloma and Leukemia is challenging for physicians. The significant contributions of the proposed research are as follows.

- Propose a practical CAD-based framework to detect the various stages of white blood cell cancers, such as Acute Myeloid Leukemia (AML), Acute lymphoblastic Leukemia (ALL), and Myeloma, along with their subcategories, using the DCNN model.
- Utilize the Gaussian distribution and K-means clustering to segment the preprocessed Image to attain the features.
- Employ the Gray Level Covariance Matrix (GLCM) technique to extract the texture features to train the proposed DCNN model from the segmented microscopic images.

Table 1. Advantages and disadvantages of SOTA methods.

Author(s)	Methods/ Classifiers	Advantages	Disadvantages/ Limitations
Kumar *et al.* [9]	k-NN	• Training is not required, and it can include new data easily. • Also, it is straightforward to implement.	• Failed to handle larger datasets and dimensions. • High computational cost.
Liu and Hu [11]	Support Vector Machine (SVM)	• Well suited for non and semi-structured input sample. • Scaling to higher dimensional data is easy.	• Training time is more, and it occupies more memory • Not suitable for multiclass classification
Mirmohammadi *et al.* [12]	Random Forest Classifier	• It is also suitable for non-structured data samples. • Easy to assign input sample features.	• It occupies more memory due to the longer training time for a larger data sample. • It is not resistant to noisy data samples.
Saidani *et al.* [18]	Neural Networks	• Suitable for both more minor and more extensive data samples. • Requires lesser training.	• Computational complexity is more. • Possibility for overfitting. • The duration of the network is not predictable.
Saini *et al.* [26]	CNN	• Lesser training is sufficient. • Suitable for multiclass classification.	• Computational Complexity is high. • It uses higher memory space.

MATERIALS AND METHODS

The Materials and Methods should be described with sufficient details to allow others to replicate and build on the published results. Please note that the publication of your manuscript implies that you must make all materials, data, computer code, and protocols associated with the publication available to readers. Please disclose any restrictions on the availability of materials or information at the submission stage. New methods and protocols should be described in detail, while well-established methods can be briefly described and appropriately cited. The overall framework of the proposed automatic classification of leukocytes in Leukemia and Myeloma using a DCNN is illustrated in Fig. (**1**).

The proposed automatic model is focused on a two-stage approach. Initially, it differentiates between M5 AML with L1 and L2 ALL, and after that, it classifies other sub-stages of white blood cells. Therefore, the diseases are grouped into two sets because of their similarity in visual features, which can confuse the doctors and cause misclassification from the recommendations provided by the doctor based on input blood samples; one of the two approaches is executed. Different features are obtained per approach. The preprocessing mechanism is applied to remove the background noise from the input images. After that, the segmentation is performed on the preprocessed images using a k-means clustering algorithm, and finally, features are mined. The various features, such as geometry, statistics, and texture, are extracted using the Gray-Level Covariance Matrix (GLCM) technique, and these features are applied to the proposed DCNN model to train and validate for efficient classification of leukocytes.

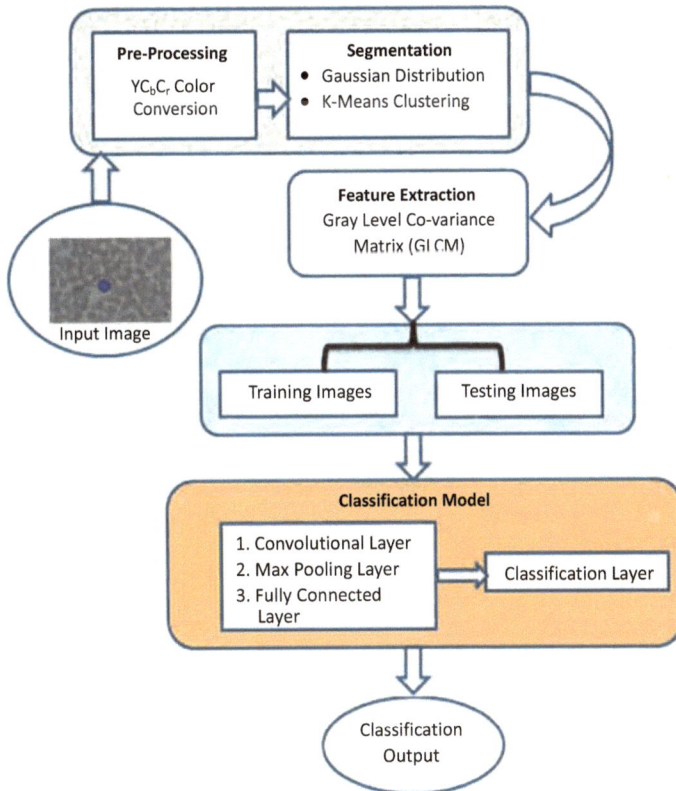

Fig. (1). CAD-based framework of the proposed model.

Dataset

The proposed CAD model is trained and tested to classify the WBC subcategories with ALL-IDB[1] dataset. The Department of Computer Science, University of Milan, Italy, provided this dataset. This dataset consists of two subsets, ALL-IDB1 and ALL-IDB2. The first dataset contains 108 non-segmented blood images containing about 39000 blood elements, and expert oncologists labeled the lymphocytes. The second dataset includes 260 segmented images, of which 130 images belong to Leukemia and 130 to the non-leukemia category.

Preprocessing and Segmentation

Preprocessing

The input blood images are converted from RGB color space to YCbCr color space for easy processing. The YCbCr color space is chosen to remove all the color contrast from the Image since it is tough to extract the region of interest from an RGB because of its color contrast. After we convert the Image to YCbCr color space, the mined C_b and Cr values are utilized for segmentation. Figs. (**2a and b**) show the actual and processed images.

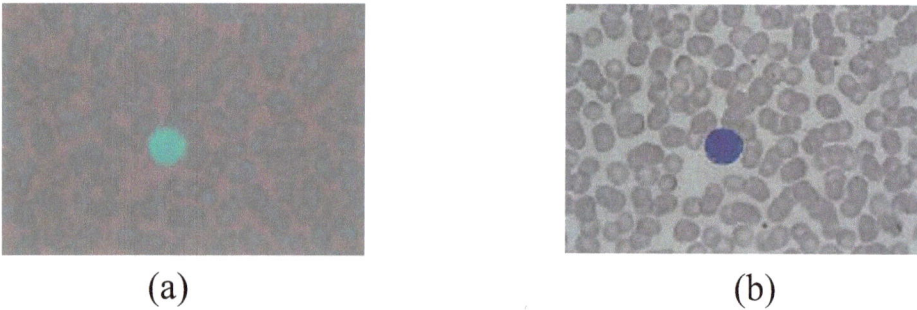

(a)   (b)

Fig. (2). (**a**) Input Image and (**b**) YC_bC_r converted image.

Segmentation

The major task of the segmentation process is to fragment the whole of the leukocytes from their background. The extracted C_b and C_r values from the training dataset are utilized to create a Gaussian Distribution, and it can be computed using the equations (1, 2 and 3).

$$x = [(C_b - C_{b_{mean}}), (C_r - C_{r_{mean}})] \qquad (1)$$

where C_{bmean} and C_{rmean} can be computed as follows.

$$C_{b_{mean}} = mean(C_b) \tag{2}$$

$$C_{r_{mean}} = mean(C_r) \tag{3}$$

This distribution extracts the valuable pixels of the test dataset required for the region of interest (RoI). Normalization and adaptive thresholding are also applied for effective segmentation. Fig. (**3**) shows the nucleus RoI, the Gaussian distribution, and the adaptive thresholding of the sample image.

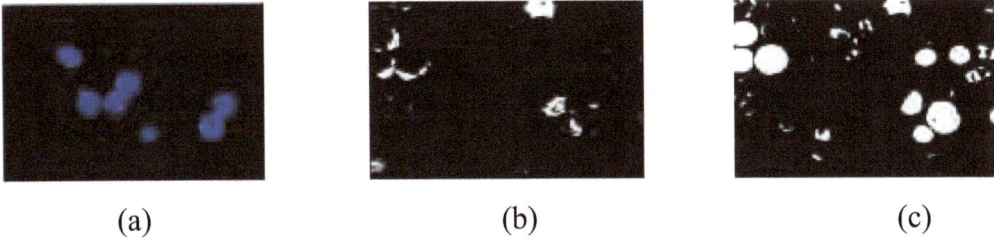

      (a)   (b)   (c)

Fig. (3). (**a**) Nucleus RoI, (**b**) Gaussian distribution, and (**c**) Adaptive thresholding image.

K-means Clustering

The clustering is used to group the multiclass dataset by measuring the correlation in their values. This proposed research uses the Partitioned clustering method to partition the input samples with k-clusters. The similar feature sample datasets are grouped, and the remaining are grouped in another cluster. The efficiency of the k-means clustering depends upon the separation and the compact size. The intensities observed have been used to form the cluster, and the sample output obtained is shown in Fig. (**4**).

K-means Cluster 1   **K-means Cluster 2**   **K-means Cluster 3**   **K-means Cluster 4**

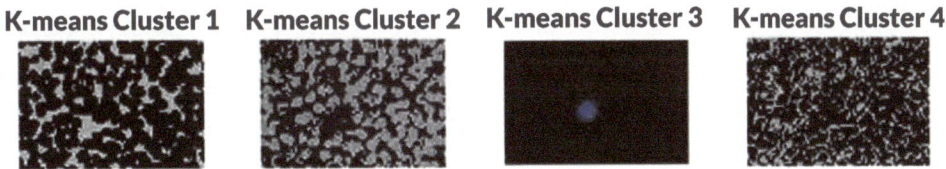

Fig. (4). k-means clustering image.

Feature Extraction

In image processing, we select the required features, also known as variable selection, for constructing the automatic model. It generates a new set of features from the primary features, and feature selection yields a subset of the features.

The feature extraction is applied to the segmented dataset containing various features. The classification between the various AML, ALL, and Myeloma categories needs different features to recompense graphic similarities. In the proposed research, multiple features, such as morphological, statistical, and texture features, have been utilized. In this feature extraction, we initially categorize L1 and L2 subtypes, and after that, M1, M2, M3, and M5 subtypes have been classified.

The GLCM is one type of feature extraction, also known as the Grey-Level Co-Occurrence Matrix technique. It is a statistical process of analyzing texture that narrates the spatial connections of each pixel. A set of features characterizes information about a picture's texture. These features include correlation, contrast, homogeneity, energy, mean, variance, standard deviation, skewness, kurtosis, and smoothness.

Deep Convolutional Neural Network Classifier

The classification process is utilized to subordinate the actual class tag to the blood test sample. During the feature extraction process, many artifacts are mined from the leukocytes, which will produce a large number of multivariate samples. This generated multivariate may reduce the classifier's performance in distinguishing between white blood cells and noise. The conventional classifier needs to improve due to the computational complexity of the features extracted. However, the CNN-based classifier model receives the input image, processes it, and categorizes it under specific subtypes (Leukemia or Myeloma). The proposed DCNN classifier model with various layer details is illustrated in Fig. (**5**).

CNN has had notable results over the previous decade in various fields identified with design acknowledgements, from voice recognition to image processing. When we compare CNN with fully connected (FC) neural networks or other architectures, CNN is widely optimal for image processing applications, and it has various technical advantages. The input image from the dataset is fed to the proposed DCNN model and the set of features. The convolutional layers and max pooling layers are utilized to attain the features from the segmented Image and train the proposed DCNN model. After this, the pooling layer section will reduce the parameters when the input image is too large. Then, we flattened the matrix into a vector and passed it into a fully connected layer. The output of the FC layer is fed into the softmax classification layer to detect and categorize the white blood cell types.

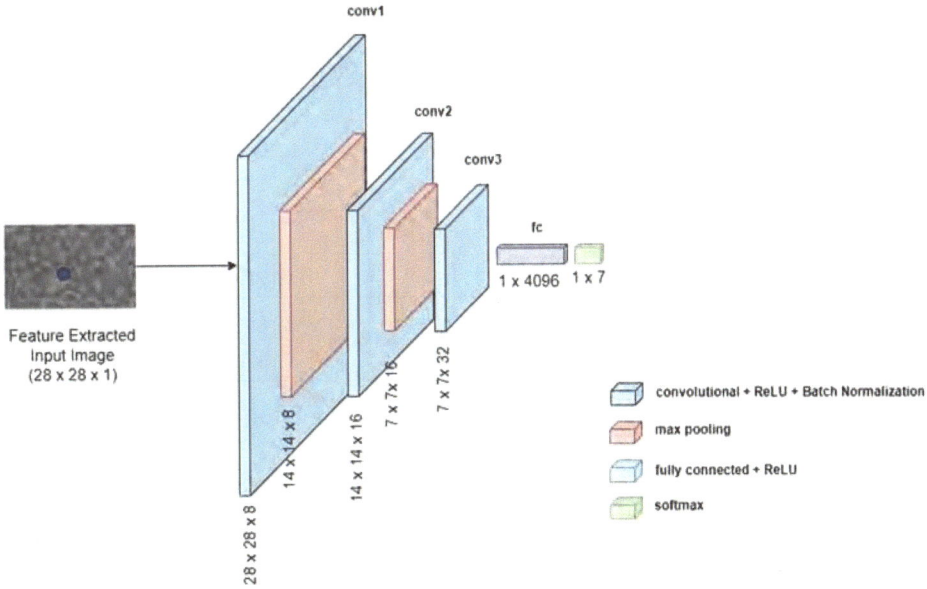

Fig. (5). The proposed DCNN classifier model.

Evaluation Metrics

The performance evaluation of the proposed CAD classifier model was carried out using accuracy, precision, recall, and F1 Score

$$Precision = \frac{TP}{TP+FP} \tag{4}$$

$$Recall = \frac{TP}{TP+FN} \tag{5}$$

$$F1\ Score = \frac{2*Recall*Precision}{Recall+Precision} \tag{6}$$

Where TP is True Positive: object present at goal location and detected as a correct class. TN is True Negative: object not present at goal location and not detected. TP is False Positive: object not present at goal location and detected as a wrong class. FN is False Negative: object present at goal location and undetected.

RESULTS

Simulation Environment

The compilation of the model is done using the Adam optimizer at the learning rate of 0.001. We compute the cross-entropy loss between the labels and predictions. One of the critical issues in CNN is overfitting, which is caused by cross entropy's scale sensitivity and constant gradient updating. A file was created to store the weights of the network during training. The proposed model uses early stopping, which monitors the value of testing accuracy, and the training automatically stops when the value of it stops increasing. The various parameters used to simulate the proposed DCNN model are tabulated in Table **2**.

Table 2. Simulation Environment/Parameters.

Parameter	Type/Range
Optimizer	Adam Optimizer
Learning Rate	1 e-03
CLR Policy	Exp. range with gamma = 0.99998
Dropout	0.5
Batch Size	128
Number of epochs	50
Loss Function	Categorical Cross Entropy

Experimental Results

The experimental analysis has been carried out using one notebook PC with specifications i7-7500U, 16GB DDR3 RAM, and NVidia 940MX processor. The proposed DCNN classifier model is verified and validated with the ALL-IDB dataset. MATLAB 2020a is utilized to train and test the proposed DCNN classifier. We obtained a classification accuracy of 92 image datasets during the testing phase of 97.8%. The accuracy and loss of the proposed DCNN model are shown in Fig. (**6**). From Fig. (**6**), the accuracy and loss stay at stable values around 50 epochs. The cross-entropy function provides a better loss value for the proposed DCNN classifier model.

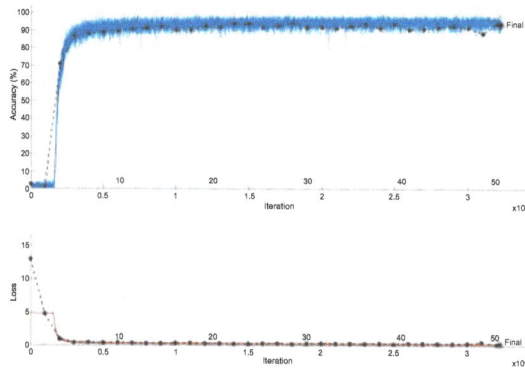

Fig. (6). The accuracy and loss of the proposed DCNN classifier model.

Fig. (**7**) represents the experimental results of the proposed DCNN classifier model for various blood samples. Figs. (**7a** and **b**) depict the detection and classification of Acute lymphoblastic Leukemia (ALL) blood cancer and its subtypes as L1 and L2, respectively. The Fig. (**7c**) depicts the detection of Myeloma blood cancer cells from the testing microscopic Image. Figs. (**7d, e,** and **7f**) represent the Acute Myeloid Leukemia (AML) blood cells and their subtypes as M2, M3, and M5, respectively. Fig. **7g** depicts the standard Image, which does not affect any blood cell cancer.

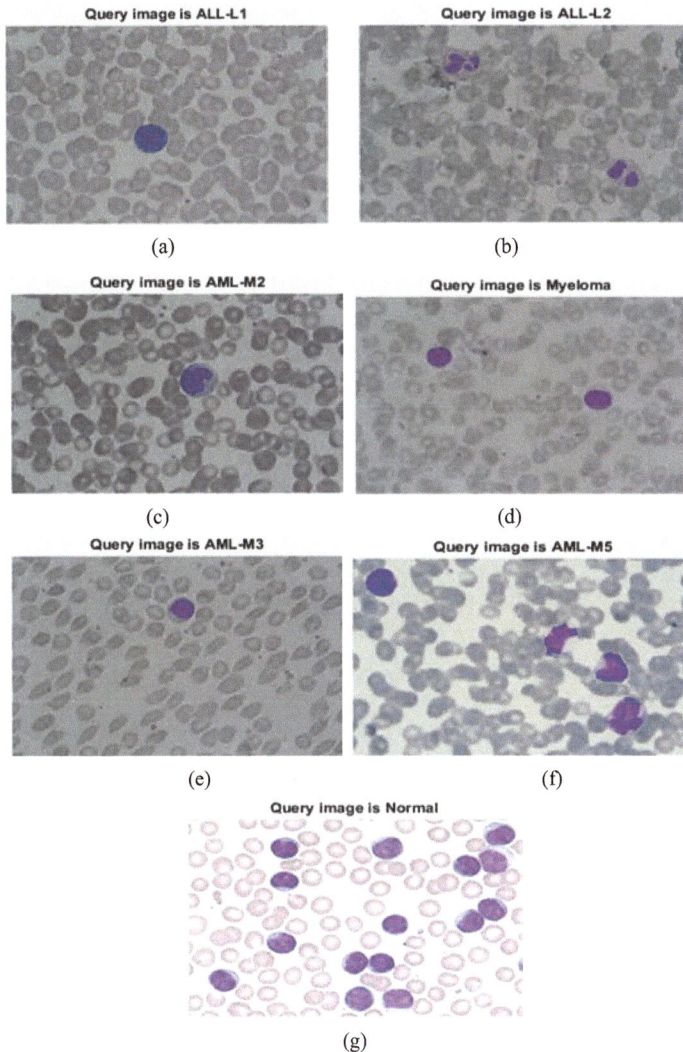

Fig. (7). Testing outcome (**a**) Acute lymphoblastic leukemia-L1 (**b**) Acute lymphoblastic leukemia-L2 (**c**) Myeloma (**d**) Acute myeloid leukemia-M2, (**e**) Acute myeloid leukemia-M3, (**f**) Acute myeloid leukemia-M5 and (**g**) Normal blood cell image.

The confusion matrix is generated to test the accuracy of the proposed DCNN classifier and is shown in Table **3**.

Table 3. The Confusion Matrix of the proposed DCNN Model.

Input Value	Leukemia	0.978	0.022
	Non-leukemia	0.022	0.978
	-	Leukemia	Non-leukemia
	Predicted Value		

Table **3** shows that the proposed DCNN classifier achieves the maximum TP and TN values due to the better feature extraction and deep learning model. It also attains the minimal FP and FN values.

DISCUSSION

This research proposed the CAD-based classifier model using DCNN to detect and classify acute lymphoblastic leukemia (ALL), Acute Myeloid Leukemia (AML), its various subcategories, and Myeloma from the microscopic blood images. The proposed CAD model was mainly focused on a two-stage approach. Initially, it differentiated between M5 AML with L1 and L2 ALL, and after that, it classified other sub-stages of white blood cells. Hence, the white blood cell diseases were grouped into two sets because of their similarity in visual features, which can confuse the doctors and cause misclassification from the recommendations provided by the doctor based on input blood samples. The microscopic Image obtained from the ALL-IDB dataset was preprocessed to remove the background noise from the input images.

We performed the Gaussian distribution and k-means clustering algorithm to segment the region of interest from the preprocessed images. The features, such as geometry, statistics, and texture, were extracted using the Gray-Level Covariance Matrix (GLCM) technique to train the proposed DCNN classifier. The validation showed that the proposed DCNN-based classification model detects the early stage of leukocyte cancer with an accuracy of 97.8%. The precision, recall, and F1 scores were achieved at 0.977, 0.955, and 0.966, respectively. The performance of the proposed automatic classification of leukocytes is compared with the existing classifier models such as CNN [18], GAN [19], Bi-CNN [20], VGGNet [22], and YOLO V4 [25]. The performance was compared using classification accuracy, precision, recall, and F1 scores. The comparison is tabulated in Table **4**. Table **4** shows that the proposed automatic classification of white blood cells using the DCNN model outperforms other models.

Table 4. Performance Comparison of the Proposed CAD model with existing method/model.

Authors	Classifier	Accuracy (%)	Precision	Recall	F1 Score
Saidani *et al*. [18]	CNN	96.89	0.967	0.966	0.970
Khaled *et al*. [19]	GAN	93.90	-	-	-
Yao *et al*. [20]	Bi-CNN	95.00	0.945	0.916	0.932
Sahlol *et al*. [22]	VGG-Net	96.10	0.956	0.936	0.946
Khandekar *et al*. [25]	YOLO V4	96.40	0.955	0.960	0.920
Proposed Model	**DCNN Model**	**97.8**	**0.977**	**0.955**	**0.966**

Table **4** shows that the proposed CAD-based classifier model using DCNN achieved a testing accuracy of 97.8%, slightly more than other existing methods. The primary reason is that the proposed DCNN classifier utilized seven texture features extracted using GLCM to train the developed Deep CNN model to classify the various white blood cancer cells. Also, the proposed classifier model achieved a higher precision and F1 score, which indicates that it will be more suitable for AML, ALL, and its subcategory classification in an efficient way.

CONCLUDING REMARKS

This paper proposes the automatic classification of white blood cell classification using the CAD-based classifier model. The white blood cell (leukocyte) diseases have been classified as Leukaemia and myeloma, and its subcategories like AML and ALL, *etc*.; it is an automated system that is used on the blood images through various stages such as pre-processing, segmentation, feature extraction, and classification. The Gaussian distribution and k-means clustering are used to obtain the features through the segmentation process. The Grey Level Covariance Matrix (GLCM) method has been used to attain the features to train the DCNN classifier model. The proposed DCNN classification model has been trained and tested using the mined features. The ALL-IDB dataset trains and tests the proposed automatic white blood cell classification model. The validation shows that the proposed DCNN-based classification model detects the early stage of leukocyte cancer with an accuracy of 97.8%. The precision, recall, and F1 scores have been achieved as 0.977, 80.955, and 0.966, respectively. It outperforms all other existing methods. In the future, more datasets can be included, and they can be deployed for real-time applications. Another improvement may be changing the parameters of the layers of the deep neural network.

REFERENCES

[1] King W, Toler K, Woodell-May J. Role of white blood cells in blood-and bone marrow-based autologous therapies. BioMed Res Int 2018; 1-8.

[http://dx.doi.org/10.1155/2018/6510842] [PMID: 30112414]

[2]   Hegde RB, Prasad K, Hebbar H, Singh BMK. Feature extraction using traditional image processing and convolutional neural network methods to classify white blood cells: a study. Australas Phys Eng Sci Med 2019; 42(2): 627-38.
[http://dx.doi.org/10.1007/s13246-019-00742-9] [PMID: 30830652]

[3]   Syeda JF, Farhath K. Blood cells, and leukocyte culture-A short review. Blood Res Transfus J 2017; 1: 555559.

[4]   Ortiz-Feregrino R, Tovar-Arriaga S, Pedraza-Ortega JC, Rodriguez-Resendiz J. Segmentation of Retinal Blood Vessels Using Focal Attention Convolution Blocks in a UNET. Technologies (Basel) 2023; 11(4): 97.
[http://dx.doi.org/10.3390/technologies11040097]

[5]   Sadeghian F, Seman Z, Ramli AR, Abdul Kahar BH, Saripan MI. Using digital image processing, a framework for white blood cell segmentation in microscopic blood images. Biol Proced Online 2009; 11: 196-206.
[http://dx.doi.org/10.1007/s12575-009-9011-2] [PMID: 19517206]

[6]   Şengür A, Akbulut Y, Budak Ü, Cömert Z. White blood cell classification based on shape and deep features. 2019 International Artificial Intelligence and Data Processing Symposium (IDAP). 1-4.
[http://dx.doi.org/10.1109/IDAP.2019.8875945]

[7]   Osowski S, Siroic R, Markiewicz T, Siwek K. Application of support vector machine and genetic algorithm for improved blood cell recognition. IEEE Trans Instrum Meas 2009; 58(7): 2159-68.
[http://dx.doi.org/10.1109/TIM.2008.2006726]

[8]   Alharbi AH, Aravinda CV, Lin M, Venugopala PS, Reddicherla P, Shah MA. Segmentation and classification of white blood cells using the UNet. Contrast Media Mol Imaging 2022; 2022(1): 5913905.
[http://dx.doi.org/10.1155/2022/5913905] [PMID: 35919503]

[9]   Kumar S, Mishra S, Asthana P. Pragya. Automated detection of acute Leukemia using a k-mean clustering algorithm. In Advances in Computer and Computational Sciences. Proceedings of ICCCCS 2016; 2018(2): 655-70. [Springer Singapore.].

[10]   Mohapatra S, Samanta SS, Patra D, Satpathi S. Fuzzy-based blood image seg-mentation for automated leukemia detection. In 2011 International Conference on Devices and Communications (ICDeCom) 1-5.
[http://dx.doi.org/10.1109/ICDECOM.2011.5738491]

[11]   Liu K, Hu J. Classification of acute myeloid leukemia M1 and M2 subtypes using machine learning. Comput Biol Med 2022; 147: 105741.
[http://dx.doi.org/10.1016/j.compbiomed.2022.105741] [PMID: 35738057]

[12]   Mirmohammadi P, Ameri M, Shalbaf A. Recognition of acute lymphoblastic leukemia and lymphocytes cell subtypes in microscopic images using random forest classifier. Phys Eng Sci Med 2021; 44(2): 433-41.
[http://dx.doi.org/10.1007/s13246-021-00993-5] [PMID: 33751420]

[13]   Gokul Kannan K, Ganesh Babu TR, Praveena R, Sukumar P, Sudha G, Birunda M. Classification of WBC cell classification using fully connected convolution neural network. J Phys Conf Ser 2023; 2466(1): 012033.
[http://dx.doi.org/10.1088/1742-6596/2466/1/012033]

[14]   Donga HV, Karlapati JSAN, Desineedi HSS, Periasamy P, Tr S. Effective framework for pulmonary nodule classification from CT images using the modified gradient boosting method. Appl Sci (Basel) 2022; 12(16): 8264.
[http://dx.doi.org/10.3390/app12168264]

[15]   Sanida T, Sideris A, Tsiktsiris D, Dasygenis M. Lightweight neural network for COVID-19 detection

from chest X-ray images implemented on an embedded system. Technologies (Basel) 2022; 10(2): 37.
[http://dx.doi.org/10.3390/technologies10020037]

[16]   Maity A, Nair TR, Mehta S, Prakasam P. Automatic lung parenchyma segmentation using a deep convolutional neural network from chest X-rays. Biomed Signal Process Control 2022; 73: 103398.
[http://dx.doi.org/10.1016/j.bspc.2021.103398]

[17]   Ramírez-Arias FJ, García-Guerrero EE, Tlelo-Cuautle E, *et al.* Evaluation of machine learning algorithms for classification of EEG signals. Technologies (Basel) 2022; 10(4): 79.
[http://dx.doi.org/10.3390/technologies10040079]

[18]   Saidani O, Umer M, Alturki N, *et al.* White blood cells classification using multi-fold pre-processing and optimized CNN model. Sci Rep 2024; 14(1): 3570.
[http://dx.doi.org/10.1038/s41598-024-52880-0] [PMID: 38347011]

[19]   Almezhghwi K, Serte S. Improved classification of white blood cells with the generative adversarial network and deep convolutional neural network. Comput Intell Neurosci 2020; 2020: 1-12.
[http://dx.doi.org/10.1155/2020/6490479] [PMID: 32695152]

[20]   Yao X, Sun K, Bu X, Zhao C, Jin Y. Classification of white blood cells using weighted optimized deformable convolutional neural networks. Artif Cells Nanomed Biotechnol 2021; 49(1): 147-55.
[http://dx.doi.org/10.1080/21691401.2021.1879823] [PMID: 33533656]

[21]   Abou Ali M, Dornaika F, Arganda-Carreras I. White blood vell classification: Convolutional neural network (CNN) and vision transformer (ViT) under medical microscope. Algorithms 2023; 16(11): 525.
[http://dx.doi.org/10.3390/a16110525]

[22]   Sahlol AT, Kollmannsberger P, Ewees AA. Efficient white blood cell leukemia classification with improved swarm optimization of deep features. Sci Rep 2020; 10(1): 2536.
[http://dx.doi.org/10.1038/s41598-020-59215-9] [PMID: 32054876]

[23]   Balasubramanian K, Ananthamoorthy NP, Ramya K. An approach to classify white blood cells using convolutional neural network optimized by particle swarm optimization algorithm. Neural Comput Appl 2022; 34(18): 16089-101.
[http://dx.doi.org/10.1007/s00521-022-07279-1]

[24]   Anand V, Gupta S, Koundal D, Alghamdi WY, Alsharbi BM. Deep learning-based image annotation for leukocyte segmentation and classification of blood cell morphology. BMC Med Imaging 2024; 24(1): 83.
[http://dx.doi.org/10.1186/s12880-024-01254-z] [PMID: 38589793]

[25]   Khandekar R, Shastry P, Jaishankar S, Faust O, Sampathila N. Automated blast cell detection for Acute Lymphoblastic Leukemia diagnosis. Biomed Signal Process Control 2021; 68: 102690.
[http://dx.doi.org/10.1016/j.bspc.2021.102690]

[26]   Saini A, Guleria K, Sharma S. A deep learning-based convolutional neural networks model for white blood cell classification. 4th IEEE International Conference for Emerging Technology (INCET).
[http://dx.doi.org/10.1109/INCET57972.2023.10170666]

[27]   Jawahar M, H S, L JA, Gandomi AH. ALNett: A cluster layer deep convolutional neural network for acute lymphoblastic leukemia classification. Comput Biol Med 2022; 148: 105894.
[http://dx.doi.org/10.1016/j.compbiomed.2022.105894] [PMID: 35940163]

## CHAPTER 15

Empowering Inclusive Communication with the Haptic-Enabled Language to Pulse Device: A Novel Assistive Technology Solution for Communicative Impairments

Bhawesh Mishra[1]**, Kavita Nampoothri**[1]**, Anushka Bukkawar**[1]**, Chandrashish Kukrety**[1] **and S. Sundar**[1,*]

[1] *School of Electronics Engineering (SENSE), Vellore Institute of Technology, Vellore, India*

Abstract: In the rapidly evolving technological landscape, inclusivity and accessibility in communication are paramount. This research paper introduces the "Haptic-Enabled Language to Pulse" device, a ground-breaking solution designed to empower individuals with speech impairments. The device represents a fusion of advanced technology and assistive healthcare technology, including Python programming, TensorFlow Lite's DeepSpeech model, and Raspberry Pi hardware, all integrated through shell scripting. This comprehensive system enables effective communication by capturing user input in spoken language, processing it into text, and translating it into Morse code for visualization. The vibration patterns from a haptic feedback device then convey the encoded Morse code to the specially-abled user. Beyond its primary function as a communication aid, this device also serves as an educational tool for Morse code learning. Its versatility accommodates diverse contexts, making it valuable for individuals with speech impairments and their communities. This research paper showcases the innovative use of Raspberry Pi alongside software components, contributing to inclusive and accessible communication solutions for speech-impaired and visually impaired individuals.

Keywords: Assistive healthcare, Deepspeech model, Haptic feedback, Morse code, Python programming, Raspberry pi, Shell scripting, Speech impairments, Speech to Text, Tensorflow lite, Visual impairments.

INTRODUCTION

In this ever-evolving world, technology has brought ease of living and the comfort of innovation to our doorstep. We have automated all parts of our lives and

---

* **Corresponding author S. Sundar:** School of Electronics Engineering (SENSE), Vellore Institute of Technology, Vellore, India; E-mail: sundar.s@vit.ac.in

Sivakumar Rajagopal, Prakasam P., Konguvel E., Shamala Subramaniam, Ali Safaa Sadiq Al Shakarchi & B. Prabadevi (Eds.)

upgraded ourselves to a better quality of life. From driverless cars to keyless locks, all are revelations of accessible and inclusive technology. The same quest for inclusivity and accessibility in communication has become more pressing than ever before.

Individuals with speech and visual impairments experience significant challenges with traditional verbal communication. So, this research endeavors to introduce an innovative solution that bridges these communication gaps.

This paper introduces the "Haptic-Enabled Language to Pulse" (HELP) device, an amalgamation of the latest software and hardware technology. At its core, the HELP device empowers users by capturing spoken language, converting it into textual form, and translating it into Morse code. Realizing this code through vibration patterns enables effective communication, not only for speech-impaired individuals but also for visually impaired people.

HELP is not just about technology. It is about inclusion. It is about breaking down the barriers to communication. It signifies a step towards the progressive pursuit of creating accessible solutions for individuals with speech and visual impairments.

OBJECTIVES

The main aims of this research are to:

- Develop a robust system that accurately converts recorded spoken language into textual form using TensorFlow Lite's DeepSpeech model, ensuring precise representation of user input.
- Create a Python script that efficiently translates the textual output into Morse code and converts it into vibration stimulus, facilitating an effective visual communication form.
- Employ Raspberry Pi hardware to enhance processing power, enabling real-time speech-to-text conversion, and Morse code encoding without compromising the device's portability and usability.

The ultimate objective is to empower individuals with hearing impairments by providing them with a reliable, easy-to-use communication tool that fosters inclusivity and independence, allowing them to participate actively in various communication contexts.

LITERATURE REVIEW

To attain our objectives, exhaustive research of the proposed and existing methodologies in a similar theme was conducted. The aim is to actuate the device in a cost-effective, faster, and accurate direction. The reference papers are listed below for the conceptualization of the research.

In 2006, Huggins-Daines *et al.* introduced "Pocketsphinx," a real-time continuous speech recognition system for hand-held devices. This system addresses the need for efficient, portable speech recognition technology, particularly for mobile and embedded applications [1]. In 2010, Cheng, Abdulla, and Salcic presented significant contributions to the field of work of embedded speech recognition. This research, published in the Proceedings of the IEEE International Symposium on Signal Processing and Information Technology, developmentally focused on the speech recognition system tailored for real-time embedded applications. The authors' work addresses the critical need to optimize speech recognition accuracy and efficiency within resource-constrained environments, further advancing the applicability of speech-based interfaces in real-time embedded systems [2]. In 2011, Pan *et al.* discussed the implementation of speech recognition systems on FPGA-based embedded systems with System-on-Chip (SOC) architecture. Their research focuses on integrated speech recognition in FPGA-based embedded systems, including its potential benefits and challenges [3]. In the same year, Qu and Li developed an embedded speech recognition module based on STM32 microcontrollers, contributing to the integrated speech interfaces in embedded systems [4]. In 2012, Reddy investigated "Text to Speech Conversion Using Raspberry Pi for Embedded System." This research delved into the practicality of implementing text-to-speech conversion on the Raspberry Pi platform and its relevance in embedded systems [5]. In 2014, Varshney and Singh contributed to the field of embedded speech recognition with their paper titled "Embedded Speech Recognition System." published in the International Journal of Advanced Research in Electrical, Electronics, and Instrumentation Energy, the research explored the development and integration of speech recognition technology into embedded systems. The research showcased practical applications and progressed in this area, providing valuable insights into the field [6]. Also in 2014, Hannun *et al.* presented "Deep Speech," a groundbreaking work in scaling up end-to-end speech recognition. This research, documented in an arXiv preprint, focused on large-scale, deep learning-based models for speech recognition. It significantly advanced the capabilities of deep neural networks in the sector of automatic speech recognition, marking a crucial milestone in the progress of speech recognition technology [7]. In 2015, Vanitha *et al.* entailed the implementation of text-to-speech for real-time embedded systems using the Raspberry Pi processor. The research explored the utilization of Raspberry Pi to create text-to-speech functionalities, enhancing the capabilities of embedded systems for various applications [8]. In 2016, Lakshmi presented the "Design and Implementation of Text to Speech conversion using Raspberry Pi." This research primarily concentrated on developing and implementing a text-to-speech conversion system using the Raspberry Pi platform, thus enhancing the accessibility of Raspberry Pi-based devices [9]. Moving ahead to 2018, Park *et al.* investigated fully neural network-based speech recognition on mobile and embedded devices. Their research explored advanced neural network approaches to speech recognition, focusing on their applicability in mobile and embedded systems and providing insights into the state-of-the-art in this field [10]. In 2019, Uko *et al.* introduced a microcontroller-based speech-to-text translation system. Their research proposes the system design and its implementation that converts spoken language into text using microcontroller technology, presenting potential applications in accessibility and assistive technology [11]. In 2020, Firmansyah *et al.* presented a study on an AI-based embedded speech-to-text system using DeepSpeech technology. Their research primarily centered on implementing DeepSpeech in embedded systems to enable efficient spoken language-to-text conversion [12]. In 2023, Wang and Nishizaki introduced a "Lightweight End-to-End Speech Recognition System on Embedded Devices." Their work continued to optimize speech recognition for embedded devices, extending its utility in various applications [13].

Proposed System

The proposed system, shown in Fig. (**1**), is explained below:

1. Start

- Begin the execution of the project.

2. Audio Monitoring

- Utilize a microphone to record audio.
- This step captures spoken language from the user using a triggered button press from the user.

3. Voice Detection

- Determine if, in the recorded audio, voice detection exists.
- If a voice is detected, proceed to the next step.
- If voice detection fails, return to Step 2 for further audio monitoring.

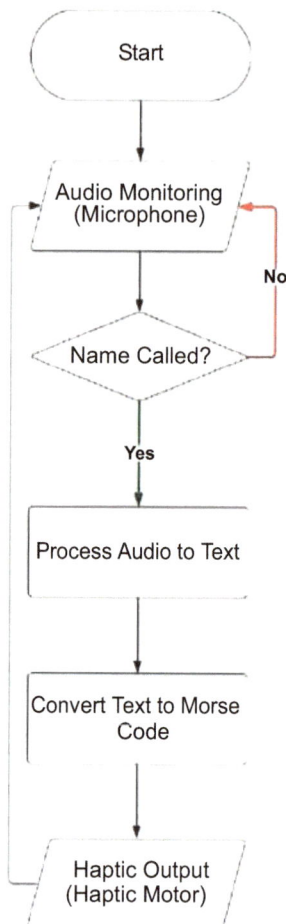

Fig. (1). Block diagram of the algorithm.

4. Process Audio to Text (using the DeepSpeech model)

- Employ DeepSpeech, a speech-to-text engine, to convert the recorded voice into text and store it on our local computer.
- This step transcribes the spoken words into a textual format, facilitating further processing

5. Convert Text to Morse Code

- We run a Python script and encode the obtained text into Morse code.
- Morse code serves as an intermediate representation between spoken language and vibration patterns.

6. Toggle the Haptic Motor

- Interpret the Morse code and control a vibration motor accordingly.
- Depending on the encoded Morse code, turn the vibration motor on or off to convey the message as a stimulus.
- After toggling the vibration motor, return to Step 2 for continuous audio monitoring.

7. Repeat

- The process continues in a loop, enabling real-time communication through voice-to-text conversion and Vibration motor-powered stimulusbased representation of the converted text.

DESCRIPTION OF THE DEVICE

Software

DeepSpeech Model

DeepSpeech is an open-source automatic speech recognition (ASR) model developed by Mozilla. It employs state-of-the-art deep learning techniques to convert spoken language into written text. The model is built upon the principles of deep neural networks and operates using the Mozilla Common Voice dataset, which includes multilingual and diverse audio recordings. The algorithm behind DeepSpeech relies on a recurrent neural network (RNN) architecture called Long Short-Term Memory (LSTM). LSTMs are ideal for sequential data like audio, as they can capture contextual information effectively. The working of DeepSpeech involves several key steps:

1. Audio Feature Extraction: The input audio is converted into a spectrogram, visually representing sound frequencies over time.
2. Acoustic Modelling: DeepSpeech uses a neural network to learn the acoustic characteristics of speech, such as phonemes and phonetic patterns.
3. Language Modelling: A language model assists in predicting the probability of a word sequence, thereby enhancing transcription accuracy.
4. Decoding: The model's output is decoded into text using post-processing techniques like Connectionist Temporal Classification (CTC).

DeepSpeech excels due to its deep neural architecture, which allows it to capture intricate acoustic and linguistic nuances in spoken language. It has demonstrated impressive accuracy and is widely used in speech-to-text applications, making it a valuable tool for transcription, voice assistants, and accessibility features.

Programming Languages

In the mentioned project, Python is crucial in several key areas. Its applications include voice recording, audio data processing, implementing the DeepSpeech model for speech-to-text conversion, and generating Morse code from the textual output. Python's flexibility and wide range of libraries make it well-suited for handling different aspects of the project.

Shell scripting is employed to integrate these Python-based processes seamlessly. It ensures efficient coordination between voice recording, text conversion, Morse code generation, and vibration motor control. This combination of Python and shell scripting enhances the project's efficiency, making it a powerful and cohesive system for accessible communication.

Hardware

Raspberry pi

As shown in Fig. (2), the core of the system design is the Raspberry Pi 3B\+ model. The Raspberry Pi board is a remarkable feat of miniaturization, packing substantial computational power into a footprint no larger than a credit card. While it boasts impressive capabilities, understanding its intricacies is essential before diving into its potential applications. At the heart of the Raspberry Pi system lies the Broadcom BCM2836 system-on-chip (SoC) multimedia processor, which integrates the system's components majorly. These components include the central and graphics processing units, audio, and communication hardware, all concealed beneath the 1 GB memory chip at the board's center.

The BCM2836 stands out for its SoC design and utilization of a distinct instruction set architecture (ISA) known as ARM. The HDMI (High-Definition Multimedia Interface) connector, positioned as the only port on the bottom of the Pi, is essential for delivering high-quality visuals. In contrast to analog composite connections, the HDMI port offers a high-speed digital link, ensuring pixel-perfect images on computer monitors and modern HDTV sets, displaying images at Full HD 1920x1080 resolution. This paper explores the hardware architecture of the Raspberry Pi and its potential applications, highlighting its unique capability to run a wide range of programs.

Fig. (2). Circuit of the system.

Other Components

In addition to the Raspberry Pi board, this project employs several essential hardware components to create a functional communication system:

Wires: Various types of wires, including jumper wires and cables, connect various components, ensuring smooth electrical connections throughout the project.

Breadboard: A breadboard provides a convenient platform for prototyping and connecting electronic components without soldering. It enables easy experimentation and modification of circuitry.

Vibration Motor: These motors act as a stimulus-based indicator in the project. When electrical current passes through them, they vibrate, conveying information using Morse code patterns.

Buttons: Buttons or switches provide user input. In this project, a button press initiates voice recording, triggering the communication process.

Cable Wires: Various types of cable wires, such as audio cables and power cables, interconnect components to ensure signal transmission and power supply.

These hardware components play crucial roles in the project's functionality, facilitating the conversion of spoken language to Morse code and its stimulus-based representation through a vibration motor, ultimately enabling effective communication for individuals with speech impairments.

Working of the Device

The project functions as a comprehensive communication system, harmonizing various hardware components and software modules to facilitate efficient communication for individuals with speech impairments. It commences with the user activating the system by pressing a designated button. At its core, the Raspberry Pi is a compact yet powerful computing platform. Pressing the button initiates the entire audio recording process.

The microphone continuously captures audio input, including the user's spoken language, which serves as the raw data for communication. An essential initial step is voice detection, where the system processes the audio input to determine the presence of a voice. This step filters out non-speech sounds, ensuring the system responds solely to the user's spoken words. Upon voice detection, the recorded voice data undergoes sophisticated processing using the DeepSpeech model. DeepSpeech, a neural network-based model from TensorFlow Lite, employs complex algorithms to accurately transcribe spoken language into textual format, effectively converting the audio signal into a more understandable form.

Subsequently, Python scripts come into play to convert the textual output from DeepSpeech into Morse code. Morse code, essentially, is a visual representation of text using dots and dashes, facilitating efficient communication through blinking patterns, which is particularly advantageous for individuals with speech impairments. The interpretation of Morse code controls a vibration motor, allowing for the stimulus-based representation of the converted text. Each dot and dash in Morse code corresponds to a short or long vibration, providing an intuitive means of communication. The vibration motor effectively conveys the translated Morse code, ensuring the user's message communication to others.

Following vibration-based communication, the system seamlessly returns to audio monitoring, ready to capture additional user input or spoken language when the button is triggered again. This entire process operates in a continuous loop, facilitating real-time communication through speech-to-text conversion and vibration-based sensory representation. Finally, the user can terminate the system when desired, effectively ending the project's execution. In this way, the project seamlessly integrates hardware components like the microphone, button, and vibration motor with the computational power of the Raspberry Pi, bridging the gap between spoken language and visual communication through Morse code. HELP has enhanced accessibility and inclusivity for individuals with speech impairments.

We assess the scalability of the employed speech-to-text model and examine the impact of processor specifications on inference time by using a set of nine standardized audio files into textual format conversion. The same model has been utilized consistently across different computing devices for comparative analysis. The reference device for this investigation was a Raspberry Pi 3B+, featuring a Broadcom BCM2836 system-on-chip processor with 1 GB of RAM. Conversely, the experimental device was a laptop incorporated with an AMD Ryzen 7 4800H Octa Core processor. We noted that the laptop processor operates at a base frequency of 2.9 GHz, whereas the BCM2836, within the Raspberry Pi 3B+, operates at a base frequency of 1.4 GHz. This research is within the context of benchmarking and evaluating the performance of the speech-to-text conversion process across these diverse computational platforms.

The nine sentences listed below were utilized for testing:

1. Hello!
2. Hello, how are you?
3. Hello, how are you doing today?
4. Hello, how are you doing on this lovely day?
5. Hello, how are you doing on this lovely, sunny day?
6. Hello, how are you doing on this lovely, sunny day in the park?
7. Hello, how are you doing on this lovely, sunny day in the park with your friends?
8. Hello, how are you doing on this lovely, sunny day in the park with your friends, playing frisbee?
9. Hello, how are you doing on this lovely, sunny day in the park with your friends, playing frisbee and having a picnic?

APPLICATIONS

Assistive communication for speech-impaired individuals: The primary application is to provide an effective means of communication for individuals with speech impairments. Translating their spoken words into Morse code and visually conveying them through vibration patterns, enables speech-impaired individuals to interact with others, promoting inclusivity.

Education and morse code learning: The project can function as an educational tool for learning Morse code. It offers an engaging and practical way for individuals to acquire this valuable skill, which can be particularly useful in emergencies or for communication in noisy environments.

Accessible communication in noisy environments: In environments with high noise levels, where verbal communication may be challenging, this system can offer an alternative means of communication. It allows individuals to convey messages visually using Morse code, ensuring effective communication even when spoken words are difficult to hear.

Emergency communication: In emergencies where speech may be hindered or impractical, such as during power outages or natural disasters, this project provides a reliable method for individuals to communicate essential information, enhancing safety and coordination efforts.

RESULTS

It is evident that the computational time for model execution on the Raspberry Pi averages 9.9 seconds for converting an audio file to a text format. In contrast, the general-purpose computer achieves the same task within 6.5 seconds, marking a significant 34.34% reduction in conversion time. This observation highlights the performance advantage of a more robust processor when using the model.

The mean processing times for the Raspberry Pi in converting audio files of varying lengths into textual output are detailed in Fig. (**3**). Additionally, the temporal discrepancy between the Raspberry Pi and the general-purpose computer is graphically represented in Fig. (**4**).

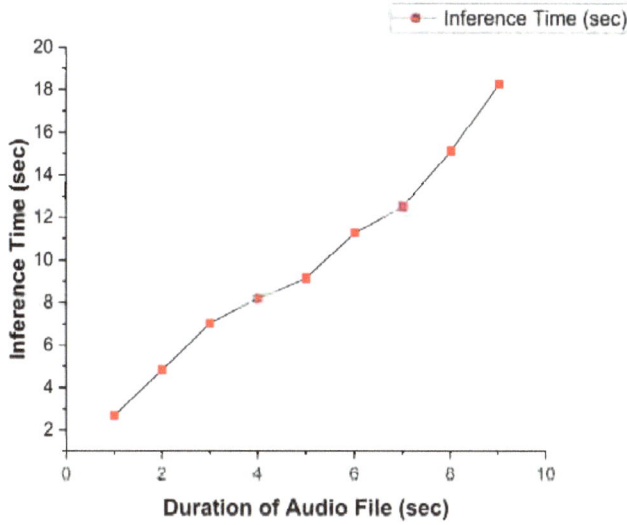

Fig. (3). Average conversion time *versus* length of audio.

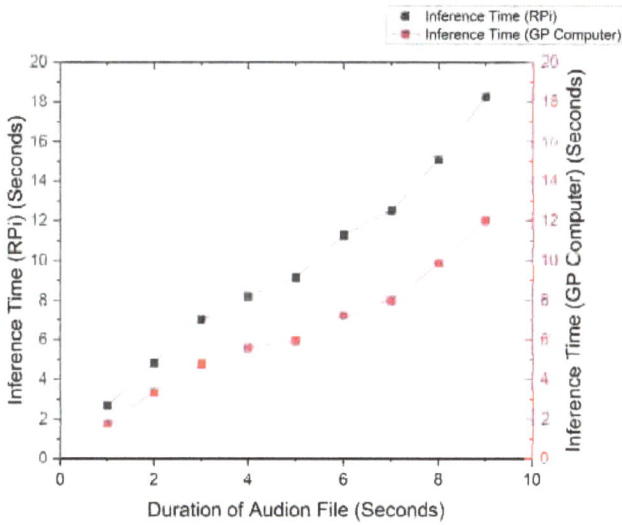

Fig. (4). Execution time comparison

A comparative analysis of the inference time between the Raspberry Pi and a general-purpose computer can be observed in Fig. (**5**).

Sentences	Duration of the Audio File	Average Time to Process on Laptop	Average Time to Process on Raspberry Pi	Time Difference
Hello!	1.568	1.7779	2.6758	0.8979
Hello, how are you?	2.624	3.3634	4.8146	1.4512
Hello, how are you doing today?	3.774	4.7597	7.0147	2.255
Hello, how are you doing on this lovely day?	4.319	5.5844	8.1833	2.5989
Hello, how are you doing on this lovely, sunny day?	4.784	5.924	9.1315	3.2075
Hello, how are you doing on this lovely, sunny day in the park?	5.45	7.202	11.2594	4.0574
Hello, how are you doing on this lovely, sunny day in the park with your friends?	6.315	7.9917	12.5116	4.5199
Hello, how are you doing on this lovely, sunny day in the park with your friends, playing frisbee?	7.477	9.86437	15.1007	5.23633
Hello, how are you doing on this lovely, sunny day in the park with your friends, playing frisbee and having a picnic?	9.277	11.9895	18.2585	6.269

Fig. (5). Tabular comparison between the execution times.

CONCLUSION

The "Haptic-Enabled Language to Pulse" (HELP) device presents an innovative solution, leveraging Python programming, TensorFlow Lite's DeepSpeech model, and Raspberry Pi hardware to empower individuals with speech impairments. The comprehensive system, converting spoken language to Morse code and delivering it through haptic feedback, demonstrates versatility and portability. Beyond its primary communication aid function, HELP serves as an educational tool, offers an alternative in noisy environments, and facilitates emergency communication. The study highlights the impact of processor specifications on performance, emphasizing the device's effectiveness while underscoring the advantages of more robust processors. Ultimately, HELP tackles communication challenges and contributes to accessible technology, promoting inclusivity and connectivity in various real-world scenarios.

ACKNOWLEDGEMENTS

The authors extend their heartfelt appreciation to the reviewers for their invaluable feedback and constructive comments on our review paper. Their insightful comments and suggestions have greatly improved the quality of the manuscript and enhanced its contribution to the field. We are thankful to the Vellore Institute of Technology for their support and resources that made this research possible. Finally, we would like to acknowledge our colleagues and collaborators for their stimulating discussions and ideas that have contributed to the development of this work. Without the support and contributions of these individuals and organizations, this paper would not have been possible. We are deeply grateful for their support and contributions to our research.

REFERENCES

[1]   Huggins-Daines D, Kumar M, Chan A, Black AW, Ravishankar M, Rudnicky AI. Pocketsphinx: A free, real-time continuous speech recognition system for hand-held devices. In2006 IEEE international conference on acoustics speech and signal processing proceedings 2006; 1.

[2]   Cheng O, Abdulla W, Salcic Z. Speech recognition system for embedded real-time applications. 2009 IEEE International Symposium on Signal Processing and Information Technology (ISSPIT). 118-22.
[http://dx.doi.org/10.1109/ISSPIT.2009.5407487]

[3]   Pan ST, Lai CC, Tsai BY. The implementation of speech recognition systems on FPGA-based embedded systems with SOC architecture. Int J Innov Comput, Inf Control 2011; 7(11): 6161-75.

[4]   Qu Q, Li L. Realization of embedded speech recognition module based on STM32. In2011 11th International Symposium on Communications and Information Technologies (ISCIT) (pp. 73-77). IEEE.
[http://dx.doi.org/10.1109/ISCIT.2011.6092186]

[5]   Reddy PVN. Text to speech conversion using raspberry-pi for embedded system. Int J Innov Res Sci Eng Technol 2012; 1(1): 144-8.
[http://dx.doi.org/10.15680/IJIRSET.2012.0101019]

[6]   Varshney N, Singh S. Embedded speech recognition system. International journal of advanced research in electrical, electronics and instrumentation energy. 2014 Apr;4:9218-27.

[7]   Hannun A, Case C, Casper J, *et al.* Deep speech: Scaling up end-to-end speech recognition. arXiv preprint arXiv:1412.5567. 2014.

[8]   Vanitha E, Kasarla PK, Kuamarswamy E. Implementation of text-to-speech for real-time embedded system using Raspberry Pi processor. International Journal and Magazine of Engineering Technology Management and Research, IISN 1995; 2015(Jul): 2348-4845.

[9]   Lakshmi K. Design and implementation of text to speech conversion using raspberry pi. Dimensions. 2016; 85: x56mm.

[10]   Park J, Boo Y, Choi I, Shin S, Sung W. Fully neural network based speech recognition on mobile and embedded devices. Adv Neural Inf Process Syst 2018; 31.

[11]   Uko VS, Dubukumah GB, Ayosubomi IK. Microcontroller based speech to text translation system. European Journal of Engineering and Technology Research 2019; 4(12): 149-54.

[12]   Firmansyah MH, Paul A, Bhattacharya D, Urfa GM. Ai based embedded speech to text using deepspeech. arXiv preprint arXiv: 12830. 2020.

[13]   Wang Y, Nishizaki H. A lightweight end-to-end speech recognition system on embedded devices. IEICE Trans Inf Syst 2023; E106(D): 1230-9.
[http://dx.doi.org/10.1587/transinf.2022EDP7221]

CHAPTER 16

Investigation of Various Transfer Learning Techniques for Classifying Alzheimer's Disease Dataset

Velswamy Karunakaran[1] and **Jain Ankur**[2,*]

[1] *Sri Eshwar College of Engineering, Coimbatore, India*

[2] *Manipal University, Jaipur, India*

Abstract: Alzheimer's disease, also called dementia, is a severe psychological disorder that affects the cerebrum, causes cognitive decline, and impairs a person's ability to reason. During the initial stage, AD patients encounter the standard adverse effects. Usually, they lose track of their everyday tasks and obligations. They find it challenging to communicate verbally with people. They also have a diminished ability to think critically, among other things. Currently, research is ongoing to determine the underlying cause of AD. We are now quite concerned about AD because the majority of its patients are older than sixty. Nervous system specialists typically perform multiple tests to differentiate AD. Human errors occur from time to time. We require high-performing deep-learning models to diagnose and forecast the illness better. This study looks at how well VGG16, InceptionResNet-V2, Resnet50, Resnet101, and Resnet152 classified the AD dataset and compares and contrasts their results. Accuracy, loss, validation accuracy, and validation loss were the performance indicators we utilized to assess the models. You may find the dataset on the Kaggle repository.

Keywords: Alzheimer's disease, Classification, Deep neural network, Healthcare, Machine learning, Transfer learning.

INTRODUCTION

Alzheimer's Disease International (ADI) states that 3/4 of AD patients do not receive the same level of care, according to the World Alzheimer's Report 2022 [1]. In underdeveloped nations, this rate is around nine out of ten. An estimated 55 million people have dementia each year. The most recent data from the World Health Organization (WHO) predicts a rise to 139 million dementia cases in 2050.

---

[*] **Corresponding author Jain Ankur**: Manipal University, Jaipur, Rajasthan, India;
E-mail: ankur.jain@jaipur.manipal.edu

Sivakumar Rajagopal, Prakasam P., Konguvel E., Shamala Subramaniam, Ali Safaa Sadiq Al Shakarchi & B. Prabadevi (Eds.)

Dementia is a condition with side effects of impeded memory, thinking, conduct, and profound control issues, leading to a deficiency of independence. Alzheimer's disease is the most widely recognized and notable type of disease. Unusual proteins disrupt synapses and nerves, disrupting the transmitters that transmit messages in the mind, particularly those responsible for storing memories.

A biomarker is any physiological, biochemical, or anatomic limit objectively assessing regular biological cycles, neurotic cycles, or responses to a corrective intervention. Clinical preliminary analyses and ongoing symptomatic interventions generally accept five promotional biomarkers as suitable. Three are cerebrum imaging measurements, and two are proteins found in the cerebrospinal fluid (CSF) [2].

Although we cannot predict AD in its early stages, we can analyze it. From the large volume of biological data sets, AI algorithms can diagnose a variety of neuropsychiatric and neurodegenerative disorders, including AD. Specialists dealing with the underlying analysis of AD can benefit from using magnetic resonance imaging (MRI) filters. To produce excellent and high-quality 2D and 3D images of the brain's architecture, magnetic resonance imaging (MRI) uses radio waves and attractive fields. However, by measuring the human essential visual cortex and identifying the topography of the cerebrum, functional magnetic resonance imaging (fMRI) provides insightful information on mental activity. Similarly, Positron Emission Tomography (PET) studies mental activity using radiotracers like fluorodeoxyglucose and amyloid [3].

Researchers in various domains have expressed interest in developing deep neural network approaches for MRI image-based AD detection [4]. Transfer learning approaches perform better than non-transfer learning-based approaches, according to Tufail *et al*. [5]. Ahmad [6] employed DNN and CAD-based techniques to achieve a 97% accuracy rate [7], while Ebrahimi used ResNet-18 and achieved 96.88% accuracy. Chao Li *et al*. [8] added two improved ResNet algorithms to the traditional ResNet residual blocks. Dong Nguyen [10] and Mingjin Liu [9] used residual network approaches to diagnose illnesses using 3D MRI images. To arrive at a final diagnosis and prognosis, we used patient demographics, cognitive exam findings, and the predictions of ResNet and XGBoost. Farheen Ramzan *et al*. [11] carefully assessed the RS-18 architecture over fMRI images. However, Wei Li *et al*. [12] worked on 4D fMRI images using deep learning methods.

M. Abdelaziz *et al*. [13] achieved an accuracy of 98.22% for NC *vs*. AD by utilizing a unique method that combined clinical score regression activity with CNN classification. Nevertheless, S. Buyrukoglu [14] found that the Random Forest method performed better when working on a predictive model. Rule induction, k-nearest neighbors (k-NN), Naive Bayes, decision trees (DT), generalized linear models (GLM), and deep learning algorithms were all used by M. Shahbaz *et al.* [15] on the ADNI dataset. Nonetheless, GLM accurately categorized the AD phases at a rate of approximately 88.24%. M. Odusami *et al.* achieved 98.86% accuracy in AD classification using ResNet18 and DenseNet201 [16]. H. S. Zaina *et al.*'s research [17] focused on four modules: pre-processing, feature extraction, categorization of AD using deep learning, and multi-layer perceptron. Their accuracy rate with both white matter and gray matter was 96.15%. Assmi *et al.* [18] looked at how well six pre-trained networks did at classifying. For VGG-19, VGG-16, ResNet-50, InceptionV3, Xception, and DenseNet169, the overall accuracy was 92.86%, 92.83%, 91.04%, 90.57%, 85.99%, and 88.64%. Nonetheless, Shukla *et al.* [19] used a variety of convolutional network models, including Alz-XceptionConvNet, Alz-DenseConvNet, Alz-ResConvNet, and Alz-VGGConvNet, in addition to Alz-MobileConvNet. Alz-MobileConvNet, Alz-MobileConvNet, and Alz-VGGConvNet achieved the best multiclass and binary classification results, with respective accuracy rates of 94% and 99%.

This chapter covered five distinct innovative deep learning-based techniques for identifying Alzheimer's disease in its initial stages. We used four stages of MRI brain images to test CNN-based transfer learning models: very mild, mild, moderate, and not dementia. The models are VGG16, InceptionResNet-V2, ResNet50, ResNet101, and ResNet152.

The Various Transfer Learning Techniques

The initial pre-processing step involves loading the model's image to perform augmentation. In this case, we divide the information into four groups: very mild, mild, moderate, and non-dementia. Fig. (**1**) displays representative brain pictures for all four categories.

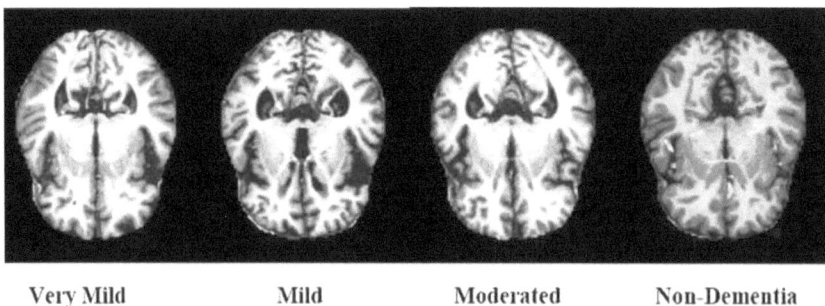

Very Mild   Mild   Moderated   Non-Dementia

Fig. (1). Categories of brain images present in the Kaggle dataset.

Next, we build a model using a CNN-based approach in the second stage. In this instance, we used the CNN-based transfer learning models InceptionResNet-V2, VGG16, ResNet50, ResNet101, and ResNet152. ResNet is a type of deep transfer learning that relies on residual learning. These models are the most efficient at mitigating the problem of gradient disappearance with a given residual block. The figure details the ResNet 50, ResNet 101, and ResNet 152 architectures.

- One layer has a convolution with a kernel (7 * 7) and 64 different kernels with a stride (size 2).
- Next, the max pooling has a stride (of size 2).
- Nine layers with a repeat of 3 thrice with 64 kernels of size 1*1, 64 kernels of size 3*3, and 256 kernels of 1*1.
- Twelve layers with a 4-time repeat of 128 kernels of 1 * 1, 128 kernels of 3 * 3, and, at last, 512 kernels of 1 * 1.
- Eighteen layers with six times repeat of a 256 kernel of 1 * 1, two more 256 kernels with 3 * 3, and 1024 kernel of 1 * 1.
- Nine layers of 3 times repeat 512 kernel of 1*1,512 with two more of 512 kernel of 3 * 3, and 2048 kernel of 1 * 1.
- Lastly, we complete a layer with an average pool, followed by a wholly linked layer with one thousand nodes and a softmax function at the top.

So, in total $1 + 9 + 12 + 18 + 9 + 1 = 50$ layers ResNet50 Deep Convolutional network.

A convolution layer appears at the beginning of ResNet101, and wholly connected layers appear at the end. Fig. (**2**) depicts ResNet101's architecture. ResNet-101 contains a total of 104 convolutional layers. It consists of 33 blocks, 29 of which require the output from the preceding block, which are considered residual connections. The remaining four blocks use the output of the preceding four blocks, which functioned in a convolution layer with a 1x1 filter and a stride of 1. The batch normalizing layer comes next. Next, we forward the output to the summation operator [20].

ResNet152 is an additional algorithm for solving the vanishing gradient problem. Deep learning can be handled with 152 convolution filters spread across multiple layers. This can guarantee a high level of categorization accuracy. Fig. (**2**) displays the number of layers for each ResNet architecture.

On the other hand, the InceptionResNet-V2 architecture is a very deep CNN that substitutes residual connections for filter concentration stages. The advantage of the inception architecture lies in its ability to adjust different filter phases without compromising the quality of the photos [21].

Visual Geometry Group, or VGG for short, is a CNN model for face recognition and image classification. Every layer of VGG uses a tiny 3x3 convolution kernel. Beginning with VGG11, it concludes with VGG19. In comparison to VGG19, which has 16 convolution layers and 3 completely linked layers, VGG16 has 13 convolution layers and 3 fully connected layers. 224x224-sized photos are the input format used by VGGNet. Next, the convolution layer contains a filter that employs receptive images in 3 by 3 dimensions. Additionally, we use a 1x1 size filter to alter the input images linearly. The ReLu (Rectified Linear Unit) activation function reduces the training time. ReLu generates a zero for negative inputs and compares the output to the input. VGGNet uses ReLu in the hidden layer instead of AlexNet because of AlexNet's longer training times and higher memory requirements. Next, the number of convolution layers in the pooling layer lowers the feature map's dimensions and parameters. The picture then moves to layers that are fully connected. VGGNet consists of one layer with 1000 channels and two with 4096 channels each. Fig. (**3**) displays the architecture diagram.

layer name	output size	18-layer	34-layer	50-layer	101-layer	152-layer
conv1	112x112	7x7, 64. stride 2				
		3x3 max pool, stride 2				
conv2.x	56x56	$\left[\begin{array}{c} 3\text{x}3, 64 \\ 3\text{x}3, 64 \end{array}\right]$x2	$\left[\begin{array}{c} 3\text{x}3, 64 \\ 3\text{x}3, 64 \end{array}\right]$x3	$\left[\begin{array}{c} 1\text{x}1, 64 \\ 3\text{x}3, 64 \\ 1\text{x}1, 256 \end{array}\right]$x3	$\left[\begin{array}{c} 1\text{x}1, 64 \\ 3\text{x}3, 64 \\ 1\text{x}1, 256 \end{array}\right]$x3	$\left[\begin{array}{c} 1\text{x}1, 64 \\ 3\text{x}3, 64 \\ 1\text{x}1, 256 \end{array}\right]$x3
conv3.x	28x28	$\left[\begin{array}{c} 3\text{x}3, 128 \\ 3\text{x}3, 128 \end{array}\right]$x2	$\left[\begin{array}{c} 3\text{x}3, 128 \\ 3\text{x}3, 128 \end{array}\right]$x4	$\left[\begin{array}{c} 1\text{x}1, 128 \\ 3\text{x}3, 128 \\ 1\text{x}1, 512 \end{array}\right]$x4	$\left[\begin{array}{c} 1\text{x}1, 128 \\ 3\text{x}3, 128 \\ 1\text{x}1, 512 \end{array}\right]$x4	$\left[\begin{array}{c} 1\text{x}1, 128 \\ 3\text{x}3, 128 \\ 1\text{x}1, 5\,12 \end{array}\right]$x8
conv4.x	14x14	$\left[\begin{array}{c} 3\text{x}3, 256 \\ 3\text{x}3, 256 \end{array}\right]$x2	$\left[\begin{array}{c} 3\text{x}3, 256 \\ 3\text{x}3, 256 \end{array}\right]$x6	$\left[\begin{array}{c} 1\text{x}1, 256 \\ 3\text{x}3, 256 \\ 1\text{x}1, 1024 \end{array}\right]$x6	$\left[\begin{array}{c} 1\text{x}1, 256 \\ 3\text{x}3, 256 \\ 1\text{x}1, 1024 \end{array}\right]$x23	$\left[\begin{array}{c} 1\text{x}1, 256 \\ 3\text{x}3, 256 \\ 1\text{x}1, 1024 \end{array}\right]$x36
conv5.x	7x7	$\left[\begin{array}{c} 3\text{x}3, 512 \\ 3\text{x}3, 512 \end{array}\right]$x2	$\left[\begin{array}{c} 3\text{x}3, 512 \\ 3\text{x}3, 512 \end{array}\right]$x3	$\left[\begin{array}{c} 1\text{x}1, 512 \\ 3\text{x}3, 512 \\ 1\text{x}1, 2048 \end{array}\right]$x3	$\left[\begin{array}{c} 1\text{x}1, 512 \\ 3\text{x}3, 512 \\ 1\text{x}1, 2048 \end{array}\right]$x3	$\left[\begin{array}{c} 1\text{x}1, 512 \\ 3\text{x}3, 512 \\ 1\text{x}1, 2048 \end{array}\right]$x3
	1x1	average pool, 1000-d fc, softmax				
FLOPs		$1.8\text{x}10^9$	$3.6\text{x}10^9$	$3.8\text{x}10^9$	$7.6\text{x}10^9$	$11.3\text{x}10^9$

Fig. (2). ResNet architecture.

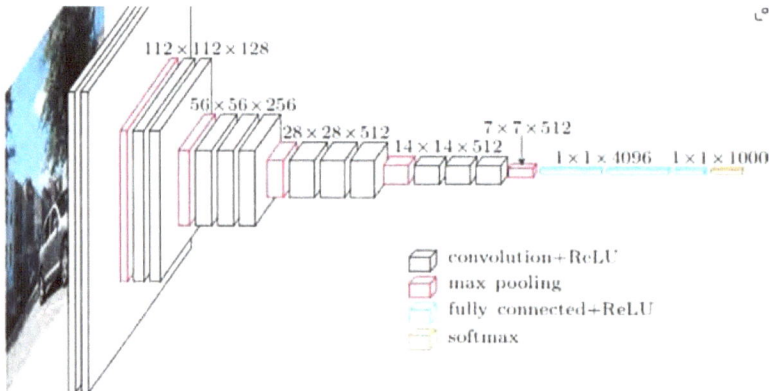

Fig. (3). VGGNet architecture.

RESULTS AND DISCUSSIONS

Using MRI brain pictures of Alzheimer's disease spanning 30 epochs, we applied five transfer learning strategies and compared the outcomes. Based on Table **1** (below), the ResNet 50 model had the highest training accuracy (93%) compared to the other models. VGG16 achieved the highest validation accuracy at 66%. Regarding losses, VGG16 had the most significant training loss (2.9%), whereas InceptionResNet-V2 had the most considerable validation loss (9.8%) compared to the other models. Figs. (**5** to **9**) display the accuracy and loss graphs for VGG16, ResNet-50, Inception ResNet V2, ResNet-101, and ResNet-152, in that order.

The output shape and the number of parameters are shown in Fig. (**4**) while training and testing the VGG16 model on the dataset.

Layer (type)	Output Shape	Param #
Input_1 (Inputaxer)	[(None, 224, 224, 3)]	0
Block1_conv1 (Conv2D)	(None, 224, 224, 64)	1792
Block1_conv2 (Conv2D)	(None, 224, 224, 64)	36928
Block1_pool (MaxPooling2D)	(None, 112, 112, 64)	0
Block2_conv1 (Conv2D)	(None, 112, 112, 128)	73856
Block2_conv2 (Conv2D)	(None, 112, 112, 128)	147584
Block2_pool (MaxPooling2D)	(None, 56, 56, 128)	0
Block3_conv1 (Conv2D)	(None, 56, 56, 256)	295168
Block3_conv2 (Conv2D)	(None, 56, 56, 256)	590080
Block3_conv3 (Conv2D)	(None, 56, 56,256)	590080
Block3_pool (MaxPooling2D)	(None, 28, 28, 256)	0
Block4_conv1 (Conv2D)	(None, 28, 28,512)	1180160
Block4_conv2 (Conv2D)	(None, 28, 28, 512)	2359808
Block4_conv3 (Conv2D)	(None, 28, 28,512)	2359808
Block4_pool (MaxPooling2D)	(None, 14, 14,512)	0
Blocks_conv1 (Conv2D)	(None, 14, 14, 512)	2359808
Block5_conv2 (Conv2D)	(None, 14, 14,512)	2359808
Block5_conv3 (Conv2D)	(None, 14, 14,512)	2359808
Block5_pool (MaxPooling2D)	(None, 7, 7,512)	0
Flatten (Flatten)	(None, 25088)	0
Dense (Dense)	(None, 4)	100356

Total params: 14,815.044 Trainable params: 100.356
Non-trainable params: 14,714.688|

Fig. (4). Model summary of VGG16.

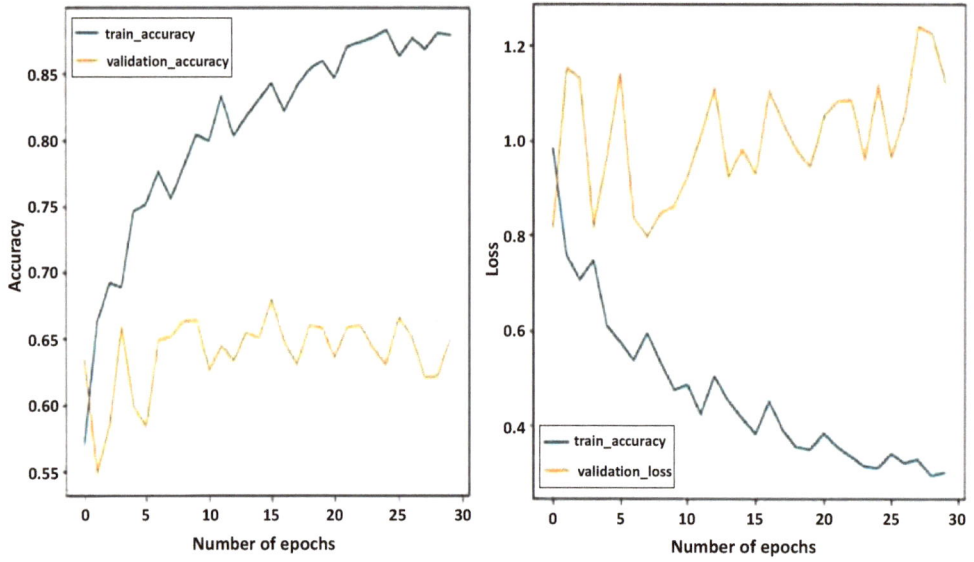

Fig. (5). VGG 16 results.

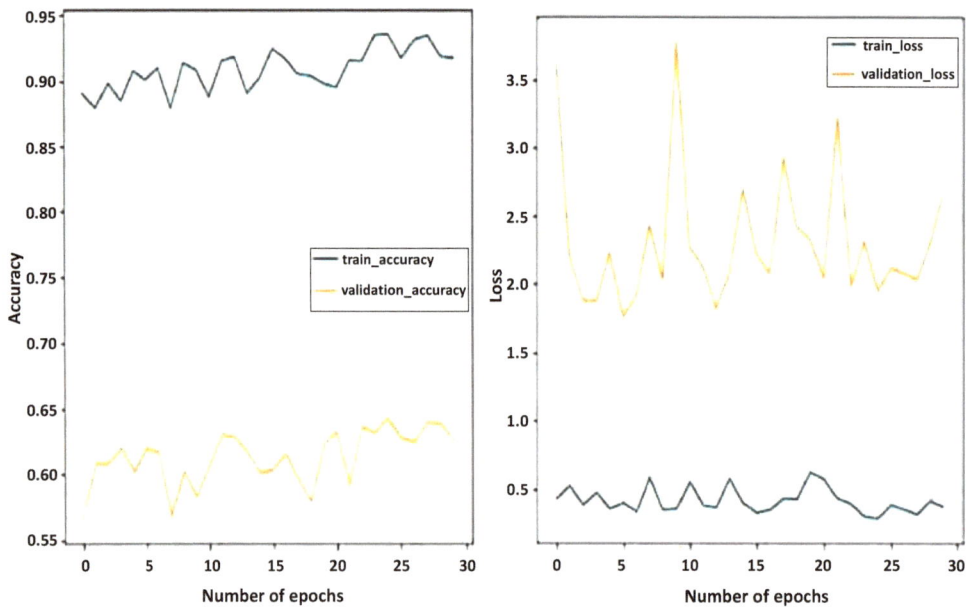

Fig. (6). ResNet 50 results.

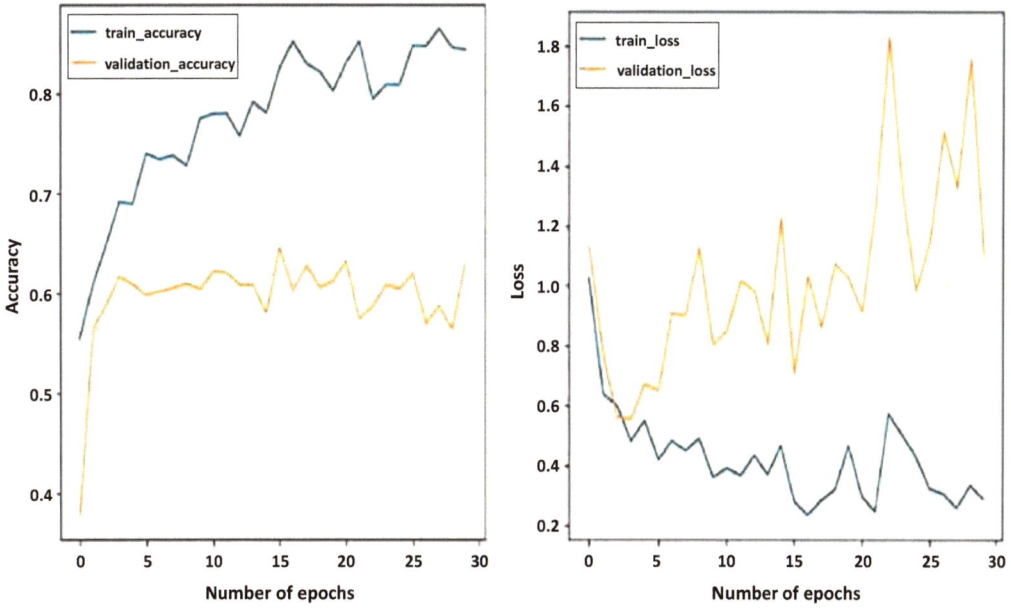

Fig. (7). IncpetionResNet-V2 results.

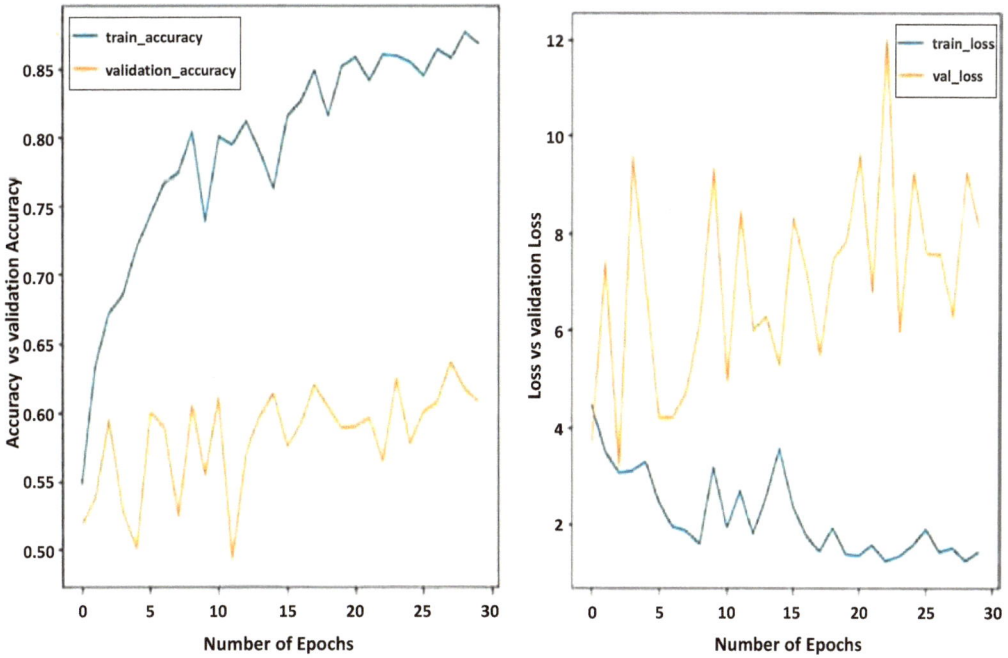

Fig. (8). ResNet 101 results.

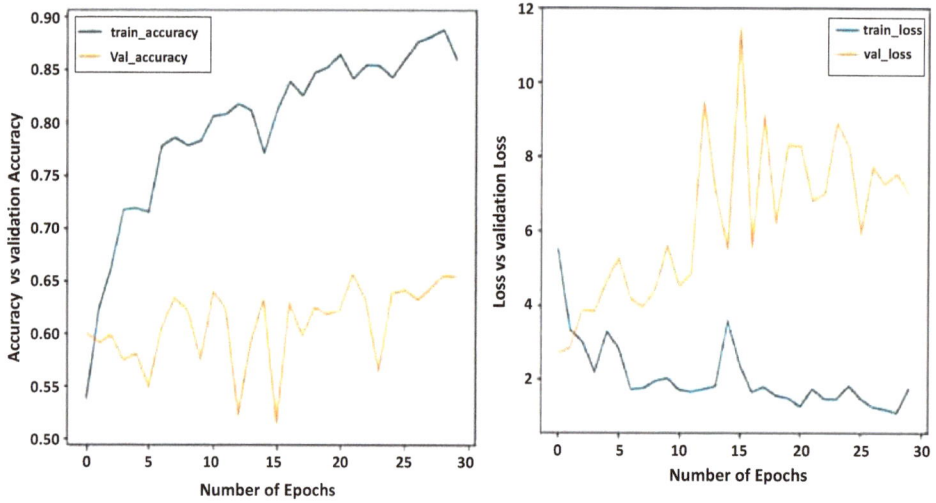

Fig. (9). ResNet 152 results.

Table 1. Validation and Training accuracy and losses for various transfer learning techniques.

Various transfer learning	Training Accuracy (Number of epochs=30)	Validation Accuracy (Number of epochs=30)	Training Loss (Number of epochs=30)	Validation Loss (Number of epochs=30)
VGG16	0.8332±0.0004	0.6661±0.00051	2.941±0.00026	7.843±0.00072
ResNet 50	0.9392±0.0002	0.6372±0.00013	2.871±0.00021	5.618±0.00031
InceptionResNetV2	0.8649±0.0002	0.6325±0.00012	2.587±0.00031	9.823±0.00016
ResNet 101	0.8764±0.0003	0.6317±0.00028	1.240±0.00024	5.570±0.00017
ResNet 152	0.8815±0.0004	0.6518±0.00019	1.244±0.00034	6.953±0.00036

CONCLUSION

Alzheimer's disease (AD) is a severe neurological disorder that damages the brain, reduces cognitive function, and makes it difficult for a person to reason. Specialists in the neurological system often employ a variety of tests to distinguish Alzheimer's disease. There should occasionally be human error. We need some reliable deep-learning models to forecast and classify diseases more accurately. The results of this study compared and looked into how the classification of the AD dataset worked with VGG16, InceptionResNet-V2, Resnet50, Resnet101, and Resnet152 as transfer learning methods. The outcome indicates that, compared to the other models, the Resnet 50 transfer learning model offers a higher accuracy of 93%. Resnet 101 provides a superior loss value at 1% for training and 5% for validation, whereas VGG 16 performs better at 66% for validation accuracy.

The appendix of this chapter is given in the end of this book in the appendix section.

CONFLICT OF INTEREST

All the authors of this article declare that there is no competing interest.

DATASET AVAILABILITY

The following link can be used to access the dataset used in this chapter: https://drive.google.com/drive/folders/1Zb3hgrapFJFoD3DOoV7-cEuX2M_ R1_g3?usp=sharing

CODES AVAILABILITY

We included the Python codes for various transfer learning techniques used in this chapter in the Appendix section for practice purposes.

REFERENCES

[1]   Yong K, Zimmermann N, Crutch S, Rossor M, Harding E. Posterior cortical atrophy, in: World Alzheimer Report 2022 Life after diagnosis: Navigating treatment, care and support, pp. 310-312.

[2]   Jack CR Jr, Holtzman DM. Biomarker modeling of Alzheimer's disease. Neuron 2013; 80(6): 1347-58.
[http://dx.doi.org/10.1016/j.neuron.2013.12.003] [PMID: 24360540]

[3]   Al-Shoukry S, Rassem TH, Makbol NM. Alzheimer's diseases detection by using deep learning algorithms: a mini-review. IEEE Access 2020; 8: 77131-41.
[http://dx.doi.org/10.1109/ACCESS.2020.2989396]

[4]   Jain R, Jain N, Aggarwal A, Hemanth DJ. Convolutional neural network based Alzheimer's disease classification from magnetic resonance brain images. Cogn Syst Res 2019; 57: 147-59.
[http://dx.doi.org/10.1016/j.cogsys.2018.12.015]

[5]   Tufail AB, Ma YK, Zhang QN. Binary classification of Alzheimer's disease using sMRI imaging modality and deep learning. J Digit Imaging 2020; 33(5): 1073-90.
[http://dx.doi.org/10.1007/s10278-019-00265-5] [PMID: 32728983]

[6]   Ebrahimi A, Luo S, Chiong R. Introducing transfer learning to 3D ResNet-18 for Alzheimer's disease detection on MRI images. In2020 35th international conference on image and vision computing New Zealand (IVCNZ) 2020 Nov 25 (pp. 1-6). IEEE.

[7]   Ahmad MF, Akbar S, Hassan SA, Rehman A, Ayesha N. Deep learning approach to diagnose Alzheimer's disease through magnetic resonance images. 2021 International Conference on Innovative Computing (ICIC). 1-6.
[http://dx.doi.org/10.1109/ICIC53490.2021.9693041]

[8]   Li C, Wang Q, Liu X, Hu B. An attention-based CoT-ResNet with channel shuffle mechanism for classification of alzheimer's disease levels. Front Aging Neurosci 2022; 14: 930584.
[http://dx.doi.org/10.3389/fnagi.2022.930584] [PMID: 35898323]

[9]   Liu M, Tang J, Yu W, Jiang N. Attention-Based 3D ResNet for Detection of Alzheimer's Disease Process. In Neural Information Processing: 28th International Conference, ICONIP 2021, Sanur, Bali, Indonesia, December 8–12, 2021. Proceedings 2021; 28(Part I): 342-53. [Springer International Publishing].

[10]   Nguyen D, Nguyen H, Ong H, *et al.* Ensemble learning using traditional machine learning and deep neural network for diagnosis of Alzheimer's disease. IBRO Neuroscience Reports 2022; 13: 255-63. [http://dx.doi.org/10.1016/j.ibneur.2022.08.010] [PMID: 36590098]

[11]   Ramzan F, Khan MUG, Rehmat A, *et al.* A deep learning approach for automated diagnosis and multi-class classification of Alzheimer's disease stages using resting-state fMRI and residual neural networks. J Med Syst 2020; 44(2): 37.
[http://dx.doi.org/10.1007/s10916-019-1475-2] [PMID: 31853655]

[12]   Li W, Lin X, Chen X. Detecting Alzheimer's disease Based on 4D fMRI: An exploration under deep learning framework. Neurocomputing 2020; 388: 280-7.
[http://dx.doi.org/10.1016/j.neucom.2020.01.053]

[13]   Abdelaziz M, Wang T, Elazab A. Alzheimer's disease diagnosis framework from incomplete multimodal data using convolutional neural networks. J Biomed Inform 2021; 121: 103863.
[http://dx.doi.org/10.1016/j.jbi.2021.103863] [PMID: 34229061]

[14]   Buyrukoğlu S. Early detection of Alzheimer's disease using data mining: Comparison of ensemble feature selection approaches. Konya Journal of Engineering Sciences 2021; 9(1): 50-61.
[http://dx.doi.org/10.36306/konjes.731624]

[15]   Shahbaz M, Ali S, Guergachi A, Niazi A, Umer A. Classification of Alzheimer's Disease using Machine Learning Techniques. InData 2019 Jul 26 (pp. 296-303).
[http://dx.doi.org/10.5220/0007949902960303]

[16]   Odusami M, Maskeliūnas R, Damaševičius R. An intelligent system for early recognition of Alzheimer's disease using neuroimaging. Sensors (Basel) 2022; 22(3): 740.
[http://dx.doi.org/10.3390/s22030740] [PMID: 35161486]

[17]   Zaina HS, Brahim Belhaouari S, Stanko T, Gorovoy V. An exemplar pyramid feature extraction based Alzheimer disease classification method. IEEE Access 2022; 10: 66511-21.
[http://dx.doi.org/10.1109/ACCESS.2022.3183185]

[18]   Assmi A, Elhabyb K, Benba A, Jilbab A. Alzheimer's disease classification: a comprehensive study. Multimedia Tools Appl 2024; 1-24.

[19]   Shukla A, Tiwari R, Tiwari S. Alz-ConvNets for classification of Alzheimer disease using transfer learning approach. SN Computer Science 2023; 4(4): 404.
[http://dx.doi.org/10.1007/s42979-023-01853-7]

[20]   Demir A, Yilmaz F, Kose O. Early detection of skin cancer using deep learning architectures: resnet-101 and inception-v3. In2019 medical technologies congress (TIPTEKNO) 2019; 3: (pp. 1-4). IEEE.
[http://dx.doi.org/10.1109/TIPTEKNO47231.2019.8972045]

[21]   Szegedy C, Ioffe S, Vanhoucke V, Alemi A. Inception-v4, inception-resnet and the impact of residual connections on learning. InProceedings of the AAAI conference on artificial intelligence 2017 Feb 12 (Vol. 31, No. 1).
[http://dx.doi.org/10.1609/aaai.v31i1.11231]

Machine Learning in Women's Health: An Insight into the Role of Machine Learning in Skin, Breast, and Ovarian Cancers and PCOS

Sweety Angela Kuldeep[1], Fatema Khusnoor[2], Salma Hashem[1], Tasnim Aktar[2] and Mosae Selvakumar Paulraj[1,*]

[1] *Environmental Sciences Program, Asian University for Women, Chittagong-4000, Bangladesh*

[2] *Social Science Program, Asian University for Women, Chittagong-4000, Bangladesh*

Abstract: Concerning the diagnostic or predictive analysis of medical data, machine learning is currently receiving a significant amount of attention. Artificial intelligence and machine learning already learn and make corrections, and feedback may help them further increase their accuracy. This approach examines structured data to group patient attributes and subsequently forecasts the likelihood that a disease would manifest. Large medical data sets are mined for insights that can be utilized to improve clinical decision-making and patient outcomes, automate daily tasks for healthcare personnel, speed up medical research, and increase operational effectiveness. Even today, many women still struggle with access to basic healthcare facilities. They are biologically more vulnerable to a variety of illnesses. As a result, AI and machine learning suggest a significant improvement in women's health. Several of these machine learning tools target the particular health problems faced by them. Following these methods, this chapter presents an insight into how these algorithms aid in detecting skin cancer, breast cancer, ovarian cancer, and PCOS.

Keywords: Breast cancer, Cancer, Cervical cancer, Machine learning, Ovarian cancer, PCOS, Skin cancer, Women's health.

INTRODUCTION

Cancer affects millions of women globally. Millions of women die each year as a result of family history, hormones, and reproductive variables. Unfortunately, a lot of people die from the seriousness of different types of cancer since doctors cannot detect them until it is too late [1]. However, through the early diagnosis of

---

* **Corresponding author Mosae Selvakumar Paulraj:** Environmental Sciences Program, Asian University for Women, Chittagong-4000, Bangladesh; E-mail: p.selvakumar@auw.edu.bd

diseases, AI, data analytics, and machine learning (ML) are laying a solid foundation for women's health. They essentially enable a computing system to learn by making predictions about the data that is fed, which has evolved powerfully into deep learning algorithms [2]. When it comes to diseases like skin cancer, breast cancer, PCOS, and ovarian cancer, it can improve survival rate, treatment, and diagnosis rates. Given the high expense of medication and the significance of the disease, early detection is the most efficient way to reduce the disease's effects on both health and the economy. Self-testing is uncommon since cancer is frequently found when it has already spread. Recently, machine learning has been extensively utilized in healthcare to forecast various diseases. It is a modeling method that depicts extracting knowledge from data and uncovering hidden associations. Some studies employed simply demographic risk indicators (lifestyle and laboratory data) to predict cancer, while others used data from patient biopsies or other characteristics [3]. Others have shown how genetic information can be used to predict cancer. For better prognostication and stratification of patients toward personalized therapy, ML specifically enables the integration or combination of several layers of data, including those from medical pictures, laboratory results, clinical outcomes, biomarkers, and biological features [4]. Applied computational methodologies and usability issues prevent these prediction models from being widely deployed despite the substantial academic interest in this area of study. As a result, the current work intends to explore and evaluate multiple machine learning techniques for identifying and diagnosing skin, breast, ovarian, and cervical cancer while taking into account diverse modeling parameters and algorithms.

METHODOLOGY

The following algorithms are commonly used in machine learning to detect and predict the discussed types of cancers:

Random Forest

A supervised machine-learning technique is employed for both classification and regression. To obtain a more precise and reliable forecast, it constructs and blends numerous decision trees.

Decision Tree

A decision tree is a supervised learning method used to visually display all potential solutions to a given problem. The classification and regression tree algorithm, or CART algorithm, was utilized.

Support Vector Classifier

Support vector classifiers can split or categorize the data by returning the best fit for the input data. The numbers are nearer to the hyperplane and alter the hyperplane's position and orientation.

Logistic Regression

Logistic regression is a form of supervised learning method used to address classification issues. It uses a set of independent factors to predict the categorical dependent variable using machine learning, and the cost function can only be between 0 and 1.

K Nearest Neighbor

The supervised learning algorithm K Nearest Neighbor, sometimes referred to as the lazy learner algorithm, is used for both classification and regression. Instead of learning the dataset right away, it initially stores it before taking action on it when it comes time to classify.

XGBRF

PCOS is categorized using an ensemble method called XGBoost with Random Forest (XGBRF). A gradient boosting approach is called XGBoost, and a bagging algorithm is called Random Forest. A modified version of the XGBoost classifier is called XGBRF. The usage of XGBRF to avoid the over-fitting issue is a benefit.

CatBoost Classifier

CatBoost is used for regression and classification. It works with a variety of data types, including audio, text, and image data, as well as historical data. This algorithm's method involves converting categorical values into numbers utilizing various statistics on categorical feature combinations and categorical feature combinations with numerical characteristics.

Skin Cancer

Skin cancer refers to abnormal growth in skin cells. It usually occurs due to excess exposure to the sun's ultraviolet rays. There are many types of skin cancer, including basal cell carcinoma (BCC), squamous cell carcinoma (SCC), melanoma, cutaneous t-cell lymphoma, dermatofibrosarcoma protuberans (DFSP), Merkel cell carcinoma (MCC), and sebaceous carcinoma (SC).

Basal cell carcinoma (BCC) is the most common kind of skin cancer. It is mainly detected in people with fair skin. People who are exposed to the sun for a long period or do indoor tanning are usually susceptible to BCC. The symptoms that arise in the case of BCC are pinkish markings and a lump the size of pearls around certain parts of the body. BCC can most commonly be seen in the neck, arms, and head region, but it can also take place in other parts of the body, such as the legs, stomach, and chest area [5]. The second most common skin cancer that people are diagnosed with is squamous cell carcinoma (SCC). People with light skin color are more susceptible to SCC [6]. However, people with darker skin are also diagnosed with SCC. The symptoms of SCC are solid red lumps, itchy patches, and aches that consistently recur in the body. Melanoma is the most dangerous among all kinds of skin cancer. While BCC and SCC stay contained in one part of the body, melanoma skin cancer tends to spread throughout the body [7]. It appears as a mole at first or a dark mark on the skin. Cutaneous T-cell Lymphoma is an unusual blood cancer. It happens because of abnormal growth in the T-cell. Though it is a blood cancer, most of the T-cells reside in the skin; hence, it can be referred to as skin cancer. The symptoms of this type of cancer are itchiness, redness of the skin, and dark red patches around certain parts of the body. Apart from these, there are other types of skin cancer consisting of DFSP, MCC, and SC that transpire due to unusual growth of cells [8].

Women before the age of fifty are more receptive to skin cancer compared to men. In the present time, advanced technology has made it easier and more feasible to detect and diagnose skin cancers at an early stage. Machine learning has made the process even more malleable. Islam *et al.* [9] evaluated the data of skin cancerous people by convolutional neural networks (CNN) using Keras Sequential API. The process of CNN was made up of three steps. The first step was convolution. Most of the computations happened in the convolution part. It produced 2D pictures of the infected area. The second step, pooling, helped detect the near-specific location of the cancer cells. Max pooling functions are among the most widely known pooling functions to proceed with this step. The third segment was to fully connect the preceding and subsequent layers. This segment connected the input and output representation. As the output is a picture, it is a non-linear representation. Hence, to deal with this issue, a non-linearity step was taken immediately after the convolution step. Several forms can be used for non-linear operations, such as Sigmoid, Tanh, ReLU, *etc*. To execute the model, the authors structured it into four layers. The Conv2D is the first layer that can go across the images. It helps minimize the image. The second layer is MaxPool2D, which helps pinpoint the maximum value from the input images to create a pool image. The third layer, Flatten, transforms the image into a one-dimensional array. It allows the input images and output values to connect. The last layer is the dense layer function. It helps to point the connection between all the layers and exclude any inputs from layers outside the Conv2D layer. The authors also used multiple machine learning algorithms in evaluation, such as ResNet50, DenseNet121, and VGG11. By using these machine learning algorithms, the author has detected cancer cells with a 97% accuracy rate [9]. Machine learning has enabled people to detect skin cancer through phone applications. The application examines the skin and forwards it to the physician. It also detects the urgency of the case in forms of benign (Not Serious), malignant (serious) or naevi (Not serious), and melanoma (Serious) [10]. Clinical images of various skin deformations are clicked *via* cell phone cameras to examine and evaluate the data. One of these datasets, SD-198, using the ABCD rule, achieved a 57.62% accuracy rate compared to a 53.35% accuracy rate using deep learning. Fujisawa *et al.* [11] evaluated machine learning algorithms with 4876 images of 14 skin diseases and found a 76.5% accuracy rate, 96.3% sensitivity rate, and 89.5% specificity rate.

Breast Cancer

The number of fatalities is dramatically rising each year due to breast cancer. One of the most prevalent diseases in women, breast cancer is brought on by a variety of clinical, lifestyle, social, and economic circumstances [12]. It is the leading cause of mortality for women. Some breast cells start to develop erratically, which leads to breast cancer. These cells continue to multiply and divide more quickly than healthy cells do, generating a bulk or lump. To reach the lymph nodes or other regions of the human body, cells can spread (metastasize) through the breasts. Any advancement in the detection and treatment of cancer is crucial for maintaining good health. To update the therapy aspect and the patient survival criterion, great accuracy in cancer prognosis is crucial. Machine learning methods are effective and have become a popular area of research. They can significantly contribute to the process of early detection and prediction of breast cancer [13, 14].

Machine-learning approaches relating to laboratory, demographic, and mammographic information can be used to predict breast cancer. For instance, using various ML algorithms, Rabiei *et al.* [3] worked on the database collected from Motamed Cancer Institute, Iran. 25% of records out of 5178 records had patients suffering from breast cancer. After the data was collected, it was preprocessed and balanced using the Synthetic Minority Oversampling Technique (SMOTE). The dataset was then subjected to Random forest (RF), Gradient Boosting trees (GBT), and Multi-layer Perceptron (MLP). Variable values were optimized using a genetic algorithm. Models were initially trained using laboratory and demographic data (20 features). To assess the efficacy of mammography features in predicting breast cancer, the models, as highlighted in Fig. (**1**), were then trained with all demographic, laboratory, and mammographic variables (a total of 24 features). It was noted that, along with other qualities, mammographic features could enhance the performance of models.

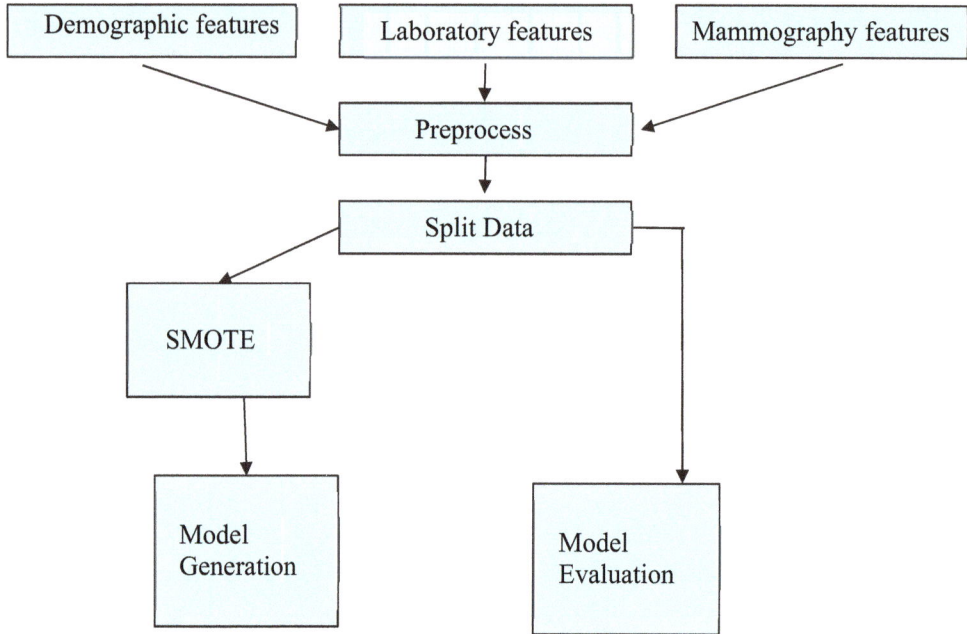

Fig. (1). Block diagram of the methods used in this approach [3].

The RF model had the highest sensitivity (95%); however, gradient-boosting models with better specificity (86%) were more effective due to the sensitivity of breast cancer diagnosis. These machine-learning algorithm approaches can predict breast cancer since timely therapeutic interventions could assist in slowing the progression of the disease and lower mortality rates through early diagnosis of this condition [3]. The performance of modeling can be enhanced by using various machine learning techniques, having access to larger datasets from several institutions (multi-center study), and taking into account important attributes from several pertinent data sources.

Additionally, using Support Vector Machine (SVM), Random Forest, Logistic Regression, Decision tree (C4.5), and K-nearest Neighbors (KNN Network) classifiers in algorithms, one can also contribute to the early diagnosis of breast cancer. Considering data gathering as their first step, Naji *et al.* [15] proceeded to the pre-processing part of their methodology, which entailed four different phases: data cleaning, attribute selection, setting the target role, and feature extraction. Building machine learning algorithms that can forecast breast cancer for a new set of measures requires prepared data. Model fresh data with labels can be displayed to see how well the algorithms function. This is typically accomplished using a training test split method to divide the labeled data that we have collected into two sections. The machine learning model using the above-mentioned classifiers deduced 75% of the data, also known as the training data or training set. 25% of the data, often known as test data or test set, was utilized to evaluate how well the model performed [15]. After putting the models to the test, the findings were compared to choose the algorithm that offered the highest level of accuracy and determine the algorithm that was most likely to predict the presence of breast cancer. Out of the four algorithms, SVM was used to classify the dataset to determine the closest data points. At training time, RF built a large number of decision trees and obtained their mean prediction. Utilizing other label points, the KNN algorithm was employed to find label points. The likelihood of an illness or health condition was evaluated using logistic regression as a function of a risk factor. In many instances, the data set was divided into multiple circumstances using the decision tree modeling tool. The Support Vector Machine outperformed all other algorithms, and it had a higher efficiency of 97.2%, Precision of 97.5%, and AUC of 96.6%. It showed an effective prediction and breast cancer diagnosis, and it delivered the greatest results in terms of accuracy and precision.

Despite significant advancements in early detection, screening, and patient care, breast cancer remains the most common disease in women. There are no malignant breast lesions; all breast lesions are benign. However, the accuracy of the diagnosis can be increased by combining preoperative testing such as mammography, physical examination, fine-needle aspiration cytology, and core needle biopsy. Machine learning techniques can be applied, particularly for the fine-needle aspiration of a breast mass. Yedjou *et al.* [4] calculated 10 real-value metrics for each cell nucleus, including its area, smoothness, concavity, symmetry, and perimeter. Additionally, the geometrical and textural characteristics of the most accurate biopsy core were taken into account and computed. From a digital image of a fine needle aspirate (FNA) of a breast mass, accurate analyses of the geometrical features and textural aspects were made. With the aid of a computerized tomography (CT) scan or ultrasound monitor, a tiny needle is injected into a region of aberrant tissue or cells during an FNA operation. These characteristics, which represent the simplest aspects of breast cancer images, are crucial for breast cancer analysis. In the field of medical science and cancer research, understanding the distinction between benign and malignant tumors is crucial. Additionally, having this knowledge can aid medical professionals in determining the most effective course of action for managing and treating disease, particularly breast cancer. According to the results produced using various feature values, 63% of the 569 patients diagnosed with breast cancer were benign, and 37% were malignant [4]. When compared to their comparable features in malignant tumors (cancerous), we discovered that the mean value for benign tumors (non-cancerous) is lower, which suggests that malignant tumors have spread to other parts of the body. To perform predictive analyses of individuals and their medical diagnoses, various classifier algorithms can be applied to medical information.

Breast cancer can be caused by several variables, including hormones, reproductive factors, and family history. One million women receive a breast cancer diagnosis for the first time each year. Despite the paucity of information regarding the causes and therapies of breast cancer, the theory contends that all cancers result from uncontrolled cell development. Experts can more easily identify certain diseases thanks to automated techniques, which also facilitate early detection. The MLP, KNN, GP, and RF algorithms were employed by Bhardwaj *et al.* [16] to classify breast cancer. According to the outcomes, the RF classifier performed better than the MLP, KNN, and GP classifiers. For instance, the RF classifier's minimum, average, and maximum classification accuracy for a 10-fold partition were 94.32%, 95.54%, and 96.24%, respectively, and the classifiers' performance metrics included sensitivity, accuracy, and specificity [16]. From a variety of inputs, the MLP produces a set of outputs. A directed graph connecting the input and output layers is made up of many layers of input nodes. To compare components in the k-nearest neighbors, the KNN was utilized. The distance between the neighbors and the newly added items for categorization was used to further weigh the neighbors. Genetic programming was useful for assessing the effectiveness of features and figuring out whether traits could withstand evolution. Using a random feature selection method, Random Forest gathered decision trees. Large-scale datasets improved in accuracy and precision thanks to their effectiveness.

Ovarian Cancer

Ovarian cancer mainly takes place in female reproductive organs. The ovaries or fallopian tubes are disturbed from their function, and when the cells become visible or noticeable more than required, it is called ovarian cancer. The outgrowing level of tumor is diverse, along with complications. Epithelial cancer forms mainly ovarian cancers. Ovarian cancer is one of the most common reasons for the death of women. Family history has a great influence on it, whereas the common genes BRAC1 and BRAC2 are responsible for ovarian cancer [17]. Research on ovarian cancer finds out that sexually transmitted infections raise the chance for benign ovarian tumors and mucinous ovarian cancer. In the fallopian tube, inflammatory damage can cause ovarian cancer. Furthermore, trachomatis salpingitis can cause this cancer. Much research has been done to prevent ovarian cancer. It found out that βC caryophyllene plant oil works on cell cycle resistance as well as controls part of an organism's growth in ovarian cancer. For that, it could be considered an anti-cancer agent [18]. Many clinical trials have been conducted to cure these cancer cells. Infertility and ovarian cancer are correlated with breast cancer. The delay in the diagnosis of ovarian cancer causes an extensive impact on women's bodies [19].

In the field of science, medical applications based on artificial intelligence have been found to be widespread. The developed computing power of machines has been used in diagnostics and clinical medicine analysis. Currently, with advances in computing and algorithms qualifying machine learning, the focus is mainly on the deep learning intersections that simulate the human brain in functional form. Ovarian cancer is one of the most common cancers in women. Till the current time, there are no medical therapies available to get a proper cure for this fatal disease. However, the primary stage of detection can disseminate the life expectancy of a cancer patient. Applying machine learning models for early diagnosis will help cancer patients.

In ovarian cancer, the human immune system, by the confines of the PD-1/PD-L1 axis, has proven to be a positive clinical trial [20]. With the assistance of HLA chemotherapy- resistance, the cancer cells can be regulated. If it is possible to be screened at a primary stage, the risk of ovarian cancer can be reduced in some portions. Hormonal factors are also involved in the cause of ovarian cancer [21].

To categorize individual cells from ovarian cancer across a machine learning basis, a connection of 2D light scattering anisotropy cytometry is used. In the support vector machine results, the machine learning algorithms are supported to delimit the model of performance for the alignment tasks. Along with that, artificial neural networks, K-nearest neighbors, Decision Trees, and image processing methods are used to analyze ovarian cancer [21]. To forecast the risk factors of ovarian cancer, numerous machine-learning approaches have been displayed. In addition, machine learning and statistical approaches contribute to the detection of the gene expression pattern of ovarian cancer. To predict and classify the risk factors of ovarian cancer, two models of machine learning have been frequently used. Those two models are the merger of a Convolution Neural Network and Relief [22]. CNN is network tectonics for deep learning. It directly acquires data and extracts the requirement for manual feature extraction. Moreover, it is effective for finding patterns in images that acknowledge the objects, scenes, and faces. The antlion-optimized convolutional neural network model is constructed to function in histopathology reflections. To generate it for better performance, the topology of each customized characteristic network is set by the antlion optimization algorithm itself. Depending on weighted linear aggregation, the result from the two prosper networks are fused. These deep fused functions are ultimately used to predict ovarian cancer stages.

In the machine learning features, there are two approaches: traditional machine learning and deep machine learning. The traditional machine learning methods mainly depend on humans to identify the characteristics to extract. On the other hand, deep machine learning convolutional neural networks recognize the significant features of the image without human input. It also knows how to perform alignment-wise during the training process. Deep learning models can interpret medical images to execute diagnosis. Furthermore, these algorithms are efficient in uncovering exceptions and risks in medical images. It is widely used in detecting cancers [23].

Ovarian cancer is mostly responsible for women's higher mortality rate. Deep learning algorithms, as shown in Fig. (**2**), are a useful component for declining the rate of these dangerous diseases. However, most of the state of deep learning models involve individual impartiality data. This can cause a low level of production because of the inadequate representation of the significant ovarian cancer characteristics. In addition, deep learning models are insufficient for the optimization of model building. This needs a high analytical cost to instruct and establish. In this process approached by Ghoniem, R. *et al*. [24], multi-modal data are used and suggested by the hybrid evolutionary deep learning model. The accepted multi-modal fusion framework fuses gene procedure with the histopathological image method. Depending on the different states and forms of every process, the author established a deep feature extraction network. This process involves a prognostic antlion-optimized long and short-term evocation model to operate gene lengthwise data. One more prognostic antlion-optimized convolutional neural network is involved in the process for pathology images. The geography of this specifically made feature network is accordingly placed by the antlion optimization algorithm for getting a better result. Based on the weighted linear aggregation, the upgraded networks are combined. These deep combined characteristics are used to anticipate the stage of ovarian cancer. The result of this study finds that the suggested model is accurate and error-free in diagnosing ovarian cancer along with other cancers.

For ovarian cancer, ultrasound is a censorious, non-invasive test for functional diagnosis. Deep learning is working to make an advanced effect on this. It is sharing the image recognition tasks. Therefore, Gao, Y. *et al*. [25] tried to grow a deep convolutional neural network (DCNN) model. This will motorize the assessment of ultrasound images along with making possible an error-free diagnostic of ovarian cancer apart from the contemporary method. In this multicenter study, from September 2003 to May 2019, the author collected pelvic ultrasound representations from ten different hospitals across China.

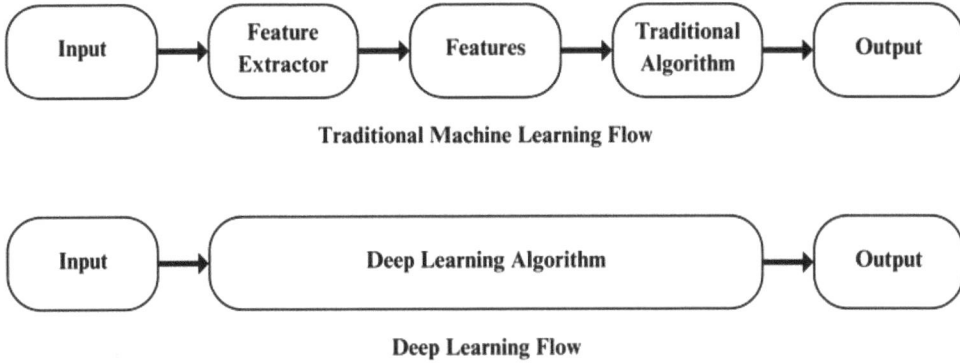

Traditional Machine Learning Flow

Deep Learning Flow

Fig. (2). Deep learning flow.

The author considered patients who are above 18 years old with adnexal noun injury in ultrasonography with healthy controls as well as patients who were out of compulsive diagnosis. In the DCNN model development, the datasets were from 34,488 images of 3,755 patients who were sufferers of ovarian cancer. Of 5,41,442 images, 101,777 were controls. To verify the model, the patients were allocated to the validation datasets. Where 3032 images of 266 were ovarian cancer patients, other 5385 images of 602 were patients with benign adnexal lesions. By using this dataset, the authors evaluated the diagnostic practicality of DCNN. The authors compared DCNNs with radiologists and tried to find out whether DCNNs are capable enough to augment diagnostic accuracy. They referred to the pathological diagnosis. In the outcome, it was observed that the DCNN model is more well-aimed than the radiologists at identifying ovarian cancer in the internal and external datasets.

Akazawa *et al.* [26] tried to utilize artificial intelligence to predict the clinical diagnosis of ovarian tumors. For that, the authors operated data from surgical examinations and patient information. To examine the method, 202 patients suffering from ovarian tumors were enrolled. Of these, 126 were benign ovarian tumor patients, 23 were borderline malignant tumor patients, and 53 were ovarian cancer patients. To examine the study, authors used 5 machine learning algorithms: support vector machines, naive Bayes, Logistic regression, Random Forest, and XGBoost, as shown in Fig. (**3**).

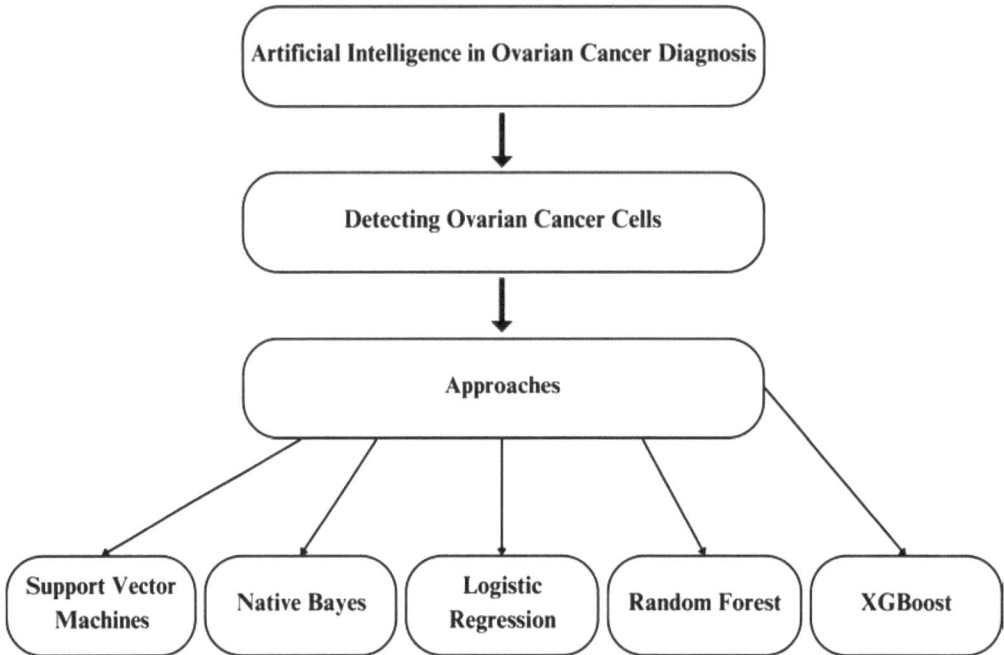

Fig. (3). Detecting ovarian cancer cells [26].

They tried to obtain the result with the help of 16 features, which were obtainable from patients' backgrounds, imaging tests, and further blood tests. In this region, they tried to examine the 16 features to predict the disease. In the result section, they found that the highest accuracy was in the XGBoost algorithm. Different results have been observed among the correlation coefficients. This work has provided evidence that artificial intelligence plays a huge role in the prediction of clinical diagnosis of ovarian cancer.

PCOS and Cervical Cancer

Technology has transformed our world and changed our lives over the last few decades, making them easier daily. There are numerous ways in which new technology is influencing humanity. These days, the healthcare industry is heavily reliant on machine learning, a field of research that enables computers to learn without being explicitly programmed. Machine learning can handle absurdly large datasets, transform analyzed data into clinical insights, and assist in the diagnosis of a variety of diseases. A medical illness called polycystic ovary syndrome (PCOS) affects hormones and affects women who are of reproductive age [27]. Hormonal imbalances are the cause of PCOS. Women with PCOS typically experience difficulties getting pregnant because the ovaries form tiny clusters of fluid called follicles (cysts) and fail to release eggs in this illness. Many women have PCOS; however, they are not first diagnosed with it. Studies claim that PCOS is typically inherited, although its exact causes are still unknown [27]. Since there is no discernible trend for this medical illness, it is unexpected and has no known remedy. Both patients and doctors struggle with the time and expense of undergoing several medical exams and scans. Early diagnosis and treatment are crucial since simple lifestyle modifications can prevent long-term health problems like type 2 diabetes and cardiovascular diseases. Period irregularities, high levels of androgen (male hormones), and polycystic ovaries are common signs. Several machine learning classifiers can be used to determine whether a woman has PCOS, including Random Forest, SVM, Logistic Regression, Gaussian Naive Bayes, K Neighbors, XGBRF, CatBoost Classifier, and so on [28].

Data preparation should come before a machine learning model is developed. The data is prepared for processing by the algorithms once it has been thoroughly cleaned and chosen. The Random Forest, Decision Tree, Support Vector Classifier, Logistic Regression, K Nearest Neighbors, XGBRF, and CatBoost Classifier techniques were used to build the model. The models can aid in the early detection of PCOS. The CatBoost Classifier has been determined to perform better than other algorithms. Doctors can use this model to identify individuals who are likely to develop this condition early on and to diagnose them.

The fourth most common cause of death in women worldwide is cervical malignant development. The prevalence of the human papillomavirus (HPV) is linked to the growth of cervical cancer [29]. The global burden of cervical cancer has been reduced as a result of early screening, making the illness avoidable. Women do not participate in enough screening programs in underdeveloped nations due to the high costs associated with routine examinations, a lack of awareness, and restricted access to medical facilities. In this way, a very high level of risk is anticipated for each patient. Numerous risk factors can lead to the development of cervical cancer. Machine learning plays an important role in predicting cervical cancer. The techniques of the decision tree, logistic regression, support vector machine (SVM), K-nearest neighbors (KNN), adaptive boosting, gradient boosting, random forest, and XGBoost can be used to identify cervical cancer.

A method known as CervDetect assessed the risk factors of malignant cervical development using machine learning algorithms. The data was pre-processed by CervDetect using Pearson correlation between the input variables and the output variable. CervDetect chose important features using the random forest (RF) feature selection method. To identify cervical cancer, CervDetect employed a hybrid strategy that combined RF and shallow neural networks [30].

SCOPES AND CHALLENGES

Although machine learning algorithms have appeared with favorable outcomes for the diagnosis of various cancers, their applicability, as well as generalizability, remain difficult to track down. Data class imbalance, dataset scale, and multitask classification are some key circulations that are faced by numerous intelligent classification algorithms in the diagnosis of cancer. It should be noted that all of the results referring to various previously-approached methods are only related to a single type of database, which can be viewed as a limitation. As a result, it is imperative to consider applying the same algorithms and methods to other databases in future works to confirm the results with databases of other machine learning algorithms using new parameters on larger data sets with more disease classes to achieve higher accuracy. Table **1** below summarizes the various approaches taken with the help of machine learning for the discussed health diagnosis.

Table 1. Various machine learning approaches for health diagnosis.

Type of Ailment	Algorithm Used	Key Features	Performance Data
Skin Cancer	CNN, ResNet50, DenseNet121, and VGG11	Pooling, 2-D images, and non-linearity operations	CNN showed 97% accuracy
Breast Cancer	RF, GBT, MLP, SVM, Logistic Regression, KNN, *etc.*	Laboratory, demographics, and mammographic data	RF and SVM showed 95% sensitivity and 97.2% efficiency respectively
Ovarian Cancer	SVM, ANN, KNN, Decision Trees, and CNN	Various gene expression patterns and clinical features	DCNN and XGBoost models perform better and have a higher accuracy
PCOS	RF, SVM, Logistic Regression, Gaussian Naive Bayes, and CatBoost Classifier	Irregular periods, androgen levels, and ovarian cysts	CatBoost classifier performed the best
Cervical Cancer	Decision Tree, Logistic Regression, SVM, KNN, RF, and XGBoost	Prevalence of HPV and other lifestyle-related factors	CervDetect used a hybrid model to assess the risk factors using RF and shallow neural networks

CONCLUSION

It is critical to create cutting-edge methods for screening, diagnosing, and treating skin cancers, breast cancers, ovarian cancers, and PCOS because these are among the major causes of death for women worldwide. The prediction, diagnosis, and categorization of breast cancer using ML algorithms are discussed in this work.

These machine-learning algorithms may be able to anticipate cancer-related illnesses because their early diagnosis may aid in slowing the progression of the sickness and lowering the death rate through timely treatment interventions. The modeling performance can be enhanced by using various machine learning techniques, having access to larger datasets from several institutions (multi-center study), and considering the important attributes from several pertinent data sources.

REFERENCES

[1]   Al-Azri MH. Delay in cancer diagnosis: causes and possible solutions. Oman Med J 2016; 31(5): 325-6.
[http://dx.doi.org/10.5001/omj.2016.65] [PMID: 27602184]

[2]   Zhang B, Shi H, Wang H. Machine learning and AI in cancer prognosis, prediction, and treatment selection: A critical approach. J Multidiscip Healthc 2023; 16: 1779-91.
[http://dx.doi.org/10.2147/JMDH.S410301] [PMID: 37398894]

[3]   Rabiei R, Ayyoubzadeh SM, Sohrabei S, Esmaeili M, Atashi A. Prediction of breast cancer using machine learning approaches. J Biomed Phys Eng 2022; 12(3): 297-308.
[http://dx.doi.org/10.31661/jbpe.v0i0.2109-1403] [PMID: 35698545]

[4]   Yedjou CG, Tchounwou SS, Aló RA, Elhag R, Mochona B, Latinwo L. Application of machine learning algorithms in breast cancer diagnosis and classification. International Journal of Science academic research. 2021 Jan; 2(1): 3081.

[5]   Al-Qarqaz F, Marji M, Bodoor K, Almomani R, Al Gargaz W, Alshiyab D, Muhaidat J, Alqudah M. Clinical and demographic features of basal cell carcinoma in North Jordan. Journal of skin cancer. 2018.
[http://dx.doi.org/10.1155/2018/2624054]

[6]   Howell JY, Ramsey ML. Squamous cell skin cancer. In StatPearls [Internet] 2022 Aug 1. StatPearls Publishing.

[7]   D'Orazio J, Jarrett S, Amaro-Ortiz A, Scott T. UV radiation and the skin. Int J Mol Sci 2013; 14(6): 12222-48.
[http://dx.doi.org/10.3390/ijms140612222] [PMID: 23749111]

[8]   Laikova KV, Oberemok VV, Krasnodubets AM, *et al.* Advances in the understanding of skin cancer: ultraviolet radiation, mutations, and antisense oligonucleotides as anticancer drugs. Molecules 2019; 24(8): 1516.
[http://dx.doi.org/10.3390/molecules24081516] [PMID: 30999681]

[9]   Islam A, Khan D, Chowdhury RA. An efficient deep learning approach to detect skin Cancer (Doctoral dissertation, Brac University).

[10]   Das K, Cockerell CJ, Patil A, *et al.* Machine learning and its application in skin cancer. Int J Environ Res Public Health 2021; 18(24): 13409.
[http://dx.doi.org/10.3390/ijerph182413409] [PMID: 34949015]

[11]   Fujisawa Y, Otomo Y, Ogata Y, *et al.* Deep-learning-based, computer-aided classifier developed with a small dataset of clinical images surpasses board-certified dermatologists in skin tumour diagnosis. Br J Dermatol 2019; 180(2): 373-81.
[http://dx.doi.org/10.1111/bjd.16924] [PMID: 29953582]

[12]   Momenimovahed Z, Salehiniya H. Epidemiological characteristics of and risk factors for breast cancer in the world. Breast Cancer: Targets and Therapy. 2019 Apr 10:151-64.
[http://dx.doi.org/10.2147/BCTT.S176070]

[13]   Khalid A, Mehmood A, Alabrah A, *et al.* Breast cancer detection and prevention using machine learning. Diagnostics (Basel) 2023; 13(19): 3113.
[http://dx.doi.org/10.3390/diagnostics13193113] [PMID: 37835856]

[14]   Yue W, Wang Z, Chen H, Payne A, Liu X. Machine learning with applications in breast cancer diagnosis and prognosis. Designs 2018; 2(2): 13.
[http://dx.doi.org/10.3390/designs2020013]

[15]   Naji MA, Filali SE, Aarika K, Benlahmar ELH, Abdelouhahid RA, Debauche O. Machine learning algorithms for breast cancer prediction and diagnosis. Procedia Comput Sci 2021; 191: 487-92.
[http://dx.doi.org/10.1016/j.procs.2021.07.062]

[16]   Bhardwaj A, Bhardwaj H, Sakalle A, Uddin Z, Sakalle M, Ibrahim W. Tree-based and machine learning algorithm analysis for breast cancer classification. Comput Intell Neurosci 2022; 2022: 1-6.
[http://dx.doi.org/10.1155/2022/6715406] [PMID: 35845866]

[17]   Ramus SJ, Gayther SA. The contribution of BRCA1 and BRCA2 to ovarian cancer. Mol Oncol 2009; 3(2): 138-50.
[http://dx.doi.org/10.1016/j.molonc.2009.02.001] [PMID: 19383375]

[18]   Arul S, Rajagopalan H, Ravi J, Dayalan H. Beta-caryophyllene suppresses ovarian cancer proliferation by inducing cell cycle arrest and apoptosis. Anti-Cancer Agents in Medicinal Chemistry (Formerly Current Medicinal Chemistry-Anti-Cancer Agents). 2020 Sep 1;20(13):1530-7.
[http://dx.doi.org/10.2174/1871520620666200227093216]

[19]   Vela-Vallespín C, Medina-Perucha L, Jacques-Aviñó C, *et al.* Women's experiences along the ovarian cancer diagnostic pathway in Catalonia: A qualitative study. Health Expect 2023; 26(1): 476-87.
[http://dx.doi.org/10.1111/hex.13681] [PMID: 36447409]

[20]   Dumitru A, Dobrica EC, Croitoru A, Cretoiu SM, Gaspar BS. Focus on PD-1/PD-L1 as a therapeutic target in ovarian cancer. Int J Mol Sci 2022; 23(20): 12067.
[http://dx.doi.org/10.3390/ijms232012067] [PMID: 36292922]

[21]   Hossain MA, Saiful Islam SM, Quinn JMW, Huq F, Moni MA. Machine learning and bioinformatics models to identify gene expression patterns of ovarian cancer associated with disease progression and mortality. J Biomed Inform 2019; 100: 103313.
[http://dx.doi.org/10.1016/j.jbi.2019.103313] [PMID: 31655274]

[22]   Ziyambe B, Yahya A, Mushiri T, *et al.* A deep learning framework for the prediction and diagnosis of oCancer in pre- and post-menopausal women. Diagnostics (Basel) 2023; 13(10): 1703.
[http://dx.doi.org/10.3390/diagnostics13101703] [PMID: 37238188]

[23]   Thilakarathne H. Deep learning *vs.* traditional computer vision. NaadiSpeaks blog, Aug. 2018;12.

[24]   Ghoniem RM, Algarni AD, Refky B, Ewees AA. Multi-modal evolutionary deep learning model for ovarian cancer diagnosis. Symmetry (Basel) 2021; 13(4): 643.
[http://dx.doi.org/10.3390/sym13040643]

[25]   Gao Y, Zeng S, Xu X, *et al.* Deep learning-enabled pelvic ultrasound images for accurate diagnosis of ovarian cancer in China: a retrospective, multicentre, diagnostic study. Lancet Digit Health 2022; 4(3): e179-87.
[http://dx.doi.org/10.1016/S2589-7500(21)00278-8] [PMID: 35216752]

[26]   Akazawa M, Hashimoto K. Artificial intelligence in ovarian cancer diagnosis. Anticancer Res 2020; 40(8): 4795-800.
[http://dx.doi.org/10.21873/anticanres.14482] [PMID: 32727807]

[27]   Dennett CC, Simon J. The role of polycystic ovary syndrome in reproductive and metabolic health: overview and approaches for treatment. Diabetes Spectr 2015; 28(2): 116-20.
[http://dx.doi.org/10.2337/diaspect.28.2.116] [PMID: 25987810]

[28]   Bharati S, Podder P, Mondal MR. Diagnosis of polycystic ovary syndrome using machine learning

algorithms. In2020 IEEE region 10 symposium (TENSYMP) 2020 Jun 5; pp. 1486-1489.
[http://dx.doi.org/10.1109/TENSYMP50017.2020.9230932]

[29]　Pimple S, Mishra G. Cancer cervix: Epidemiology and disease burden. Cytojournal 2022; 19: 21.
[http://dx.doi.org/10.25259/CMAS_03_02_2021] [PMID: 35510109]

[30]　Mehmood M, Rizwan M, Gregus ml M, Abbas S. Gregus ml M, Abbas S. Machine learning assisted cervical cancer detection. Front Public Health 2021; 9: 788376.
[http://dx.doi.org/10.3389/fpubh.2021.788376] [PMID: 35004588]

## CHAPTER 18

An Insight into the Mathematical Modeling of Physiological Systems

Suvendu Ghosh[1], Sonia Mondal[2], Partha Sarathi Singha[3] and Debosree Ghosh[4,*]

[1] *Department of Physiology, Hooghly Mohsin College, Chinsura, Hooghly, Pin 712101, West Bengal, India*

[2] *Department of Mathematics, Government General Degree College, Kharagpur II, Paschim Medinipur, Pin 721149, West Bengal, India*

[3] *Department of Chemistry, Government General Degree College, Kharagpur II, Paschim Medinipur, Pin 721149, West Bengal, India*

[4] *Department of Physiology, Government General Degree College, Kharagpur II, Paschim Medinipur, Pin 721149, West Bengal, India*

Abstract: Mathematics is extensively used in designing physiological modeling. There exists a long and rich history of mathematical modeling in physiology. Mathematical modeling refers to creating a mathematical representation of a real-life condition. Physiological modeling is creating a physiological system's representation in mathematical form. Mathematical models for many aspects of human pathology and physiology have been produced in recent decades. Understanding the connections between the parts of a complicated system may be accomplished with the use of mathematical models. In the biological context, mathematical models aid in our understanding of the intricate web of relationships among the various components (signaling molecules, DNA, enzymes, proteins, *etc.*) in a biological system. This improved interpretation allows us to predict and understand the behavior of the system in a diseased state. Understanding of several intricate biological systems, including metabolic networks, gene regulatory networks, enzyme kinetics, signal transduction pathways, and electrophysiology, has improved because of mathematical modeling. The study of biological systems has grown even more reliant on computational approaches and mathematical modeling as a result of recent developments in high throughput data production techniques.

Keywords: Enzymes, Mathematics, Physiological modeling, Proteins, Signaling molecules.

* **Corresponding author Debosree Ghosh:** Department of Physiology, Government General Degree College, Kharagpur II, Paschim Medinipur, Pin 721149, West Bengal, India; Tel: +919830320757; E-mail: ghoshdebosree@gmail.com

Sivakumar Rajagopal, Prakasam P., Konguvel E., Shamala Subramaniam, Ali Safaa Sadiq Al Shakarchi & B. Prabadevi (Eds.)

INTRODUCTION

In physiology, mathematical modeling has a long and illustrious history [1]. It is helpful to quickly explain the general modeling technique before describing how models have advanced our understanding. Mathematical modeling starts with a well-formulated hypothesis based on prior findings, like experimental research. There are broadly four different types of mathematical modeling (Fig. **1**). The mathematical model is a quantitative depiction of the main idea. For instance, Otto Frank developed a mathematical model of the arterial pulse in the late nineteenth century [2]. Over the years, the same types of mathematical techniques to comprehend the mechanical features and characteristics of the circulatory system have persisted, as Bunberg and colleagues recently reviewed [3]. Hodgkin and Huxley's foundational study on the propagation and development of neural action potentials was documented around the middle of the last century [4], from which cardiac electrophysiology models quickly developed and spread [5]. Mathematical modeling in physiology quickly transitioned from analytical techniques to computational applications of governing equations and simulation to utilize the newly emerging capability of the first analog and digital computers. The problems of larger size may be addressed and examined due to this progress. For instance, Arthur Guyton and his collaborators created an elaborate representation of the balance of fluid and electrolyte in the late 1960s that is still significant and relevant today due to the range of physiology it covers [5].

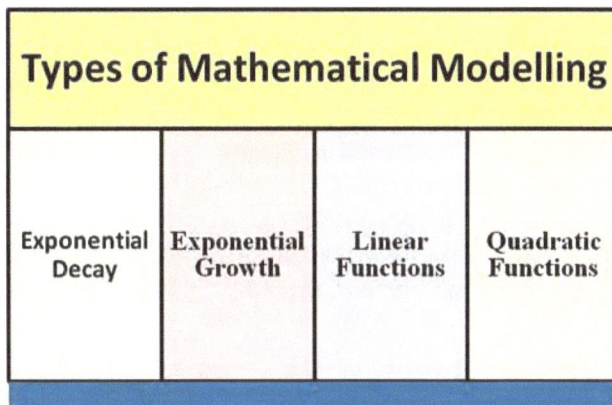

Fig. (1). Types of mathematical models.

Since the period of original research by Guyton, physiological modeling has moved from specialized and frequently single-purpose computers to the regular computer of the researcher. It is possible to assemble small-scale computer clusters at an affordable cost. This is due to the extensive availability of high-performance computing power, which is relatively cost-effective and has a high storage capacity. Numerous quantities of biomedical, biological, and even clinical data may be gathered and saved as a part of specialized research initiatives or during clinical patient treatment, thanks to technical developments in computer power and digital storage media. The need to simultaneously link observed data stream characteristics technically to the system and its properties under study, and potentially in real-time, as often needed by some clinical applications [6], is even more urgent. With the help of this, the vast amount of biomedical data is converted to give a better interpretation of the biological systems themselves is possible. It links the mathematical, mechanistic, and computational modeling of biological systems at all lengths, breadths, and time scales of physiological systems of the human body, as revealed by the Physiome project [1, 7, 8].

Mechanistic Mathematical Physiological System Models

Mechanistic mathematical models represent the knowledge we now have about the functional relationships that control the entire behavior of the system that is under study. By putting our understanding of physiology into the context of dynamical systems (stochastic or deterministic), we make it possible to make exact quantitative predictions and compare them to the outcomes of carefully selected trials. Mechanistic mathematical models frequently enable us to investigate a system in much greater depth than is feasible in experimental research and can thus help to determine the reason behind any specific discovery [6]. Mathematical models and experiments work very well when integrated into a scientific program, as the presence of one increases the value of the other. In addition to illuminating experimental findings, allowing for discrimination between competing scientific hypotheses, and aiding in experimental design, models depend on experiments for defining and improving parameter values [6]. The application of mathematical models sheds light on the molecular and ionic processes underlying inherited and acquired diseases. Dr. Yoram Rudy's lab's work has been essential to this attempt. Notably, the Luo-Rudy dynamic model and its variations continue to rank among the most often mentioned models of cardiac action potentials and are frequently used to explore the fundamentals of cardiac electrophysiology [9 - 13]. Additionally, investigations utilizing these models have shown the effectiveness of computational methods in producing fresh mechanistic insights into cardiac arrhythmia [14 - 17].

Mathematical Modeling of the Reproductive System

Men and women differ significantly in many organ systems, including the architecture of the brain, the functioning of the immunological and stress systems, and the metabolic and cardiovascular systems. They also differ in the reproductive system and reproductive behaviors [18]. The creation of successful sex-based treatments depends on a thorough knowledge of how these sex variations affect health and disease. A renewed focus on computational physiology will be necessary to enable the integration and analysis of enormous amounts of life-science data. It is because of the recent explosion of biological, biomedical, and clinical data at all levels of organization, from cellular to organismic. The knowledge of sex variations in health and illness has the potential to be facilitated and advanced by mathematical modeling. Many people are also aware of the disparities in physical composition between the sexes: men usually have proportionally higher bone mass, muscular mass, and body fat percentage than women.

The structural/morphological variations between adult males and females for most (if not all) organ systems are underappreciated yet can have a significant influence on physiological function. The condition of the differential impact of genders on various physiological functions is mathematically represented. The mathematical model helps in research and studies. Mathematical models are the new tools for enriching our knowledge of the reproductive system and simplifying its complexities [19]. Modeling using mathematics also helps us simplify, compare, analyze, and understand the variations in the physiology of the male and female physiological systems [20]. Studies show that the Laplace's equation applied to the event of follicle rupture is beneficial in interpreting the simultaneous impact of several factors that lead to the bursting of the follicle in females. The dynamics of the gamete transfer by peristaltic analysis applied to the beat of the cilia lining the female oviduct is interpreted. The metachronal wave generated in the wall of the vas deferens in males is known to contradict peristalsis and becomes the dominating factor in spermatozoa transport. Similarly, mathematical modeling is applied to understand the biomechanical significance of the forces associated with the mechanics of sperm-egg interactions and fertilization [20]. Von Foerster's equation is used to formulate a fertilization index that accounts for the epididymal reserve of spermatozoa and the number of spermatozoa present in a specific ejaculation [20, 21].

Mathematical Modeling of Developmental Biology

Studies show that mathematical modeling is beneficial to analyze and understand the various complex aspects of developmental biology [22]. The application of generalized Hook's law to get the displacements for specific load conditions at the time of molding of the fetal head is known [20]. The use of mathematics in developmental biology is evolving daily. Mathematical principles help understand the basic principles that drive morphogenesis in the embryo developmental process. Studies report the use of mathematical modeling of gastrulation in the embryo of a chick in order to understand the events of morphogenesis. The study reports the use of mathematical formulations in addressing the problem of the interplay between the dynamics of the gradients of morphogenesis and cellular movements [22]. Various mathematical models have also been reported in studying the developmental cycle of *Dictyostelium discoideum* [22]. Studies show that differential equation models are the best for studying the development of embryos [23]. Segment polarity network in the early development of the embryo of *Drosophila melanogaster* is interpreted using mathematical modeling. Simple models are designed and used to understand the complex process of segment polarity networks [24]. Whole embryo modeling of early segmentation in the Drosophila melanogaster has been used to identify crucial robust and fragile expression domains [24]. Diffusion models are also in extensive use and help in forming hypotheses about the properties of morphogens in embryos of various species [22].

Mathematical Modeling of the Nervous System

The development of neurons is a very complex process. Mathematical modeling and numerical simulation are widely used to interpret the complex processes involved in the initiation, elongation, axon formation, and branching of the neurites [25]. Thus, mathematical modeling is highly beneficial in studying the process of morphological development of neurons. Differential equations are widely applied for understanding various neural events, including the morphological development of the nervous system [25]. Some of the mathematical models that are used in studying neuroscience and in understanding the nervous system are models of neurons and synapses, models of memory, models of sleep and wakefulness, models of disorders and diseases, *etc.*

Mathematical modeling of the neural network has been an interesting topic of research for many years. The model helps us not only to understand and study the electrical activities of the neural network but also the application of these interpretations, which has helped us to develop advanced tools like neurorobotics for better and more efficient exploration of the nervous system. Various mathematical models have been designed for different types of neural disorders. Some of the models mimic a particular neural disease condition and are used to understand the underlying mechanism of the disease. They are also used to design the treatment regimes and therapy for the same [26]. Mathematical models of the nervous system have also been used to address critical conditions of the nervous system and the brain, such as hydrocephalous [27].

Mathematical Modeling of the Endocrine System

Mathematical models of an endocrine system coupled with carefully designed experiments help us to understand in detail the complex processes and endocrine regulations at various levels of the organization. Using novel mathematical models in which the pancreas is considered a network of beta cells, the rhythmic secretion of insulin is studied [28]. Complex regulatory mechanisms underlying various events of stress, metabolism, and reproductive axes have been studied in detail using mathematical modeling. Mathematical modeling has revealed the mechanism of an exhibition of normal ultradian pulsatility by glucocorticoid. The mechanism of how glucocorticoid responds to stressors like inflammation is also revealed using mathematical modeling [29]. Mathematical modeling of the immune regulation by glucocorticoid is used to understand various aspects and mechanisms involved in the immune regulation by glucocorticoid [29]. A possibility of dose-dependent suppression of the anti-tumor immune response by glucocorticoid is revealed by the mathematical modeling study [29]. New strategies for ovarian stimulation are also revealed by using mathematical modeling and simulations [30]. Statistical models have also been designed and are in use to address different outcome criteria of IVF [31].

Mathematical Modeling of the Digestive System

Mathematical modeling is of extensive use in understanding and interpreting the various processes involved in digestion. A mathematical model is used to study the process of digestion and absorption in pigs [32]. A mathematical model is also used to describe the regulation of digestion. The model is based on data acquired from the analysis of more than twelve hundred sources of experimental observations reported on digestion performed on dogs [33]. A mathematical model of digestion of the small intestine is also reported. The model is used to understand and describe the different aspects of digestion. These include primarily the transport of food bolus along the digestive tract and degradation of the food depending on the enzymes, the physical conditions, *etc.* Various numerical computations are used for this purpose [34]. Mathematical modeling helps to understand the mechanism and details of food hydrolysis [35 - 37].

CONCLUSION

To summarize the topic, physiological phenomena have specific inputs, central integrators, and outputs. These throughputs can be computed by mathematical expressions. Based on these expressions, laboratory-based computational physiology models can be devised, having a wide range of applications in clinical and nonclinical studies that provide insight into the complex working mechanisms in *in vivo* systems. These models have fascinating approaches to understanding the various patterns of developmental aspects in humans, sexual dimorphic patterns in brain functioning, metabolism, energy homeostasis, cardiovascular functions, stress, and immune responses. Robotics and artificial intelligence coupled with mathematical modeling may be adopted for better understanding and detailed study of various physiological and pathological conditions [38, 39]. Finally, mathematical modeling thus explores the clues to the mechanisms of development of diseases and pathophysiological conditions in males and females by integrating the copious amount of observed data, simulation, exploration of hypothesis, and parameter estimation. Various physical and mathematical laws like differential equations, Laplace's transformations, and computer-assisted techniques contribute a lot to the development of drugs, therapeutic regimens, and management of diseases, which is a great strategy towards the satisfactory living of mankind in the 21st generation.

ACKNOWLEDGEMENTS

Dr. SG acknowledges the Department of Physiology, Hooghly Mohsin College, Chinsurah, Hooghly, West Bengal, India. Dr . PSS acknowledges the Department of Chemistry, Dr. SM acknowledges the Department of Mathematics, and Dr. DG acknowledges the Department of Physiology of the Government General Degree College, Kharagpur II, West Bengal, India.

REFERENCES

[1]   Bassingthwaighte JB. Strategies for the physiome project. Ann Biomed Eng 2000; 28(8): 1043-58.
      [http://dx.doi.org/10.1114/1.1313771] [PMID: 11144666]

[2]   Frank O. Die Grund form des arteriellen Pulses. Erste Abhandlung. Mathematische Analyse. Z Biol (Münch) 1899; 37: 483-526.

[3]   Brunberg A, Heinke S, Spillner J, Autschbach R, Abel D, Leonhardt S. Modeling and

[4]   Brunberg A, Heinke S, Spillner J, Autschbach R, Abel D, Leonhardt S. Modeling and simulation of the cardiovascular system: a review of applications, methods, and potentials / Modellierung und Simulation des Herz-Kreislauf-Systems: ein Überblick zu Anwendungen, Methoden und Perspektiven. Biomed Tech (Berl) 2009; 54(5): 233-44.
      [http://dx.doi.org/10.1515/BMT.2009.030] [PMID: 19807287]

[5]   Hodgkin AL, Huxley AF. A quantitative description of membrane current and its application to conduction and excitation in nerve. J Physiol 1952; 117(4): 500-44.
      [http://dx.doi.org/10.1113/jphysiol.1952.sp004764] [PMID: 12991237]

[6]   Noble D. The surprising heart: a review of recent progress in cardiac electrophysiology. J Physiol 1984; 353(1): 1-50.
      [http://dx.doi.org/10.1113/jphysiol.1984.sp015320] [PMID: 6090637]

[7]   Heldt T, Verghese G, Long W, Szolovits P, Mark R. Integrating data, models, and reasoning in critical care. Proceedings of the 28th IEEE EMBC International Conference 2006; 350-3.
      [http://dx.doi.org/10.1109/IEMBS.2006.259734]

[8]   Crampin EJ, Halstead M, Hunter P, *et al.* Computational physiology and the physiome project. Exp Physiol 2004; 89(1): 1-26.
      [http://dx.doi.org/10.1113/expphysiol.2003.026740] [PMID: 15109205]

[9]   Hunter PJ, Borg TK. Integration from proteins to organs: the Physiome Project. Nat Rev Mol Cell Biol 2003; 4(3): 237-43.
[http://dx.doi.org/10.1038/nrm1054] [PMID: 12612642]

[10]   Faber GM, Rudy Y. Action potential and contractility changes in [Na(+)](i) overloaded cardiac myocytes: a simulation study. Biophys J 2000; 78(5): 2392-404.
[http://dx.doi.org/10.1016/S0006-3495(00)76783-X] [PMID: 10777735]

[11]   Livshitz LM, Rudy Y. Regulation of Ca^{2+} and electrical alternans in cardiac myocytes: role of CAMKII and repolarizing currents. Am J Physiol Heart Circ Physiol 2007; 292(6): H2854-66.
[http://dx.doi.org/10.1152/ajpheart.01347.2006] [PMID: 17277017]

[12]   Luo CH, Rudy Y. A dynamic model of the cardiac ventricular action potential. I. Simulations of ionic currents and concentration changes. Circ Res 1994; 74(6): 1071-96.
[http://dx.doi.org/10.1161/01.RES.74.6.1071] [PMID: 7514509]

[13]   Viswanathan P, Rudy Y. Pause induced early afterdepolarizations in the long QT syndrome: a simulation study. Cardiovasc Res 1999; 42(2): 530-42.
[http://dx.doi.org/10.1016/S0008-6363(99)00035-8] [PMID: 10533588]

[14]   Zang Y, Dai L, Zhan H, Dou J, Xia L, Zhang H. Theoretical investigation of the mechanism of heart failure using a canine ventricular cell model: Especially the role of up-regulated CaMKII and SR Ca^{2+} leak. J Mol Cell Cardiol 2013; 56: 34-43.
[http://dx.doi.org/10.1016/j.yjmcc.2012.11.020] [PMID: 23220154]

[15]   Kurata Y, Tsumoto K, Hayashi K, *et al.* Dynamical mechanisms of phase-2 early afterdepolarizations in human ventricular myocytes: insights from bifurcation analyses of two mathematical models. Am J Physiol Heart Circ Physiol 2017; 312(1): H106-27.
[http://dx.doi.org/10.1152/ajpheart.00115.2016] [PMID: 27836893]

[16]   Clancy CE, Rudy Y. Linking a genetic defect to its cellular phenotype in a cardiac arrhythmia. Nature 1999; 400(6744): 566-9.
[http://dx.doi.org/10.1038/23034] [PMID: 10448858]

[17]   Silva JR, Pan H, Wu D, *et al.* A multiscale model linking ion-channel molecular dynamics and electrostatics to the cardiac action potential. Proc Natl Acad Sci USA 2009; 106(27): 11102-6.
[http://dx.doi.org/10.1073/pnas.0904505106] [PMID: 19549851]

[18]   Ai X, Curran JW, Shannon TR, Bers DM, Pogwizd SM. Ca^{2+}/calmodulin-dependent protein kinase modulates cardiac ryanodine receptor phosphorylation and sarcoplasmic reticulum Ca^{2+} leak in heart failure. Circ Res 2005; 97(12): 1314-22.
[http://dx.doi.org/10.1161/01.RES.0000194329.41863.89] [PMID: 16269653]

[19]   Balhara YPS, Verma R, Gupta C. Gender differences in stress response: Role of developmental and biological determinants. Ind Psychiatry J 2011; 20(1): 4-10.
[http://dx.doi.org/10.4103/0972-6748.98407] [PMID: 22969173]

[20]   Clark AR, Kruger JA. Mathematical modeling of the female reproductive system: from oocyte to delivery. Wiley Interdiscip Rev Syst Biol Med 2017; 9(1): e1353.
[http://dx.doi.org/10.1002/wsbm.1353] [PMID: 27612162]

[21]   Sharma S, Guha SK. Mathematical modeling In reproductive biomedicine. Biomathematics, 2006; 305-314.

[22]   Vasan SS. Semen analysis and sperm function tests: How much to test? Indian J Urol 2011; 27(1): 41-8.
[http://dx.doi.org/10.4103/0970-1591.78424] [PMID: 21716889]

[23]   Vasieva O, Rasolonjanahary MI, Vasiev B. Mathematical modelling in developmental biology. Reproduction 2013; 145(6): R175-84.
[http://dx.doi.org/10.1530/REP-12-0081]

[24]   Tomlin CJ, Axelrod JD. Biology by numbers: mathematical modelling in developmental biology. Nat Rev Genet 2007; 8(5): 331-40.
[http://dx.doi.org/10.1038/nrg2098] [PMID: 17440530]

[25]   Bieler J, Pozzorini C, Naef F. Whole-embryo modeling of early segmentation in Drosophila identifies robust and fragile expression domains. Biophys J 2011; 101(2): 287-96.
[http://dx.doi.org/10.1016/j.bpj.2011.05.060] [PMID: 21767480]

[26]   Graham BP, van Ooyen A. Mathematical modelling and numerical simulation of the morphological development of neurons. BMC Neurosci 2006. 30; 7(Suppl 1): S9.
[http://dx.doi.org/10.1186/1471-2202-7-S1-S9]

[27]   Green HS, Triffet T. Mathematical modelling of nervous systems. Math Model 1980; 1(1): 41-61.
[http://dx.doi.org/10.1016/0270-0255(80)90006-8]

[28]   Availabel from: https://www.uoguelph.ca/ceps/events/2023/01/brain-biomechanics-mathematical-modelling-central-nervous-system-compartment

[29]   Zavala E, Wedgwood KCA, Voliotis M, *et al.* Mathematical modelling of endocrine systems. Trends Endocrinol Metab 2019; 30(4): 244-57.
[http://dx.doi.org/10.1016/j.tem.2019.01.008] [PMID: 30799185]

[30]   Yakimchuk K. Mathematical modeling of immune modulation by glucocorticoids. Biosystems 2020; 187: 104066.
[http://dx.doi.org/10.1016/j.biosystems.2019.104066] [PMID: 31734335]

[31]   Fischer S, Ehrig R, Schäfer S, *et al.* Mathematical modeling and simulation provides evidence for new strategies of ovarian stimulation. Front Endocrinol (Lausanne) 2021; 12: 613048.
[http://dx.doi.org/10.3389/fendo.2021.613048] [PMID: 33790856]

[32]   Fischer-Holzhausen S, Röblitz S. Mathematical modelling of follicular growth and ovarian stimulation. Curr Opin Endocr Metab Res 2022; 26: 100385.
[http://dx.doi.org/10.1016/j.coemr.2022.100385]

[33]   Bastianelli D, Sauvant D, Rérat A. Mathematical modeling of digestion and nutrient absorption in pigs. J Anim Sci 1996; 74(8): 1873-87.
[http://dx.doi.org/10.2527/1996.7481873x] [PMID: 8856442]

[34]   Kuznetsov VL, Troitskaia VB, Vershinina EA, Polenov SA, Kucher VI. Mathematical model of the digestion process regulation and its computer presentation. Usp Fiziol Nauk 2002; 33(4): 53-64.
[PMID: 12449807]

[35]   Taghipoor M, Lescoat P, Licois JR, Georgelin C, Barles G. Mathematical modeling of transport and degradation of feedstuffs in the small intestine. J Theor Biol 2012; 294: 114-21.
[http://dx.doi.org/10.1016/j.jtbi.2011.10.024] [PMID: 22085739]

[36]   Le Feunteun S, Verkempinck S, Floury J, *et al.* Mathematical modelling of food hydrolysis during *in vitro* digestion: From single nutrient to complex foods in static and dynamic conditions. Trends Food Sci Technol 2021; 116: 870-83.
[http://dx.doi.org/10.1016/j.tifs.2021.08.030]

[37]   Valiente Fernández M. Models that link physiology with outcomes. Am J Respir Crit Care Med 2023; 208(1): 111.
[http://dx.doi.org/10.1164/rccm.202304-0718LE] [PMID: 37159945]

[38]   Ghosh D, Singha PS, Ghosh S. Neurorobotics: Artificial intelligence in neuroscience. Published in the book futuristic trends in biotechnology by IIP series 2024: 3.

[39]   Zhao, Ting. Artificial intelligence in mathematical modeling of complex systems. EAI endorsed transactions on e-learning. 2024. 10.
[http://dx.doi.org/10.4108/eetel.5256]

CHAPTER 19

Diverse Disease Prognostication through Machine Learning Models

M. Vanitha[1,*] and **R. Charanya**[1]

[1] School of Computer Science Engineering and Information Systems, Vellore Institute of Technology, Vellore, India

Abstract: Today's generation faces various diseases due to the current atmosphere, pollution, poor quality of food, and their living habits. It is difficult for doctors to predict and analyze all the diseases manually; in most cases, the prediction of diseases goes wrong due to the large number of samples. The work aims to build a healthcare web application for identifying and predicting multiple diseases like heart disease, diabetes prediction, liver disease, breast cancer, kidney disease, *etc.*, and machine learning models such as Decision trees, SVM, KNN, Random Forest, *etc.*, used to accomplish this. To improve the accuracy level, datasets were gathered for every condition and trained them. we created an end-to-end web application using Flask framework where the user enters data to view the outcomes of various diseases' predictions. The drawbacks of the existing system are the users have to go to different sites to get different disease predictions, it becomes difficult for the user to move from one site to another, and in many cases, there is no proper user-friendly web application for disease prediction with no proper accuracy level mentioned. The proposed system focuses on developing a web application that offers users a variety of disease predictions based on their preferences. Multiple models were taken into consideration for training and testing the data. The evaluation results of each model were collected and then compared using a box plot.

Keywords: Decision tree, KNN, Random forest, SVM.

INTRODUCTION

The WHO has published an updated list of diseases that cause death in this generation, which includes diabetes, heart disease, kidney disease, *etc*. According to a WHO statistic, there are more than 20 million fatalities worldwide each year as a result of various diseases. As per the statistics, some of the leading causes of illness and mortality among the global population are heart disease, kidney disease, diabetes, liver disease, and breast cancer.

---

*Corresponding author M. Vanitha : School of Computer Science Engineering and Information Systems, Vellore Institute of Technology, Vellore, India; E-mail: mvanitha@vit.ac.in

Sivakumar Rajagopal, Prakasam P., Konguvel E., Shamala Subramaniam, Ali Safaa Sadiq Al Shakarchi & B. Prabadevi (Eds.)

In recent years, the current generation has seen a gradual rise in the worldwide disease burden. Several studies have been conducted in an attempt to identify the primary risk factors for this kind of disease and precisely estimate the total risk. It causes death without evident signs; this type of sickness is even called a silent killer. Making decisions on lifestyle modifications for high-risk patients depends on early detection of these diseases, which lowers complications and death rates.

Here, machine learning repeatedly demonstrated its ability to help with decision-making and forecasting using the vast amounts of data generated by various healthcare-related companies. The purpose of this study is to make predictions for several kinds of diseases, such as diabetes, heart disease, renal disease, liver illness, and others. Through analysis, the patient data uses machine learning models to categorize whether or not the patient has this type of condition. Machine learning models have been used in this situation. Even though diseases can present in various ways, there is a common set of basic risk factors that can establish whether an individual is ultimately at risk by obtaining data from various sources and classifying it appropriately.

While numerous websites and platforms exist to forecast diseases, none exist for the simultaneous prediction of multiple diseases. When the user wants to predict one disease, he must go to a different location to diagnose another. In some situations, there are multiple disease prediction techniques available, although the forecast accuracy varies greatly. The major goal of this study is to anticipate diseases in advance and save time by using existing cases, which can predict various diseases with higher accuracy. One can recover easily from diseases in advance. The problem will be addressed with the help of machine learning approaches. With the help of machine learning, one can build a model and import that model into the web application with the help of the Python Flask framework. Here, the user can diagnose the type of disease and view the current health status.

Motivation

Building an intuitive web application that can be used to anticipate many diseases without visiting different websites is the primary objective of this study, keeping in mind the existing circumstances. In this study, we have built a multiple-disease web application for diabetes, heart disease, liver disease, kidney disease, *etc*. Predicting diseases in advance can effectively lower mortality rates and optimize time by conducting simultaneous checks for multiple conditions. In some cases, disease accuracy levels are lower, leading to potential future issues. In these instances, machine learning techniques can help predict diseases with higher accuracy. The first step here is analyzing the data. This work has some advanced features of machine learning techniques like one hot encoding and feature scaling.

With the help of these two techniques, anyone can easily pre-process the 100% clean data, and better accuracy results can be achieved.

Literature Survey

The authors eloquently highlighted the machine-learning techniques utilized for diabetes-related disease classification, early detection, and prediction. As an additional feature, it also provides an IoT-based method for monitoring diabetes, enabling both healthy and affected individuals to monitor their blood glucose levels [1]. They predicted diabetes using various machine learning classification techniques such as LR, KNN, Naive Bayes, SVM, RF, and DT. Comparing the models, SVM showed the highest accuracy of 81.21% [2].

Eight different algorithms were considered, and each model's results were compared with the other to see which would yield the highest accuracy prediction [3]. The author suggested a remote monitoring system using advanced-level machine learning for diabetes risk prediction and management. This end-to-end application utilizes personal health devices like smartwatches and smartphones. By developing a support vector machine (SVM) model using the Pima Indian Diabetes dataset, the author achieved an accuracy of 93.23%, a sensitivity of 97.21%, and an F-score of 88% [4]. An application for accurately diagnosing several diseases was created using machine learning classification models. Users gain from the work since it makes it accessible to them to remotely check their status, which increases life expectancy and saves time [5].

To predict cardiac disease, a machine learning classification model and a user-friendly online application were developed. The prediction process takes into consideration 13 factors in total, such as BMI, sex, and age. The author achieves 80% accuracy using random forest and implements the model in a web page using the Python Streamlit framework for single-site web applications. Users can enter the necessary fields and click the prediction button to check their disease status [6]. The author aims to achieve 100% accuracy by comparing and training multiple algorithms. They conduct a comparative study on these algorithms, predicting and ranking their accuracy levels. The results are visualized using bar charts and ROC curve graphs [7].

The author investigates preprocessing and training techniques to improve the accuracy of diabetes prediction. They employ supervised learning algorithms, including DT, KNN, LR, SVM, and random forest. Random forest achieves the highest accuracy of 97%. Utilizing the Python Flask framework, the author created a web application to enable result checking and uses one-hot encoding to further improve accuracy [8].

The author has used a centralized system approach in a paper to address both prediction recommendations and the analysis of previous medical records to determine the current state, both of which are necessary for an efficient monitoring system [9]. Moreover, to develop a web application for recommendation systems utilizing machine learning algorithms so that the required safety measures can be adopted earlier.

The authors proposed an improved version of the KNN (k-nearest neighbor) classifier that achieves higher accuracy compared to the original KNN [10]. The paper outlines the implementation process and highlights the differences of this improved algorithm. With 18 demonstrated steps for disease prediction, the improved KNN achieves an impressive accuracy of 99%. This work presented a unique genetic algorithm-based approach for predicting liver disease. Different features were found and used in machine learning algorithms, including random forest, SVM, KNN, and DT. Additionally, a neural network with backpropagation is employed for binary operation and improved accuracy [11].

The author introduces a method using neural networks to increase disease prediction accuracy. The paper focuses on early disease detection to reduce mortality rates. A user-friendly web application was developed using Python frameworks, allowing users to easily predict diseases using imported machine learning classification models such as Support Vector Machine, Logistic Regression, KNN, and random forest [12].

The author of [13] suggests a framework for handling missing values and variable feature ranges in medical data preprocessing. Eight classification algorithms—Random Forests, Decision Trees, SVM, Naive Bayes, KNN, Logistic Regression, C4.5, and MLP—are used to test the method on the PIMA Indian Diabetes dataset. The outcomes show that the suggested preprocessing method performs better for accurate classification than earlier methods and algorithms. The application utilizes past patient data to monitor and pre-emptively manage health conditions, enhancing life expectancy and saving time [14]. The author delved into the realms of neural networks and classification methods to foresee the presence of diabetes. Adequate technique choice and forecast accuracy are crucial [15]. Artificial neural networks (ANNs) and artificial intelligence systems were employed. The study utilized ANNs to identify diabetes, utilizing a neural network error function during training.

The author of this work has used glucose and ECG monitoring sensors to improve performance. In this study, the author suggested an ensemble learning technique method to identify the disease with the help of algorithms like a random forest-one example of ensemble learning by using the majority of voting the model predicts the disease here. Authors achieved accuracy in multiple disease prediction [17]. A predictive method has been proposed for identifying diseases, including diabetes, breast cancer, and heart disease, through the utilization of the Flask API. Flask API and TensorFlow can be used to identify more diseases.

The authors used a hybrid-decision-support system for predicting heart disease, achieving 100% clean clinical data. A multiverse chained approach and feature selection techniques improve accuracy [18]. Various prediction methods using seven classifier algorithms were employed [19]. Techniques for feature selection were used for every classifier.

To predict breast cancer, a comprehensive analysis was carried out encompassing a variety of machine learning classification algorithms, including Decision Tree, SVM, Random Forest, Logistic Regression, and KNN. The performance of the classifiers was evaluated and compared [20].

Rani *et al.* introduced a hybrid decision support system for early heart disease detection, utilizing a combination of Genetic Algorithm and Recursive Feature Elimination based on clinical parameters. Data pre-processing involved SMOTE (Synthetic Minority Oversampling Technique) and standard scalar methods. The final stage of system development incorporated a support vector machine, naive Bayes, logistic regression, random forest, and AdaBoost classifiers. Notably, the random forest classifier yielded the most accurate results in the evaluation [21].

Chittora *et al.* used a dataset from the UCI repository to predict Chronic Kidney Disease. Seven different classifier algorithms were utilized for the analysis, including the artificial neural network, C5.0, Chi-square Automatic Interaction Detector, logistic regression, linear support vector machine (with penalty L1 & L2), and random tree. Several feature selection methods were employed, including the wrapper method, least absolute shrinkage, correlation-based, and selection operator regression. Results for each classifier were computed using various feature selection methods, ultimately achieving accurate predictions [22].

With the help of software applications and networked biomedical devices, the Internet of Medical Things (IoMT) has developed into a tool for next-generation bioanalysis. This technology effectively supports healthcare responsibilities. In reality, a wide range of conditions can cause a person to acquire one or more chronic or non-chronic diseases. Because of this, it is anticipated that AI and IoMT will enable the early detection of any health concerns by individuals and the appropriate action to be taken. By analyzing various aspects, the authors conducted a systematic literature review (SLR) of AI-based clinical decision support systems and Internet of Medical Things (IoMT) techniques for multi-disease forecasting [23].

An approach for forecasting seven diseases -heart disease, diabetes, renal disease, liver disease, breast cancer, kidney problems, malaria, and pneumonia was suggested by this study [24]. Authors have proposed a multi-disease prediction web app with prediction models for the range of clinical conditions mentioned above using Flask. The authors employed a variety of algorithms, including SVM, Naïve Bayes, Random Forest, K-Nearest Neighbors, K-means, Adaboost Classifiers, and others, to treat conditions like diabetes, liver illness, and heart disease. Table **1** discusses algorithms that are producing improved results for specific medical conditions.

Table 1. Comparison of strengths and limitations of various algorithms.

Algorithm Used	Key Findings	Pros	Cons
Support Vector Machine, Naive Bayes, Logistic Regression, Random Forest, K-Nearest Neighbors, K-means and Adaboost Classifiers [10, 21].	Random forest classifier gave the most accurate results in the evaluation.	Due to its ability to produce estimates of feature relevance, handle high-dimensional datasets efficiently, and be resistant to noise and outliers, Random Forest is a popular option for numerous real-world applications.	It can be computationally expensive to use a lot of trees in the forest or to train a Random Forest model on a large dataset.
Naïve Bayes (NB) classifier, C4.5 classifier, and Artificial Neural Network (RNN)-Back Propagation (BP) methods are used [2].	Results demonstrate that the Naive Bayes has higher classification accuracy.	Effectively works in multi-class predictions.	Conditional independence assumption does not always hold.
K-Nearest Neighbors, Decision Trees, Support Vector Machine, Random Forest [3, 10, 11].	K-Nearest Neighbour gives an accuracy of 87% for heart disease prediction.	There is no training phase; therefore, adding new data to the model will not have any adverse impacts.	Due to the significant cost of determining the distances between individual data instances, it is not advised for large datasets.
K-Nearest Neighbors, Logistic regression, Support Vector Machine [4].	With SVM, non-linear data outcomes can be predicted by mapping input features into a higher-dimensional feature space. The radial basis kernel is applied to yield the SVM-RBF.	SVM is more effective in high-dimensional spaces.	SVM does not perform very well when the data set has more noise.

Consequently, diseases are proliferating in the modern world and are even being called "silent killers" since they cause mortality without presenting any outward signs. To satisfy the needs of contemporary health, a variety of tools and techniques are always being evaluated. These machine-learning techniques are well suited to the prediction of multiple diseases. Early disease prediction and control can help to prevent and reduce disease-related deaths. In many existing systems, the approach is limited to analyzing a single disease. Users are limited to using one website to study diabetes; they must visit a different website to check for other diseases. The accuracy level is low, but with all this in mind, a multiple-disease web application was proposed for users to check multiple diseases at a time without traversing to different websites. Also, in many of the existing cases, the accuracy levels are too low to overcome this; feature scaling was used to improve the accuracy levels.

Gaps identified in the Existing System

In today's world, diseases are spreading rapidly, leading to significant efforts in forecasting and controlling them. Machine learning techniques can effectively predict multiple diseases, aiding early detection and reducing mortality rates. Existing systems often analyze a single disease, requiring users to navigate across the websites. To solve this problem, a web application was created that enables users to check numerous diseases at once, utilizing feature scaling and one-hot encoding approaches to improve accuracy.

Proposed System

The recommendation we make is to offer a web application for multiple disease prediction that allows users to select based on their personal preferences, therefore reducing the need for the user to browse between many websites to predict diseases. To examine the results, the user must enter the parameters of the disease. The web application will then execute the appropriate model and present the data. In this instance, several types of models were considered for training and testing the data.

The procedures for training and deploying the model are depicted in Fig. (1). The evaluation results for each model were gathered, and then the results were plotted using a box plot. The web application will use the model that provides the highest level of accuracy for predictions.

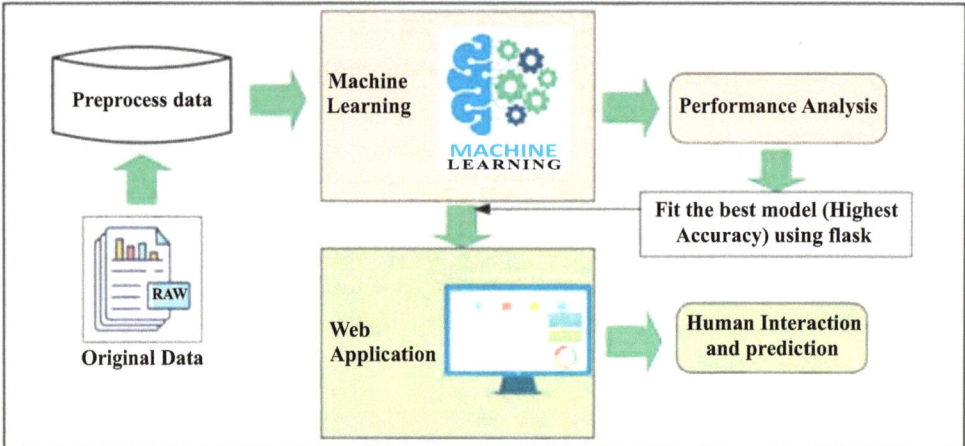

Fig. (1). Steps involved in training and deploying the model.

STEP 1: Start

STEP 2: Import the required libraries and dataset of diseases.

STEP 3: Pre-processing of the data is required. Data pre-processing is necessary. The data was converted to the required format during the preprocessing stage. It was applied to address missing, noisy, and duplicate datasets.

STEP 4: Divide the dataset into two different forms: one is a trained set, and the other one is a test set, where training takes 80% of data and testing takes 20% of data.

STEP 5: Apply machine learning algorithms for evaluation.

STEP 6: After evaluation of accuracy results, compare the accuracy results by using a box plot.

STEP 7: At last, choose the best accuracy model.

STEP 8: Fit that model into a web application for multiple disease prediction.

Fig. (2) shows the steps involved in the prediction of disease using machine learning algorithms. The first step is importing required libraries from Python like pandas, NumPy, and matplot. The second step is to import datasets like heart disease, diabetes, liver disease, kidney disease, *etc*. Once the dataset is imported, then visualization of each imported data takes place. To do away with the need for the user to switch between multiple websites to forecast illnesses, our suggested solution aims to provide a web application for multiple disease prediction that allows users to decide based on their needs. We then used the random forest, XGBoost, and KNN algorithms on the training dataset and applied our expertise to the classified method by using the testing dataset. We compared the accuracy levels of each machine learning model. After applying the data, we selected the optimal algorithm; the model with the highest accuracy was our final prediction model.

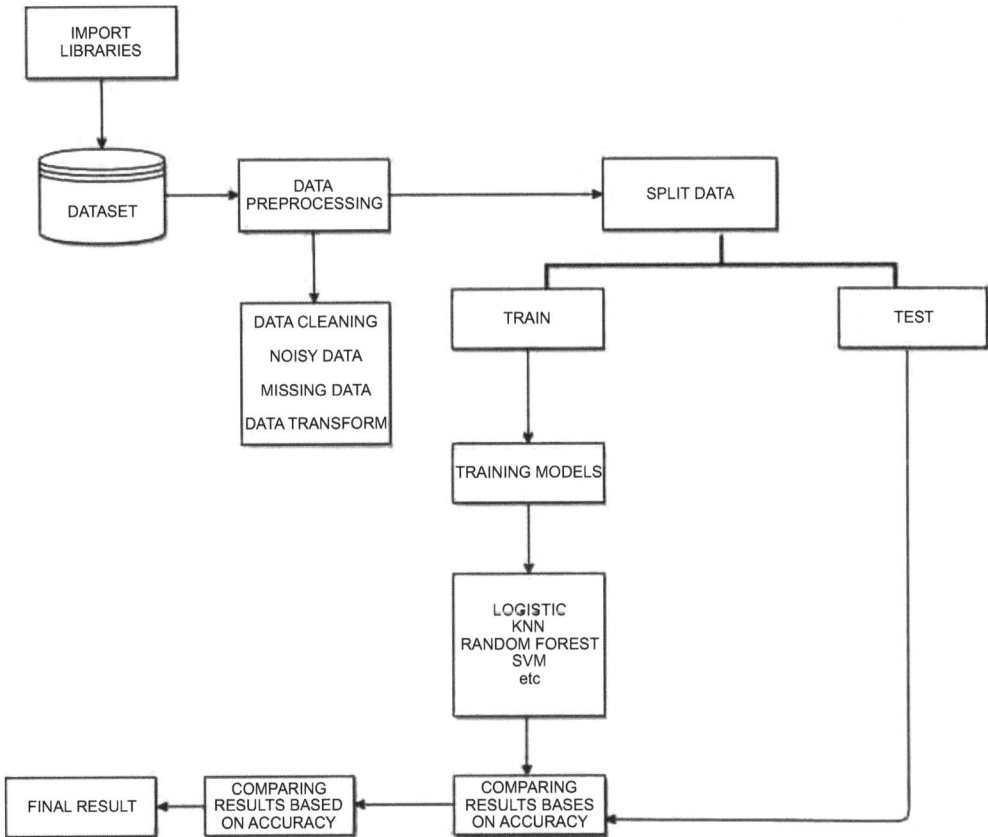

Fig. (2). Stages in the prediction of disease.

Fig. (**3**) shows the stages in a web application to predict the diseases. The user's credentials were first used to log into the disease prediction website. After logging into the system, the user may select the disease type from a list of diseases on this website. The list contains diseases like Liver diabetes, heart disease, and kidney disease. After selecting the type of disease, the user has to enter the details of that particular disease. Once all the data has been entered, the disease can be predicted using ML algorithms. A particular disease will show as "positive" if the person has it; if not, it will show as "negative". The detailed procedure is depicted in Fig. (**4**).

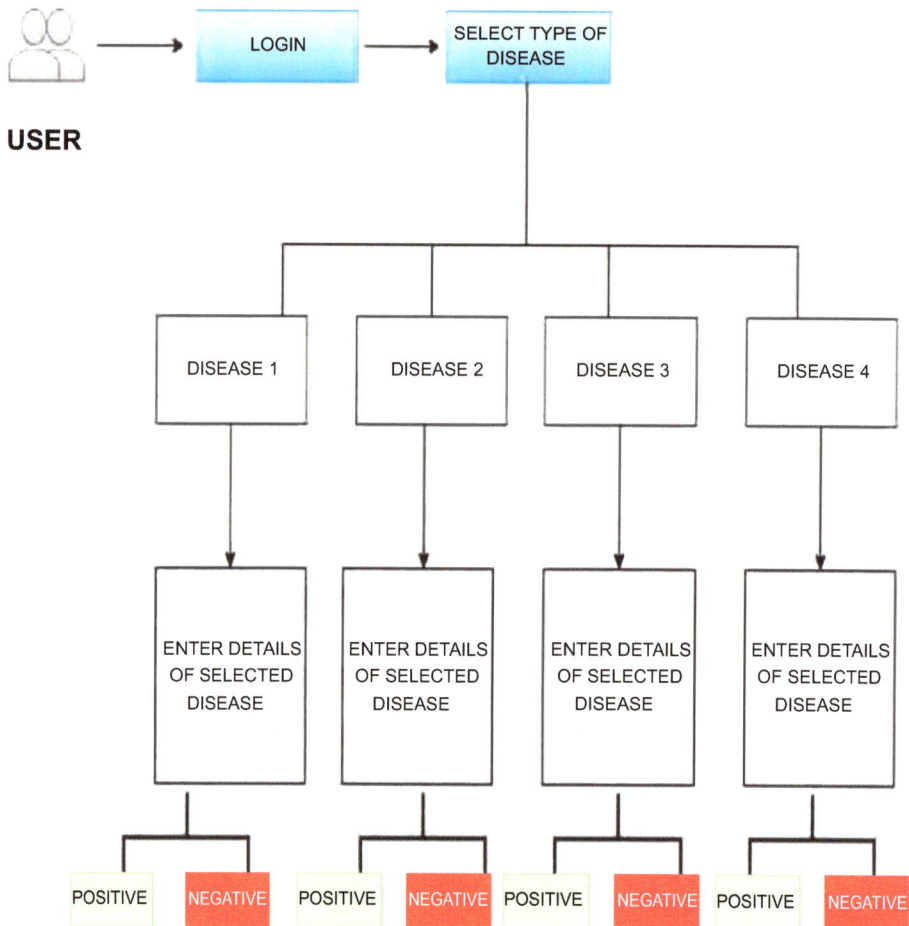

Fig. (3). Stages in a web application to predict the disease.

Implementation

Importing libraries like pandas, NumPy, seaborn, matplotlib, *etc.*, was necessary for the prediction of algorithms. Fig. (**5**) shows the Python libraries involved in the ML model.

Data Collection

The disease's datasets were collected from many websites that deal with healthcare. For diabetes prediction, the Pima Indian Diabetes Dataset was used. The testing information is used to assess the model's prediction, whereas the training dataset is used to anticipate how the model will learn. Fig. (**6**) shows the example of the train and test model for this work; 70% of the data is used for training data, and 30% of the data is used for testing.

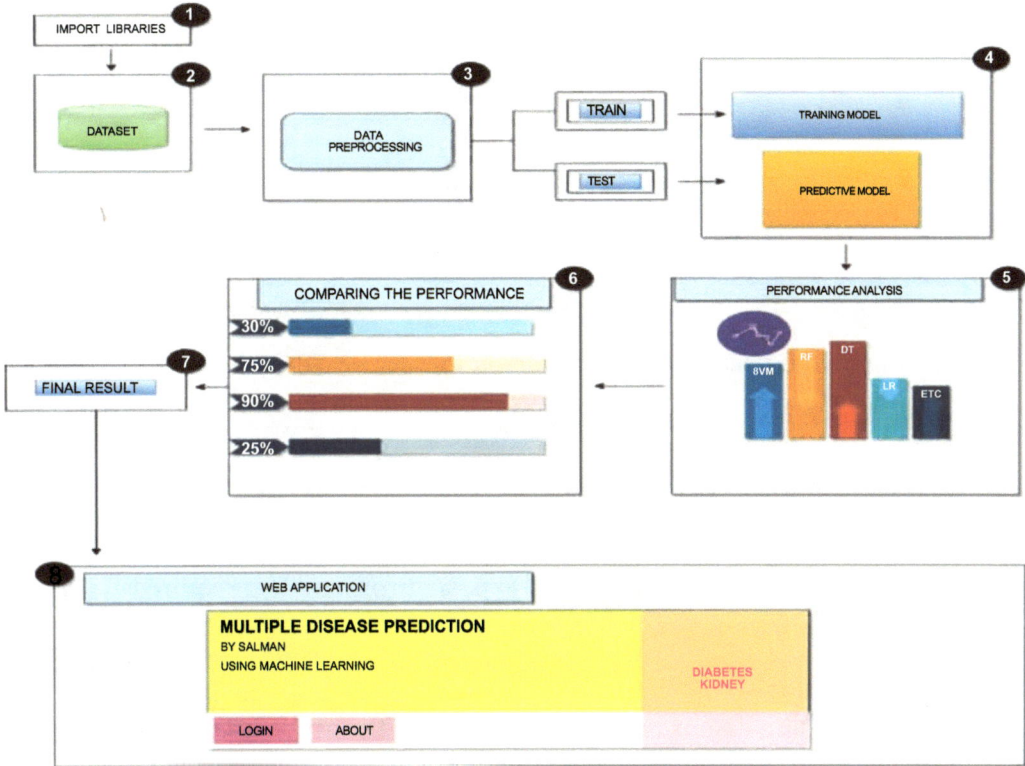

Fig. (4). Architecture for multiple disease prediction.

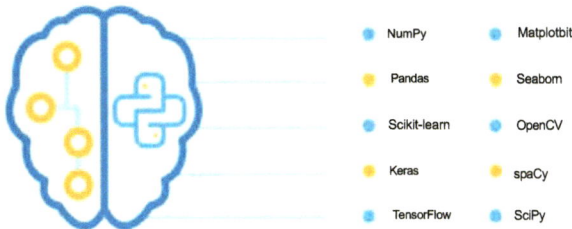

Fig. (5). Python libraries involved in ML model.

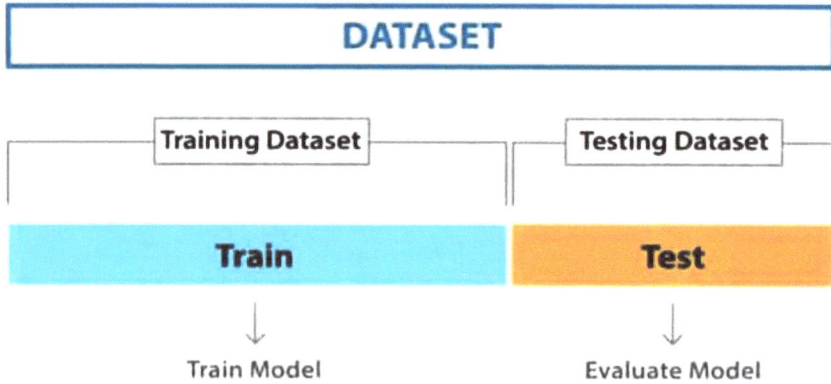

Fig. (6). The example of the train and test model.

Data Analysis and Data Preprocessing

Data pre-processing is the most important step for predicting the disease. Initial data quality issues could include noisy data, missing data, or data not in the proper format, which could lead to erroneous findings and conclusions. In pre-processing, data is transformed into the expected format. The required format is used to deal with noisy data, duplicates in the data, and missing values of the dataset. Fig. (7) shows the steps involved in data Preprocessing. Data pre-processing involves several steps, such as attribute scaling, dataset splitting, and import. Preprocessing data is necessary to improve the accuracy of the model and make sure that correct results are predicted more often.

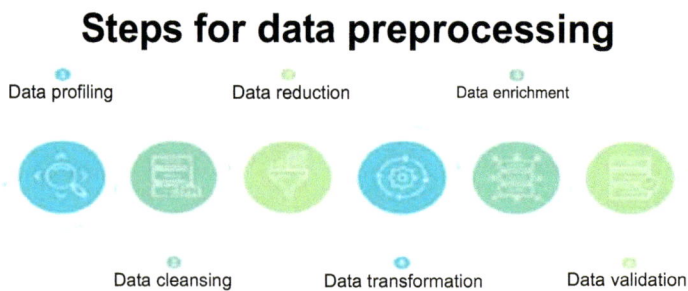

Fig. (7). Steps involved in data preprocessing.

Prediction of Disease

Various machine learning algorithms like Support vector machine (SVN), Naive Bayes, Decision Tree (DT), Random Forest (RF), Logistic Regression (LR), Adaboost, and Xg-boost are used for classification. We compared various algorithms and applied the one that provides the best accuracy for our multiple disease prediction application.

Fig. (**8**) shows the prediction result of the ML Model.

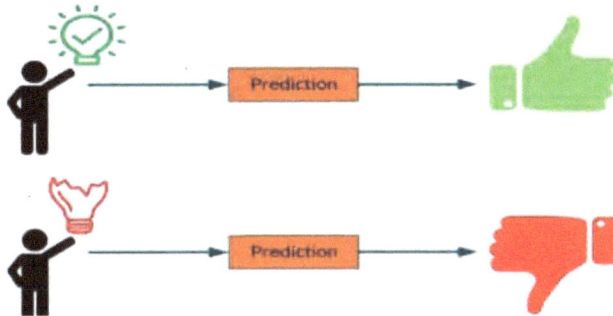

Fig. (8). The prediction result of the ML model.

Different Algorithms

The predictive classifier models were developed to accurately identify the disease. The classification model for predicting the Disease is Random Forest (RF), Decision Tree, K Nearest Neighbour, Naïve Bayes, Logistic Regression, and SVM Algorithm.

Logistic Regression

Machine Learning (ML) classification algorithms like Logistic Regression (LR) are used to forecast the probability of specific classes based on dependent variables. The output of logistic regression always lies between (0 and 1), which is suitable for a binary classification task. Fig. (**9**) below illustrates the accuracy levels of training and testing data, with 84% accuracy reached in training data and 88% accuracy achieved in testing data. Here, 70% of the data were used for training and 30% for testing.

KNN

One of the most significant machine learning algorithms is KNN, or k-nearest neighbor, which uses historical data to enable supervised learning approaches understandably. It divides the categories of datasets into different formats; for example, if we give a dataset like a mango and an apple, it separates both the datasets in different, creates a boundary line, and draws a circle in that. When we enter a new dataset, it checks which algorithm meets near the line to see whether it is an apple or a mango. Fig. (**10**) shows the accuracy levels achieved in train and test sets.

LR

```
# fitting data to model

from sklearn.linear_model import LogisticRegression

log_reg = LogisticRegression()
log_reg.fit(X_train, y_train)

· LogisticRegression
LogisticRegression()

# model predictions

y_pred = log_reg.predict(X_test)

# accuracy score

from sklearn.metrics import accuracy_score, confusion_matrix, classification_report

print(accuracy_score(y_train, log_reg.predict(X_train)))

log_reg_acc = accuracy_score(y_test, log_reg.predict(X_test))
print(log_reg_acc)

0.8402255639097744
0.8815789473684211
```

Fig. (9). Classification report and accuracy score of the LR model.

KNN

```
from sklearn.neighbors import KNeighborsClassifier

knn = KNeighborsClassifier()
knn.fit(X_train, y_train)

· KNeighborsClassifier
KNeighborsClassifier()

# model predictions

y_pred = knn.predict(X_test)

# accuracy score

print(accuracy_score(y_train, knn.predict(X_train)))

knn_acc = accuracy_score(y_test, knn.predict(X_test))
print(knn_acc)

0.8665413533834586
0.8333333333333334
```

Fig. (10). Classification report and accuracy score of the KNN model.

SVM

SVM chooses the vectors or extreme points that can help in creating the hyperplane. Fig. (**11**) shows the accuracy levels of train and test datasets using SVM, with 89% train data and 84% test data.

```
Out[70]:    -GridSearchCV
            -estimator: SVC
                - SVC

In [73]:  #best parameters
          grid_search.best_params_
Out[73]:  {'C': 1, 'gamma': 0.13}

In [74]:  #best score
          grid_search.best_score_
Out[74]:  0.8665843766531477

In [75]:  svc = SVC(C = 1, gamma = 0.1, probability=True)
          svc.fit(X_train, y_train)
Out[75]:             SVC
          SVC(C-1, gamma=0.1, probability-True)

In [76]:  #model predictions
          y_pred = svc.predict(X_test)

In [77]:  #accuracy score
          print(accuracy_score(y_train, svc.predict(X_train)))
          svc_acc = accuracy_score(y_test, svc.predict(X_test))
          print(svc_acc)
          0.8947368421052632
          0.8421052631578947
```

Fig. (11). Classification report and accuracy score of the SVM model.

The main goal of SVM is to build the best boundary line so that one can easily insert a new dataset into a model by ##DIViding them into two categories where it can easily put the data according to the category and form the result.

Decision Tree

A decision tree is like a tree data structure, where it has a root node, parent node, child node, and leaf node. The first node, or top node of the tree, is known as the root node, and the last node is known as the leaf node. A decision tree is a combination of multiple if else conditions where its first step is to find a correct in machine learning, say feature selection. Feature selection is done using some log probability formulas where we have to find the probability that matches all 0s in one place and all 1s in another place. After feature selecting a model, we have to pass our new datasets to check the result the constructed tree will pass each data's one by one and checks for which condition satisfies the result. Fig. (**12**) shows an example of the accuracy levels attained using the decision tree model, showing 99% in train data and 98% in test data following model evaluation.

DT

```
from sklearn.tree import DecisionTreeClassifier

dtc = DecisionTreeClassifier()
dtc.fit(X_train, y_train)

# accuracy score, confusion matrix and classification report of decision tree

dtc_acc = accuracy_score(y_test, dtc.predict(X_test))

print(f"Training Accuracy of Decision Tree Classifier is {accuracy_score(y_train, dtc.predict(X_train))}")
print(f"Test Accuracy of Decision Tree Classifier is {dtc_acc} \n")

print(f"Confusion Matrix : \n{confusion_matrix(y_test, dtc.predict(X_test))}\n")
print(f"Classification Report : \n {classification_report(y_test, dtc.predict(X_test))}")

Training Accuracy of Decision Tree Classifier is 1.0
Test Accuracy of Decision Tree Classifier is 0.8201754385964912
```

Fig. (12). Classification report and accuracy score of the DT model.

Random Forest

As part of the ensemble learning technique, the random forest makes predictions by utilizing the majority vote. If there are 100 data in ##DIVides data into 10 sets and passes each set of data to each without repeating values, then the random forest is a combination of multiple decision trees. It forms multiple decision trees and imports some set of data to each tree in bootstrap format. Then, as a new algorithm is inserted into those trees, it checks each one individually and outputs a result in the format of 0s and 1s. After receiving each result, it then checks the majority output again, determining which has the highest number—for instance, out of 10, if 1s are repeated six times and 0s repeated four times—then our is `1`. Like this, machine learning forecasts the outcome. Random forest is one of the best machine-learning algorithms and is mostly used for large datasets. The accuracy scores obtained in the train and test sets are displayed in Fig. (13).

```
RF
from sklearn.ensemble import Random Forestclassifier

rand_clf = RandomForestClassifier(criterion = 'entropy', max_depth=15 = max_features = 'auto', min_samples_leaf = 2, min_samples
rand_clf.fit(X_train, y_train)

                    RandomForestclassifier
RandomForestClassifier(criterion-'entropy', max_depth-15, max_features='auto',
                min_samples_leaf-2 , min_samples_split-3,
                n_estimators=130)

y_pred = rand_clf.predict(X_test)

# accuracy score
print(accuracy_score(y_train, rand_clf.predict(X_train)))

ran_clf_acc = accuracy_score(y_test, y_pred)
print(ran_clf_acc)
0.9830827067669173
0.9035087719298246
```

Fig. (13). Classification report and accuracy score of the RF model.

RESULT ANALYSIS

After evaluating the models with various algorithms, the optimal algorithm with the highest degree of accuracy can be found by contrasting such algorithms. The user may enter data according to the disease to check the results.

The above accuracy levels are achieved using various machine-learning algorithms. Algorithms may vary for each disease according to its accuracy levels to get higher accuracy levels from SVM, decision tree, and random forest. The accuracy achieved using the ML algorithm for each disease is shown in Table **2**.

Table 2. Accuracy achieved using ML algorithms for each disease.

DISEASE	ACCURACY
Diabetes	98%
Heart Disease	99%
Kidney Disease	96%
Liver Disease	80%
Malaria	95.65%
Pneumonia	91.35%

Performance Analysis Scores for the Disease

A high-accuracy random forest algorithm is used to predict the classification report. A classification report shows the performance evaluation metrics and classification scores in machine learning. It is used to display the visual table and scores of the precision score, recall score, and F1 Score, and it supports the score of your trained classification model of algorithms. Table **3** shows the accuracy achieved by each algorithm, and Tables **4** and Table **5** show the accuracy levels achieved in each algorithm for kidney disease and liver disease.

Table 3. Accuracy scores achieved by each algorithm.

Model	Score
Random Forest Classifier	90.35
Logistic Regression	88.16
SVM	84.21
KNN	83.33
Decision Tree Classifier	82.02

Table 4. Accuracy levels achieved in each algorithm for kidney disease.

Model	Score
Random Forest Classifier	0.991667
Gradient Boosting	0.975000
xgboost	0.966667
Decision Tree Classifier	0.941667
Logistic Regression	0.908333
KNN	0.700000
SVM	0.700000

Table 5. Accuracy levels achieved in each algorithm for liver disease.

Model	Score
SVM	71.18
Gradient Boosting Classifier	70.59
Logistic Regression	69.41
xgboost	69.41
Random Forest Classifier	68.82
Decision Tree Classifier	67.06
KNN	62.94

To further evaluate the data and predict the disease, machine learning classifier algorithms like random forest, logistic regression, decision tree, SVM, and KNN are used. These algorithms produced accuracy scores of 91% for random forest, 87% for logistic regression, 77% for Decision Tree, 81% for SVM, and 81 for KNN. After obtaining the accuracy levels of each algorithm, the accuracy levels were compared to determine which algorithm achieved the highest level of accuracy. Fig. (**14**) shows the ROC Curve graph on accuracy levels obtained by each algorithm.

The performance evaluation is shown in Fig. (**15**), with random forest having the highest accuracy level after comparing with all the algorithms. The accuracy level of each algorithm m for heart disease is shown in Fig. (**16**).

The performance evaluation is shown in Fig. (**17**), with random forest having the highest accuracy level after comparing with all the algorithms.

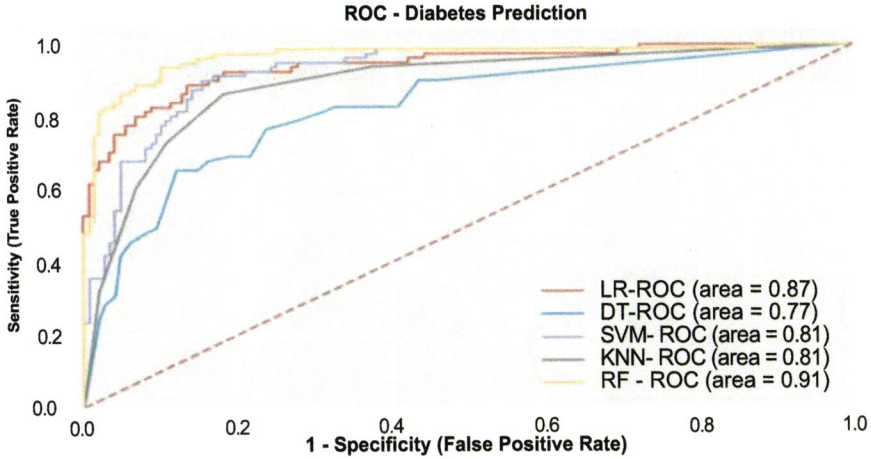

Fig. (14). Diabetes ROC curve graph on accuracy levels obtained by each algorithm.

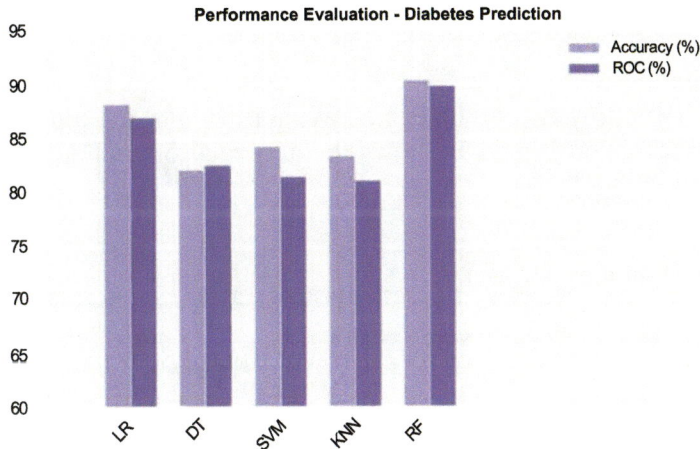

Fig. (15). Performance evaluation of each algorithm for diabetes.

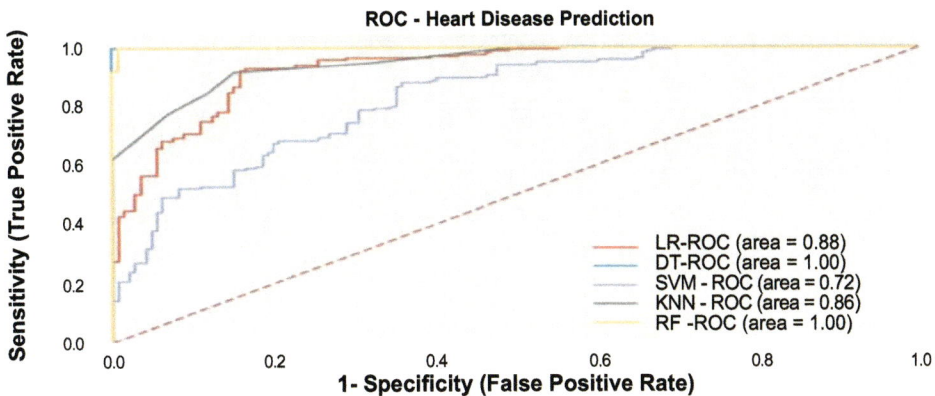

Fig. (16). Accuracy levels obtained by each algorithm for heart disease.

[88.31, 99.35000000000001, 71.75, 86.04, 98.7]
[88.63999999999999, 97.50999999999999, 71.89999999999999, 86.16, 98.61999999999999]

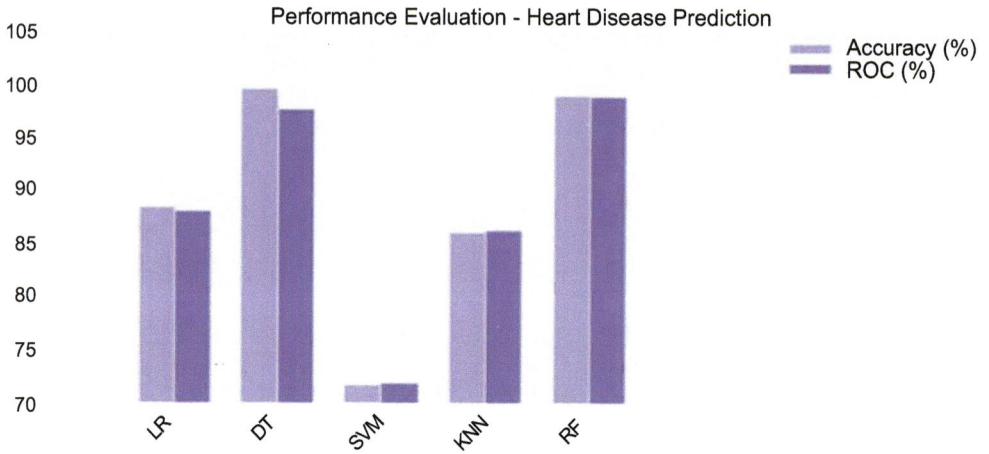

Fig. (17). Performance evaluation of each algorithm for heart disease.

Figs. (17 and 18) show the performance evaluation of each algorithm for heart and kidney disease, and Fig. (19) shows the accuracy of each algorithm for liver disease.

[90.83,94.17,70.0,70.0,96.67, 99.17, 97.5]
[90.28, 95.83, 64.92999999999999, 69.78999999999999, 95.83, 98.96000000000001, 96.88]

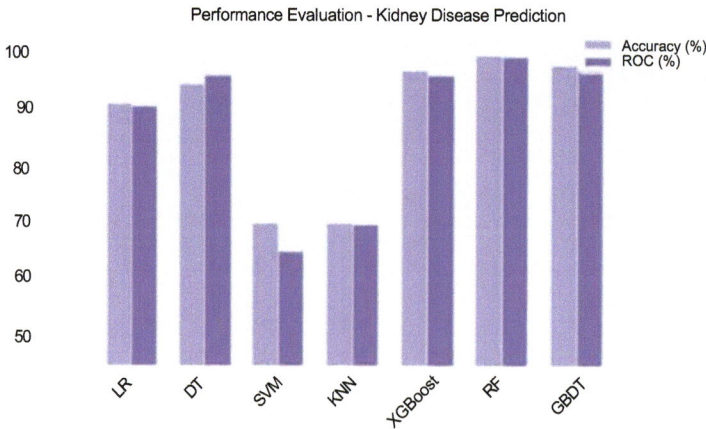

Fig. (18). Performance evaluation of each algorithm for kidney disease.

[69.41000000000001, 67.06, 71.17999999999999, 62.94, 69.41000000000001, 68.82000000000001, 70.59]
[57.26, 55.46, 50.0, 55.75, 60.9, 58.86, 61.919999999999995]

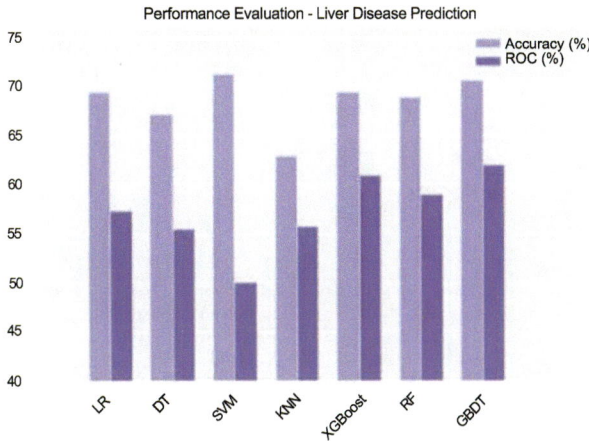

Fig. (19). Accuracy level obtained by each algorithm for liver disease.

CONCLUSION

The major objective of this study is to build a web application that can correctly predict multiple diseases. Here, the disease prediction is made based on the user's input in the required fields. The user will be offered the option. It also saves time for the end user because they do not have to go through numerous websites. Early disease detection can prevent financial problems and increase your life expectancy. To attain the highest level of accuracy, decision tree, random forest, SVM, and K closest neighbor (KNN) models were used. A web-based application for multi-disease prediction helps users predict when specific diseases are likely to manifest, which can reduce the death rate. With the aid of smartwatches, blood pressure monitors, heart rate monitors, and other devices, this approach may eventually be used to forecast the disease using real-time statistics.

REFERENCES

[1]   Yaganteeswarudu A. Multi-disease prediction model by using machine learning and Flask API. In2020 5th International Conference on Communication and Electronics Systems (ICCES) 2020, Jun 10; pp. 1242-1246.
[http://dx.doi.org/10.1109/ICCES48766.2020.9137896]

[2]   MO RR. Novel Implementation of Cardio-vascular disease (CVD) using machine learning techniques.

[3]   Srivastava K, Choubey DK. Heart disease prediction using machine learning and data mining. International Journal of Recent Technology and Engineering (IJRTE) 2020; 9(1): 212-9.
[http://dx.doi.org/10.35940/ijrte.F9199.059120]

[4]   Ramesh J, Aburukba R, Sagahyroon A. A remote healthcare monitoring framework for diabetes prediction using machine learning. Healthc Technol Lett 2021; 8(3): 45-57.
[http://dx.doi.org/10.1049/htl2.12010] [PMID: 34035925]

[5]   Kalshetty JN, Achyutha Prasad N, Mirani D, Kumar H, Dhingra H. Heart health prediction using web

application. Int J Health Sci 2022; 6(S2): 5571-8.
[http://dx.doi.org/10.53730/ijhs.v6nS2.6479]

[6] Ali MM, Paul BK, Ahmed K, Bui FM, Quinn JMW, Moni MA. Heart disease prediction using supervised machine learning algorithms: Performance analysis and comparison. Comput Biol Med 2021; 136: 104672.
[http://dx.doi.org/10.1016/j.compbiomed.2021.104672] [PMID: 34315030]

[7] Ahmed N, Ahammed R, Islam MM, *et al.* Machine learning based diabetes prediction and development of smart web application. International Journal of Cognitive Computing in Engineering 2021; 2: 229-41.
[http://dx.doi.org/10.1016/j.ijcce.2021.12.001]

[8] Rajora H, Punn NS, Sonbhadra SK, Agarwal S. Machine learning equipped web-based disease prediction and recommender system.Advanced Machine Intelligence and Signal Processing. Singapore: Springer Nature Singapore 2022; pp. 313-24.
[http://dx.doi.org/10.1007/978-981-19-0840-8_23]

[9] Waris SF, Koteeswaran S. Heart disease early prediction using a novel machine learning method called improved K-means neighbor classifier in python. Mater Today Proc 2021.

[10] Poonguzharselvi B, Ashraf MM, Subhash VV, Karunakaran S. Prediction of liver disease using machine learning algorithm and genetic algorithm. Ann Rom Soc Cell Biol 2021; 2347-57.

[11] Nandy S, Adhikari M, Balasubramanian V, Menon VG, Li X, Zakarya M. An intelligent heart disease prediction system based on swarm-artificial neural network. Neural Comput Appl 2023; 35(20): 14723-37.
[http://dx.doi.org/10.1007/s00521-021-06124-1]

[12] Butt UM, Letchmunan S, Ali M, Hassan FH, Baqir A, Sherazi HHR. Machine learning-based diabetes classification and prediction for healthcare applications. J Healthc Eng 2021; 2021: 1-17.
[http://dx.doi.org/10.1155/2021/9930985] [PMID: 34631003]

[13] Bhuiyan MF, Rahman MT, Anik MA, Khan M. A Framework for Type-II Diabetes Prediction Using Machine Learning Approaches. In 2021 12th International Conference on Computing Communication and Networking Technologies (ICCCNT) 2021 Jul 6; pp. 1-6.
[http://dx.doi.org/10.1109/ICCCNT51525.2021.9580158]

[14] Gupta S, Karanth G, Pentapati N, Prasad VB. A web-based framework for liver disease diagnosis using combined machine learning models. 2020 International Conference on Smart Electronics and Communication (ICOSEC) 2020; 421-8.
[http://dx.doi.org/10.1109/ICOSEC49089.2020.9215454]

[15] Trivedi NK, Maheswari S, Sharma H, Jain S, Agarwal S. Early Detection & Prediction of Heart Disease using Various Machine Learning Approaches. In 2022 9th International Conference on Computing for Sustainable Global Development (INDIACom) 2022, Mar 23; pp. 793-797.

[16] Rady M, Moussa K, Mostafa M, Elbasry A, Ezzat Z, Medhat W. Diabetes Prediction Using Machine Learning: A Comparative Study. 2021.
[http://dx.doi.org/10.1109/NILES53778.2021.9600091]

[17] Kumar J, Tiwari RK, Pandey V. Diabetes prediction using machine learning tools. In 2021 4th International Conference on Recent Trends in Computer Science and Technology (ICRTCST) 2022 Feb 11; pp. 263-267.
[http://dx.doi.org/10.1109/ICRTCST54752.2022.9781963]

[18] Thilagavathi G, Priyanka S, Roopa V, Shri JS. Heart disease prediction using machine learning algorithms. 2022 International Conference on Applied Artificial Intelligence and Computing (ICAAIC) 2022; 494-501.
[http://dx.doi.org/10.1109/ICAAIC53929.2022.9793107]

[19] Kavitha M, Gnaneswar G, Dinesh R, Sai YR, Suraj RS. Heart disease prediction using hybrid machine

learning model. In 2021 6th International Conference on inventive Computation Technologies (ICICT) 2021, Jan 20; pp. 1329-1333.
[http://dx.doi.org/10.1109/ICICT50816.2021.9358597]

[20] Mohanty S, Gantayat PK, Dash S, Mishra BP, Barik SC. Liver Disease Prediction Using Machine Learning Algorithm. InData Engineering and Intelligent Computing. Proceedings of ICICC 2020; 2021: 589-96. [Springer Singapore.].

[21] Rani P, Kumar R, Ahmed NMOS, Jain A. A decision support system for heart disease prediction based upon machine learning. J Reliab Intell Environ 2021; 7(3): 263-75.
[http://dx.doi.org/10.1007/s40860-021-00133-6]

[22] Chittora P, Chaurasia S, Chakrabarti P, *et al.* Prediction of chronic kidney disease-a machine learning perspective. IEEE Access 2021; 9: 17312-34.
[http://dx.doi.org/10.1109/ACCESS.2021.3053763]

[23] Merabet A, Saighi A, Laboudi Z, Ferradji MA. Multiple Diseases Forecast Through AI and IoMT Techniques Systematic Literature Review In International Conference on Intelligent Systems and Pattern Recognition. Cham: Springer 2024; pp. 189-206.

[24] Chunduru A, Kishore AR, Sasapu BK, Seepana K. Multi chronic disease prediction system using CNN and random forest. SN Computer Science 2024; 5(1): 157.
[http://dx.doi.org/10.1007/s42979-023-02521-6]

Advanced Computing Solutions for Healthcare, 2025, 354-369

CHAPTER 20

Patch Antenna Design for 2.4 GHz for On-Off Body Communication

J. Joselin Jeya Sheela[1], N. Duraichi[1], B. Jeyapoornima[1] and M. Logeshwaran[2,*]

[1] *Department of Electronics and Communication Engineering, Saveetha School of Engineering, Saveetha Institute of Medical and Technical Sciences, Saveetha Nagar, Thandalam, Chennai-602105, Tamil Nadu, India*

[2] *Department of Electronics and Communication Engineering, R.M.K. Engineering College, RSM Nagar, Kavaraipettai, Gummidipoondi Taluk, Tiruvallur-601206, Tamil Nadu, India*

Abstract: On-body communication is a crucial technology, finding applications ranging from healthcare monitoring to IoT connectivity. Using antennas, on-body communication technology exploits specific frequency bands and tailored antenna designs to facilitate efficient data exchange between devices on or near the human body. This study delves into patch antenna designs customized for on-off body communication at 2.4GHz frequencies. As the demand surges for seamless connectivity with diminutive, inconspicuous form factors, an in-depth exploration of antenna design becomes indispensable. The antenna, crafted with FR-4 (a lossy material) possessing a dielectric constant of 4.4, employs copper for the patch and ground layers. Simulation is carried out utilizing the CST Studio Suite, confirming and documenting key antenna parameters: Gain of 5.65dB, bandwidth of 96.8MHz, return loss of -41.83dB, and VSWR of 1.05. Through simulation, a thorough analysis of the antenna's performance in both free space and on the body is conducted, unveiling its exceptional performance in these scenarios. Furthermore, the proposed design optimizations exhibit resilience against environmental factors, making them well-suited for practical on-body communication applications.

Keywords: Antenna, Antenna parameters, Bandwidth, Biomedical application, Gain, ISM band, Microstrip antennas, Patch antennas, Resonant frequency, Radiation pattern, Return loss, Specific absorption rate, VSWR.

---

* **Corresponding author M. Logeshwaran:** Department of Electronics and Communication Engineering, R.M.K. Engineering College, RSM Nagar, Kavaraipettai, Gummidipoondi Taluk, Tiruvallur-601206, Tamil Nadu, India; E-mail: logeshwaran25092@gmail.com

Sivakumar Rajagopal, Prakasam P., Konguvel E., Shamala Subramaniam, Ali Safaa Sadiq Al Shakarchi & B. Prabadevi (Eds.)

INTRODUCTION

In an era marked by escalating interconnectivity, the demand for effective and dependable on-body communication systems has experienced a notable upsurge, propelled by a spectrum of applications spanning from wearable health monitoring to the Internet of Things (IoT). Body-centric communication (BCC) delineates into three primary classifications predicated on interaction modes: in-body, off-body, and on-body communications [1]. Central to the architecture of body-centric wireless systems are antennas and signal propagation, which are pivotal components in developing compact, high-efficiency RF sensor nodes, optimizing both spectrum utilization and power efficiency. Additionally, they play a pivotal role in ensuring the robustness of communications within, on, and off the body [2].

Microstrip patch antennas are assuming an increasingly pivotal role in the advancement of contemporary wireless communication systems. This significance stems from the burgeoning demand for diverse wireless applications among end-users. Researchers and scientists are actively immersed in this domain, driven by the myriad opportunities afforded by wireless technology [8]. A microstrip patch antenna comprises a radiating patch positioned on one facet of a dielectric substrate, with a ground plane situated on the opposing side. Typically fashioned from conductive materials like copper or gold, the patch can adopt various geometries. Both radiating patches and feed lines are conventionally fabricated *via* photolithography on the dielectric substrate [9]. To facilitate analysis and performance projection, patches are often configured in standardized shapes such as square, rectangular, circular, triangular, elliptical, or other commonly employed forms [10].

Furthermore, a considerable number of antennas presently in operation are undergoing significant size reduction [7]. This research paper delves into the domain of on-off body communication using patch antennas, intending to investigate their design, performance, and optimization for robust and efficient communication within the intricate on-body environment. In typical healthcare monitoring scenarios, ensuring effective radio communication is paramount. This involves the antenna emitting signals uniformly in all directions across the body's surface (omnidirectional) while also directing its signal towards devices situated off the body. The goal is to optimize radio channel performance both on and off the body, primarily focusing on minimizing signal attenuation to enhance power efficiency [3].

Patch antennas present numerous advantages over traditional antenna platforms, encompassing a low profile, compact size, facile manufacturing process, cost-effectiveness, and seamless integration with "monolithic microwave integrated circuits (MMICs)" [4]. Moreover, they demonstrate inherent resonance capabilities and commendable narrow bandwidth performance [5]. Consequently, the design and optimization of patch antennas have emerged as a focal area of research, given their direct impact on the caliber and resolution of microwave images. This paper undertakes a comprehensive investigation, delving into the meticulous design of patch antennas customized for 2.4GHz on-off body communications.

In the dynamic realm of wireless communication and remote sensing, the "Industrial, Scientific, and Medical (ISM)" band operating at 2.4GHz has emerged as a pivotal cornerstone for on-off body communications. The ISM band presents a favorable equilibrium between resolution and penetration depth, aligning well with numerous applications necessitating non-ionizing radiation. This research unfolds as a pioneering endeavor in the pursuit of harnessing the potential of patch antenna designs meticulously crafted for the 2.4GHz ISM band within the domain of on-off body communications.

Related Works

A study [11] amplifies the directivity of a compact ultra-wideband (UWB) antenna by incorporating a reflector inspired by the Yagi-Uda design. The resulting reflector-loaded antenna (RLA) exhibited significant enhancements, such as a 4 dB rise in realized gain and a notable 14.26 dB increase in transmitted field strength within a human breast model. Moreover, it sustained high signal fidelity, boasting a 94.86% correlation factor. Comparable enhancements in directivity were observed when a validated head imaging antenna was similarly outfitted with a reflector.

A study [12] presents a novel design for an aperture-fed annular ring (AFAR) microstrip antenna, emphasizing simplification in fabrication by leveraging 3D-printed and solderless 2D materials. Comprising three layers-an aperture-fed patch, a ground plane slot for power transmission, and a microstrip line for feeding-the antenna underwent optimization *via* the finite element method (FEM) in four iterative steps, targeting distinct parameters. The optimized 3D AFAR antenna achieved an S11 of approximately 17 dB, a front-to-back ratio exceeding 30 dB, and a gain of around 3.3 dBi, making it conducive for streamlined manufacturing and deployment across antenna technologies.

A study [13] elucidates the development and experimental validation of a deeply implanted conformal printed antenna. In this innovative design, the hip implant serves as the ground plane for a trapezoidal radiator, which is fed by a coaxial cable and specifically tailored to transmit biological signals captured within the body by specialized biosensors. The system comprises a metallic (or equivalent) hip implant, a biocompatible gypsum-based dielectric, and a conformal radiator. Experimental validation entailed immersing the 3D-printed plastic bone housing the setup in a tissue-like liquid within a plastic container to mimic internal body conditions, closely resembling human leg dimensions. The study assesses the matching and radiation characteristics within the industrial, scientific, and medical (ISM) frequency band (2.4–2.5 GHz), thereby validating the feasibility of the proposed configuration.

A study [14] intricately explores the development of a miniature dual-band implantable antenna tailored for communication within the human body, specifically situated beneath the skin of the arm. Operating seamlessly within the MICS and ISM 2.4 GHz bands, this antenna facilitates telemetry while offering versatility in application. Inspired by the Koch fractal structure, the antenna boasts a compact profile and straightforward design, featuring adjustable resonant frequencies and obviating the necessity for additional matching components. Performance assessment reveals commendable outcomes, including broad bandwidths and observed gains in radiation patterns. Furthermore, the study delves into various skin-mimicking phantom formulations to authenticate the antenna's functionality, with an emphasis on precision and simplicity in fabrication.

A study [15] introduces a dual-band button sensor antenna customized for on-body monitoring in Wireless Body Area Network (WBAN) systems, tackling design complexities posed by body proximity. This resilient system seamlessly integrates the antenna with a wireless sensor module on a PCB, delivering compact dimensions, exceptional radiation properties, and heightened efficiency. Measured and simulated outcomes exhibit robust correlation, highlighting the antenna's efficacy in both free space and on-body contexts, with negligible susceptibility to bending and motion. Moreover, Specific Absorption Rate (SAR) values adhere to safe thresholds. Range assessments illustrate coverage spanning up to 40 meters across diverse environments.

An article [16] introduces circularly polarized sensors and antennas tailored for applications in biomedical systems, energy harvesting, IoT, and 5G devices. It tackles the challenge of effectively evaluating wearable antennas. Through the printing of the microstrip antenna and sensor feed network on the same substrate, a cost-effective methodology is realized. The study juxtaposes simulated and measured outcomes of compact circularly polarized sensors, underscoring their efficacy, adaptability, and cost-efficiency. Both passive and active metamaterial antennas augment system performance, with those integrating Circular Split-Ring Resonators (CSRRs) showcasing a broader frequency range and reduced gain, albeit with amplified bandwidth. Collectively, these antennas contribute to heightened system efficacy across diverse applications.

A manuscript [17] delineates a compact antenna array meticulously designed for incorporation into medical masks, seamlessly integrating antennas onto the mask's structure and shielding. Leveraging a composite amalgamation of materials comprising polycarbonate (PC), polyethylene terephthalate (PET), and FR4 substrate, these antennas proficiently cover three pivotal frequency bands: 2.38–2.62 GHz, 3.38–3.74 GHz, and 5.14–8 GHz. Empirical assessments corroborate its efficacy, consistently exhibiting reflection coefficients below −10 dB. This adaptability facilitates uninterrupted wireless transmission both indoors and outdoors, spanning the spectrum of 5G FR1 and Wi-Fi 7 frequencies. Furthermore, the system's flexibility fosters facile integration with masks of diverse compositions, ensuring steadfast connectivity across heterogeneous environments.

A manuscript [18] presents a dual-band microstrip patch antenna engineered for applications in wireless body area networks (WBANs), operating within the 2.45 GHz and 5.8 GHz ISM bands. Fabricated on a substrate composed of jean material, with dimensions of mm^2, the antenna incorporates corner cuts and a circular slot to augment frequency activation and radiation pattern enhancement. Experimental evaluations reveal an impedance bandwidth of 3.68% (2.40 GHz–2.49 GHz) and 3.81% (5.67 GHz–5.89 GHz) for the respective bands, achieving maximum gains of 3.08 dBi at 2.45 GHz and 2.15 dBi at 5.8 GHz. Moreover, the antenna demonstrates consistent performance on both planar and curved body surfaces, coupled with low specific absorption rate (SAR) values, underscoring its suitability for WBAN applications.

Table 1 presents a comparative analysis of various antenna designs. The initial design introduced a reflector and Ultra-Wideband (UWB) antenna, enhancing directivity albeit with potential complexity. Another design centered around an active electronically scanned array (AESA) microstrip antenna, reducing footprint but facing challenges with back lobes and impedance mismatch. One configuration utilized a coaxial-cable-fed radiator with minimal spatial demands but encountered bio-compatibility concerns. Fractal geometries were harnessed for dual-band operation, addressing compactness while contending with frequency shifts. A button sensor antenna offered a diminutive form factor and multi-band functionality, albeit with a restricted communication range. Circularly polarized antennas exhibited improved performance albeit requiring intricate design methodologies. An antenna system tailored for multiple resonance points delivered broad bandwidth albeit with specific material requisites. Lastly, a dual-band antenna demonstrated stable patterns and low Specific Absorption Rate (SAR) values, notwithstanding fabrication complexities and sensitivity to errors.

Table 1. Comparison of prior research.

Reference	Author	Methodology	Advantages	Drawbacks
[11]	Awan, Dawar, *et al.*	Yagi-Uda-inspired reflector design, Compact UWB antenna.	Enhanced directivity and gain, High signal integrity.	Potential design complexity, Size, and placement constraints.
[12]	Alhassoon, Khaled, *et al.*	Design of aperture-fed annular ring (AFAR) microstrip antenna, Finite element method for parameter investigation.	Reduced footprint by 56% compared to rectangular antenna, Practical solderless approach, Enhanced bandwidth due to dual resonance.	A thin substrate of the feeding line causes unwanted back lobe, Mismatching due to the increased thickness of the feeding substrate.
[13]	Matekovits, Ladislau, *et al.*	The design involved a coaxial-cable-fed trapezoidal radiator with the hip implant as the ground plane.	No additional space is required beyond the implant itself, and no additional surgery is needed for antenna insertion.	Challenges in bio-compatibility and miniaturization, Limited options for bio-compatible dielectrics.
[14]	Bahrouni, Majdi, *et al.*	The antenna design is based on fractal structures, specifically derived from the first iteration of Koch's fractal curve.	Dual-band operation, compact size, simplified design.	ISM 2.4 GHz band resonant frequencies may shift due to parasitic effects, impacting higher frequency performance.
[15]	Ali, Shahid Muhammad, *et al.*	Designing a wearable button sensor antenna for WBAN applications.	Small form factor resembling a button for wearer comfort, multi-band operation compatible with WIFI standards, high stability in resonance frequency, and impedance matching.	Limited range of communication within 40m, potential performance variations based on body locations, substrate tolerances, and user movements.
[16]	Sabban, Albert.	Design and optimization of circular polarized sensors and antennas, including CSRRs integration.	Enhanced performance, wider bandwidth, improved efficiency.	Complexity in design, potential cost, sensitivity to environmental factors.
[17]	Chung *et al.*	Design antenna system to operate at multiple resonance points (2.45 GHz, 3.5 GHz, 5.5 GHz, 6.8 GHz).	Wide operating bandwidth. High efficiency and gain. Good isolation in MIMO configuration.	Requires specific materials and fabrication techniques. Complex integration into medical masks.

(Table 1) cont.....

Reference	Author	Methodology	Advantages	Drawbacks
[18]	Thaiwirot, Wanwisa, *et al.*	Design on a jean substrate with specific dimensions. Optimize rectangular patch and microstrip feed line. Cut corners and add circular slots for dual-band operation.	Dual-band operation, stable radiation patterns, low SAR values.	Fabrication complexity, potential sensitivity to fabrication errors, slight degradation in performance under bending.

ANTENNA DESIGN

This paper focuses on the customization of a patch antenna to resonate specifically within the ISM band at 2.4 GHz. The selected substrate material is FR-4, renowned for its characteristics as a lossy material, possessing a dielectric constant (ε_r) of 4.4. Pure copper is utilized as the conducting material for the antenna's patch. The physical dimensions of the antenna are fixed at 38.6 x 48.7mm, as depicted in Fig. (**1**). The fundamental design of this antenna pivots on three critical parameters: the operating frequency (f_o), the dielectric constant of the substrate (ε_r), and the height of the substrate (h), with equations derived from a study [19].

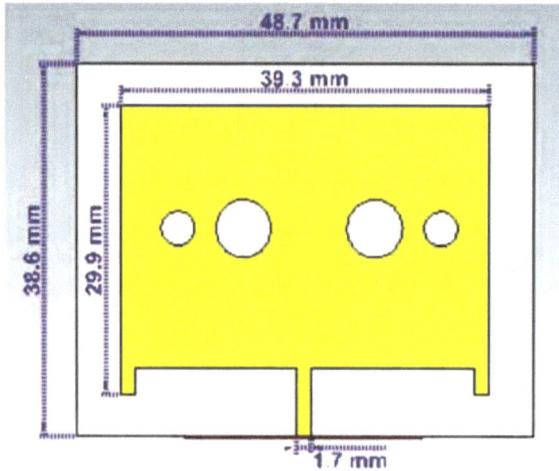

Fig. (1). Antenna design for 2.4 GHz.

Width – W,

$$W = \frac{c}{2f_o \sqrt{\frac{(\varepsilon_r + 1)}{2}}}$$

(1)

Effective Dielectric Constant (ε_{eff}),

$$\varepsilon_{eff} = \frac{\varepsilon_r + 1}{2} + \frac{\varepsilon_r - 1}{2} \left[1 + 12\frac{h}{W}\right]^{\frac{1}{2}} \tag{2}$$

L_{eff} – Effective length,

$$L_{eff} = \frac{c}{2f_o\sqrt{\varepsilon_{eff}}} \tag{3}$$

ΔL-Extension of Length,

$$\Delta L = 0.412 \times h \; \frac{(\varepsilon_{eff} + 0.3)\left(\frac{W}{h} + 0.264\right)}{(\varepsilon_{eff} - 0.258)\left(\frac{W}{h} + 0.8\right)} \tag{4}$$

L- Actual patch's Length,

$$L = L_{eff} - 2\Delta L \tag{5}$$

$$L = \frac{c}{2f_o\sqrt{\varepsilon_{eff}}} - 0.824h\left(\frac{(\varepsilon_{eff} + 0.3)\left(\frac{W}{h} + 0.264\right)}{(\varepsilon_{eff} - 0.258)\left(\frac{W}{h} + 0.8\right)}\right) \tag{6}$$

RESULTS AND DISCUSSION

The Patch Antenna is meticulously designed and evaluated utilizing the advanced software CST Studio Suite, distinguished for its prowess in three-dimensional electromagnetic analysis. Comprising three fundamental elements – the ground plane, substrate, and patch – this antenna is intricately engineered, with precise dimensions calculated within the software to meet specific criteria. To complete the antenna's design, a feed point is strategically positioned at predefined coordinates. Subsequently, the antenna undergoes a comprehensive assessment, measuring various performance parameters such as return loss, Voltage Standing Wave Ratio (VSWR), gain, and bandwidth. This exhaustive analysis equips designers with the tools to fine-tune the antenna, aligning it with desired specifications and optimizing its overall performance.

The return loss of an antenna denotes the extent of power reflected from the antenna due to impedance mismatches within the system, typically expressed in decibels (dB). It serves as a metric for assessing the efficiency of power transfer from the transmitter to the antenna. A higher return loss (greater in dB) signifies that the majority of the power is radiated by the antenna, indicating minimal reflection. A return loss of -10 dB or better is commonly deemed acceptable for a wide array of applications [20]. At 2.4 GHz, a return loss of -41.83 dB is achieved, as illustrated in Fig. (**2**).

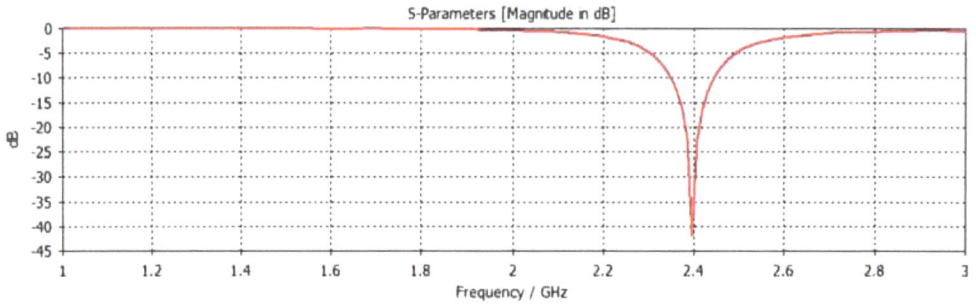

Fig. (2). Return loss (simulation).

VSWR (Voltage Standing Wave Ratio) stands as another pivotal performance metric utilized in antenna and RF systems. It gauges the effectiveness of power transfer between a source (*e.g.*, a transmitter) and an antenna or any other RF component. VSWR shares a close relationship with return loss and serves to evaluate the impedance match between components. A VSWR of 1:1 delineates a perfect match, signifying efficient power transfer devoid of any reflection. As VSWR escalates, it indicates a poorer match and increased power loss due to reflection. At 2.4 GHz, a VSWR of 1.05 is attained, as depicted in Fig. (**3**).

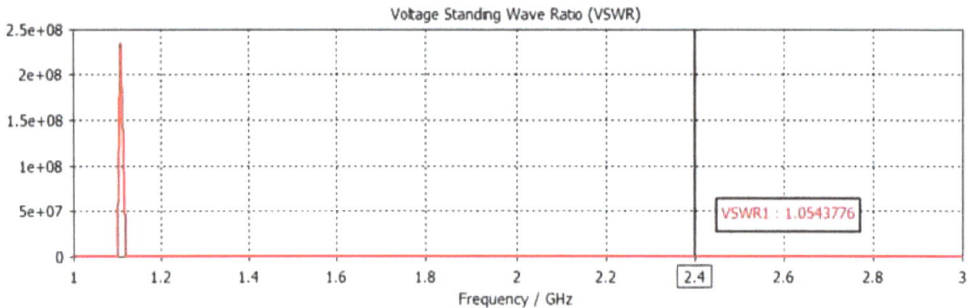

Fig. (3). VSWR (simulation).

"Bandwidth" denotes the spectrum of frequencies within which an antenna can effectively transmit or receive electromagnetic signals while upholding acceptable performance parameters. This parameter holds significant importance in antenna design as it governs the adaptability and applicability of the antenna across different scenarios. The bandwidth of a patch antenna is contingent upon its specific design and construction, showcasing notable variability between different antennas. At 2.4 GHz, a bandwidth of 96.8 MHz is achieved.

"Antenna gain" represents an antenna's capacity to concentrate electromagnetic energy in a specific direction during signal transmission or reception, in contrast to an isotropic radiator, which disperses radiation uniformly in all directions. This gain parameter holds pivotal significance in comprehending an antenna's directivity and operational attributes. Antenna gain is intricately linked to the radiation pattern, whereby antennas with higher gain demonstrate heightened directionality, focusing their radiation into distinct areas. This characteristic finds widespread application across various domains. At 2.4 GHz, a gain of 5.65 dB is achieved, as illustrated in Fig. (**4**).

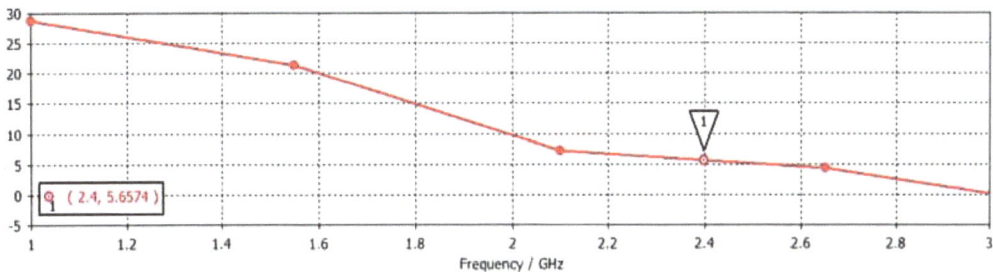

Fig. (4). Gain (simulation).

The radiation pattern of an antenna serves as a critical characteristic, offering a visual or descriptive depiction of how the antenna emits or receives electromagnetic signals within space. It essentially delineates the spatial distribution of radiation intensity emitted by the antenna. Omnidirectional antennas emit radiation uniformly in all directions, whereas directional antennas concentrate energy in specific directions. The width of the primary radiation lobe, known as beamwidth, indicates the coverage range. Understanding radiation patterns is imperative for optimizing antenna performance across various applications, encompassing wireless communication and radar systems. In this design, an omnidirectional pattern is attained, as depicted in Fig. (**5**). Lastly, the antenna parameters are summarized in Table **2**.

Signal propagation through the human body in the context of on-body communication entails the transmission of electromagnetic waves emitted by an antenna situated on the body's surface, as illustrated in Fig. (**6**). These waves traverse body tissues *via* several intricate mechanisms. These include conduction through electrolytes and ions present in bodily fluids, dielectric absorption, which converts energy into heat, and reflection and refraction at tissue interfaces. Additionally, scattering due to tissue irregularities and multipath propagation arising from reflections and refractions play pivotal roles. Collectively, these mechanisms significantly affect the signal's strength, quality, and reliability as it propagates through the body, thereby influencing the performance of on-body communication systems.

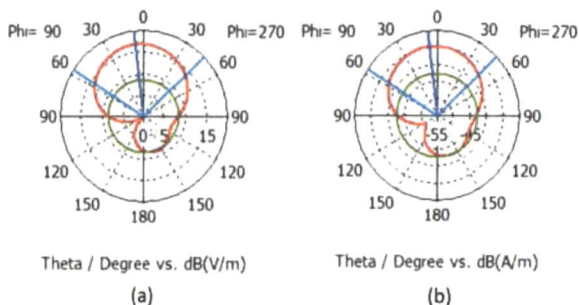

Theta / Degree vs. dB(V/m) (a) Theta / Degree vs. dB(A/m) (b)

Fig. (5). (**a**) E-Field (**b**) H-field (Simulation).

Table 2. Antenna parameters (simulation).

Antenna Parameters	
Description	**Values**
Frequency	2.4 GHz
Return loss	-41.83dB
VSWR	1.05
Bandwidth	96.8MHz
Gain	5.65dB
Radiation Pattern	Omnidirectional Pattern

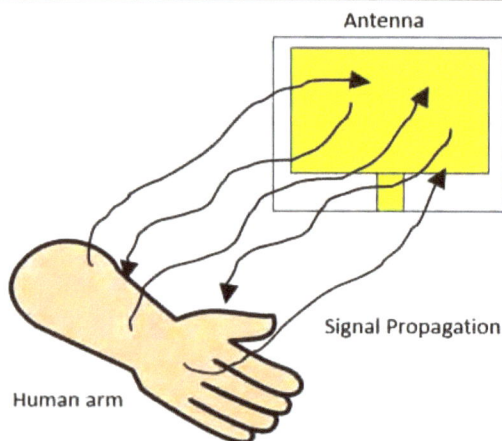

Fig. (6). Signal propagation (on-off body).

The "specific absorption rate (SAR)" is a critical parameter in evaluating the absorption of electromagnetic energy within the human body. This energy concentration predominantly occurs in the near-field region, where the user's body is near the communication device's antenna. The calculation of energy absorbed by tissue is determined using Kivekäs' method, which employs a model that accounts for both homogeneous and layered tissue characteristics.

Efficiency is paramount in antenna design to minimize losses, especially considering the human body's propensity to absorb electromagnetic (EM) waves, thereby converting them into heat and energy. Given that the human body acts as a lossy medium for EM waves, optimizing antenna efficiency is imperative [21]. Particular attention must be paid to the design to reduce backward radiation, specifically addressing the SAR, which can be detrimental to human tissues [22]. Simultaneously, it is crucial to maintain the antenna's overall dimensions as compact as possible [23].

The investigation of the SAR involved positioning the designed antenna at a 5mm distance from a simplified model representing a human arm. This model comprises four distinct layers that mimic the composition of a real human arm, as illustrated in Fig. (7). These layers consist of skin (2.5mm), fat (3mm), muscle (8mm), and bone (10mm) within the cylindrical structure [6].

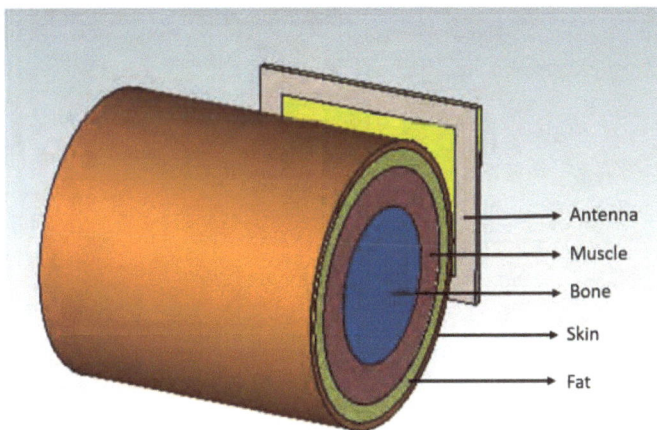

Fig. (7). Phantom of bone.

The calculation of the SAR is conducted for the ISM band, specifically at a frequency of 2.4 GHz. It is important to note that SAR values vary across different frequencies due to changes in the dielectric constant of the tissues involved. Typically, SAR values are expressed in units of watts per kilogram (W/kg) for any 1g or 10g of tissue.

The notion of SAR within the human body is contingent upon frequency fluctuations, owing to the heterogeneous dielectric characteristics inherent in distinct tissues. These characteristics, chiefly encapsulated by relative permittivity (εr) and electrical conductivity (σ), govern the interplay between tissues and electromagnetic fields. As the frequency spectrum of electromagnetic waves shifts, so does the absorption propensity of various tissues, with higher frequencies exhibiting a proclivity for superficial absorption, influencing tissues proximal to the body's surface. Conversely, lower frequencies possess a penetrative capability, reaching deeper tissue layers. Such frequency-selective behavior underpins the imperative of evaluating the safety parameters of radiofrequency (RF) apparatus and discerning the intricate dynamics of RF radiation within the multifarious strata of the human anatomical framework.

Following the European Specification ES 59005, adherence to a stringent SAR threshold of 2 W/kg for 10g of tissue is deemed optimal. The calculation of SAR entails the utilization of a standardized power level of 0.5 watts (W) uniformly applied across all frequencies. This regulatory norm is indispensable in safeguarding against excessive electromagnetic exposure to human tissue. As depicted in Fig. (8), the SAR during stimulation attains a commendable value of 1.21 W/kg, firmly residing below the prescribed threshold, thereby affirming compliance with safety standards.

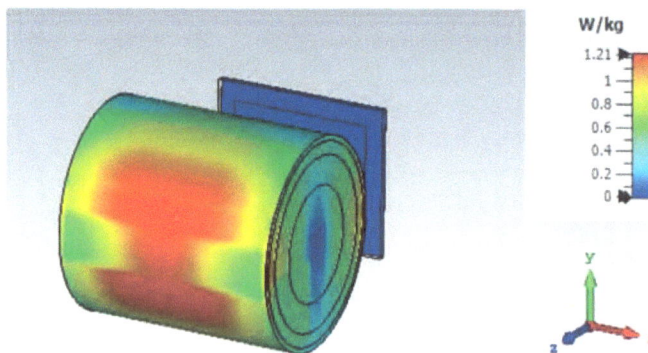

Fig. (8). Simulated SAR value.

SAR measurements are essential for evaluating the safety of electronic devices, particularly those emitting RF radiation, such as mobile phones, wireless routers, and medical equipment like MRI machines. Ensuring that SAR levels remain within established safety limits helps protect individuals from potential adverse health effects associated with RF radiation exposure. This is critical for the well-being of patients and healthcare professionals who may use or be exposed to RF-emitting medical equipment.

Understanding the safety of RF-emitting devices is crucial for patient safety and informed medical decision-making. Researchers, regulatory bodies, and manufacturers collaborate to establish and enforce safety standards, conduct thorough testing, and provide accurate information to the public. This collaborative effort promotes the responsible use of electronic devices and safeguards public health.

CONCLUSION

In this extensive investigation, a 2.4 GHz patch antenna specifically tailored for on-off body communication has been meticulously designed and optimized. The primary focus was on minimizing SAR levels to ensure user safety. The resulting antenna configuration exhibits impressive parameters, including a gain of 5.65 dB, a wide bandwidth of 96.8 MHz, a return loss of -41.83 dB, and a minimal VSWR of 1.05. By addressing SAR concerns through rigorous design enhancements and safety measures, this antenna emerges as a robust solution for practical on-body communication applications in healthcare, wearables, and the Internet of Things (IoT). This achievement underscores the utmost importance placed on user well-being. Furthermore, this work not only advances on-body communication technology but also highlights the critical role of responsible antenna design in promoting safer and more efficient data transmission in our increasingly interconnected world.

REFERENCES

[1]   Shahzad MA, Paracha KN, Naseer S, *et al.* An artificial magnetic conductor-backed compact wearable antenna for smart watch iot applications. Electronics (Basel) 2021; 10(23): 2908.
        [http://dx.doi.org/10.3390/electronics10232908]

[2]   Hao Y, Hall PS. On-body antennas and propagation: Recent development. IEICE Trans Commun 2008; E91-B(6): 1682-8.
        [http://dx.doi.org/10.1093/ietcom/e91-b.6.1682]

[3]   Khan MM, Rahman MA, Talha MA, Mithila T. Wearable antenna for power efficient on-body and off-body communications. Journal of Electromagnetic Analysis and Applications 2014; 6(9): 238-43.
        [http://dx.doi.org/10.4236/jemaa.2014.69024]

[4]   Mohammed AS, Kamal S, Ain MF, *et al.* Microstrip patch antenna: A review and the current state of the art. Journal of Advanced Research in Dynamical and Control Systems 2019; 11(7): 510-24.

[5]   Wong KL. Compact and broadband microstrip antennas. John Wiley & Sons; 2004 Apr 7.

[6]   Abdullah M, Khan A. Multiband wearable textile antenna for ISM body center communication systems. In2015 XXth IEEE International Seminar/Workshop on Direct and Inverse Problems of Electromagnetic and Acoustic Wave Theory (DIPED) 2015; Sep 21; pp. 90-96.

[7]   Mishra R, Mishra RG, Chaurasia RK, Shrivastava AK. Design and analysis of microstrip patch antenna for wireless communication. Int J Innov Technol Explor Eng 2019; 8(7): 663-6.

[8]   Rana MS, Islam SI, Al Mamun S, Mondal LK, Ahmed MT, Rahman MM. An s-band microstrip patch antenna design and simulation for wireless communication systems. Indonesian Journal of Electrical Engineering and Informatics (IJEEI) 2022; 10(4): 945-54.
[http://dx.doi.org/10.52549/ijeei.v10i4.4141]

[9]   Orban D, Moernaut GJ. The basics of patch antennas, updated. Orban Microwave Products, RF Globalnet. 2009 Sep 29.

[10]   Pozar DM. Microstrip antennas. Proc IEEE 1992; 80(1): 79-91.
[http://dx.doi.org/10.1109/5.119568]

[11]   Awan D, Bashir S, Khan S, Al-Bawri SS, Dalarsson M. UWB antenna with enhanced directivity for applications in mMedical imaging. Sensors (Basel) 2024; 24(4): 1315.
[http://dx.doi.org/10.3390/s24041315] [PMID: 38400473]

[12]   Alhassoon K, Malallah Y, Alsunaydih FN, Alsaleem F. Three-dimensional printed annular ring aperture-fed antenna for telecommunication and biomedical applications. Sensors (Basel) 2024; 24(3): 949.
[http://dx.doi.org/10.3390/s24030949] [PMID: 38339666]

[13]   Matekovits L, Mir F, Dassano G, Peter I. Deeply implanted conformal antenna for real-time bio-telemetry applications. Sensors (Basel) 2024; 24(4): 1170.
[http://dx.doi.org/10.3390/s24041170] [PMID: 38400327]

[14]   Bahrouni M, Houzet G, Vuong TP, *et al.* Modeling of a compact, implantable, dual-band antenna for biomedical applications. Electronics (Basel) 2023; 12(6): 1475.
[http://dx.doi.org/10.3390/electronics12061475]

[15]   Ali SM, Sovuthy C, Noghanian S, *et al.* Design and evaluation of a button sensor antenna for on-body monitoring activity in healthcare applications. Micromachines (Basel) 2022; 13(3): 475.
[http://dx.doi.org/10.3390/mi13030475] [PMID: 35334779]

[16]   Sabban A. Wearable circular polarized antennas for health care, 5G, energy harvesting, and IoT systems. Electronics (Basel) 2022; 11(3): 427.
[http://dx.doi.org/10.3390/electronics11030427]

[17]   Chung MA, Lee MC, Hsiao CW. Antenna systems in medical masks: Applications for 5G FR1 and Wi-Fi 7 wireless systems. Electronics (Basel) 2022; 11(13): 1983.
[http://dx.doi.org/10.3390/electronics11131983]

[18]   Thaiwirot W, Hengroemyat Y, Kaewthai T, Akkaraekthalin P, Chalermwisutkul S. A dual-band low SAR microstrip patch antenna with jean substrate for WBAN applications. Int J RF Microw Comput-Aided Eng 2024; 1-12.
[http://dx.doi.org/10.1155/2024/5076232]

[19]   Sheela JJ, Logeshwaran M, Kumar KU, Vamsi M, Kumar NC. Design of ultra-wideband of rectangular shaped emoji designed microstrip patch antenna of 4.5 GHz for military applications. In 2022 3rd International Conference on Smart Electronics and Communication (ICOSEC) 2022; Oct 20; pp. 71-75.

[20]   Punith S, Praveenkumar SK, Jugale AA, Ahmed MR. A novel multiband microstrip patch antenna for 5G communications. Procedia Comput Sci 2020; 171: 2080-6.
[http://dx.doi.org/10.1016/j.procs.2020.04.224]

[21]   Basir A, Bouazizi A, Zada M, Iqbal A, Ullah S, Naeem U. A dual-band implantable antenna with wide-band characteristics at MICS and ISM bands. microwave and optical technology letters. 2018 Dec; 60(12): 2944-9.

[22]   Bouazizi A, Zaibi G, Iqbal A, Basir A, Samet M, Kachouri A. A dual-band case-printed planar inverted-F antenna design with independent resonance control for wearable short range telemetric systems. Int J RF Microw Comput-Aided Eng 2019; 29(8): e21781.
[http://dx.doi.org/10.1002/mmce.21781]

[23]   Michel A, Colella R, Casula GA, *et al.* Design considerations on the placement of a wearable UHF-
         RFID PIFA on a compact ground plane. IEEE Trans Antenn Propag 2018; 66(6): 3142-7.
         [http://dx.doi.org/10.1109/TAP.2018.2811863]

Performance Evaluation of Syringe Control Systems: Servo Motors *versus* Stepper Motors

S. Sundar[1,*], Jaswanth K.[1], D. Ravi Teja[1] and Vinyas Shetty[1]

[1] *Vellore Institute of Technology, Vellore, India*

Abstract: This academic paper presents a comparative analysis of 8051-controlled syringes utilizing servo and stepper motors for precise fluid injection. The study thoroughly investigates the design, performance, functionality, and other relevant aspects of both systems. It evaluates the servo motor-driven syringe pump for its continuous and smooth fluid delivery capabilities, while the stepper motor-driven syringe pump is recognized for its discrete and precise steps. Various experiments are conducted to assess each system's capabilities and limitations, including fluid dispensing accuracy, response time, and resilience to external disturbances. The findings of this comparative research offer valuable insights for selecting the most appropriate system based on specific application requirements, such as accuracy and speed. Ultimately, this study contributes to the optimization of fluid delivery systems across various industries.

Keywords: Healthcare, Precise injection, Ring and pinion, Rack and pinion, Syringe, Stepper motor, Servo motor, 8051 controller.

INTRODUCTION

The accurate and controlled administration of medication is crucial in healthcare. The effectiveness and safety of medical treatments depend on precise dosing and delivery. Advanced technological solutions have led to innovative approaches to medication administration, particularly in cases where precision and customization are essential. Our research introduces a groundbreaking development: an 8051 microcontroller-based syringe control system that uses a servo motor for the meticulous and controlled dispensing of medications in medical applications.

Medication management in healthcare is complex, with different patients, conditions, and treatments requiring diverse dosages and delivery profiles. Manual administration of medication, while common, is prone to errors and lacks

---

* **Corresponding author S. Sundar:** Vellore Institute of Technology, Vellore, India; E-mail: sundar.s@vit.ac.in

Sivakumar Rajagopal, Prakasam P., Konguvel E., Shamala Subramaniam, Ali Safaa Sadiq Al Shakarchi & B. Prabadevi (Eds.)

the precision needed in modern medical settings. In response to these challenges, our research explores the integration of cutting-edge technology to provide a more accurate and adaptable solution for a variety of medical scenarios. At the heart of this innovation is the 8051 microcontroller, a versatile and programmable platform that serves as the core control unit for our syringe control system. Combined with a precision servo motor, this system offers an unprecedented level of control over the syringe's plunger movement, maintaining the exact dosage and delivery rate throughout treatment. The integration of an intuitive user interface empowers healthcare professionals to customize medication delivery to the unique needs of each patient, thereby enhancing the safety and effectiveness of treatment plans.

The implications of this research are significant. The 8051 microcontroller-based syringe control system, enhanced by a servo motor, offers a novel approach to improving patient care, addressing the need for precision and customization in medication administration across various medical settings. By ensuring accurate and reliable medication delivery, this research contributes to the progress of medical technology, furthering the safety and effectiveness of patient treatment.

The remainder of this paper is organized as follows:

Section 2 provides a literature review of various techniques proposed for driver drowsiness detection. Sections 3 and 4 describe the execution principle and experimental setup in detail. Section 5 provides a detailed workflow of the system built. Section 6 presents the experimental results and the evaluation of the proposed method.

Literature Review

The research paper authored by Ashmi M., S. Jayaraj, and K. S. Sivanandan addresses the development of a control system for motorized assistive devices, particularly prosthetic legs. The primary aim is to achieve more natural and coordinated limb movements, recognizing the increasing demand for advanced artificial limbs due to the rising number of individuals with lower limb weakness or amputations. The paper highlights the importance of replicating human gait patterns, focusing on the coordination of muscles, bones, and joints in prosthetic legs. The authors detail a control system that employs an 8051 microcontroller to govern 12V DC series motors responsible for controlling knee and hip joint movements. The direction of motor rotation is managed using an H-bridge converter, specifically the L293D. The study demonstrates the system's functionality *via* simulations using Keil μVision and Proteus software, illustrating its capacity to control DC motors for actions like locking/unlocking legs and moving knee and hip joints. The research suggests that for further improvement, incorporating feedback mechanisms like optical encoders to transition to a closed-loop system would enhance the precision and functionality of prosthetic leg movements. This work contributes to the advancement of prosthetic limb technology, potentially improving mobility and overall quality of life for individuals with limb impairments [1].

In the article "Development of a Microcontroller-Based Motor Speed Control System Using Intel 8051", researchers delve into the development of a motor speed control system employing the Intel 8051 microcontroller. This system is designed to regulate the speed of Direct Current (DC) motors, which are integral to numerous applications. The article introduces various speed control methods, including Phase-Locked-Loop (PLL) control and Pulse-Width Modulation (PWM) [2]. It emphasizes the microcontroller's role in processing feedback from optical encoders and adjusting the motor's voltage supply based on user-defined speed settings. Detailed descriptions of the hardware components, such as the motor drive and optical encoder, are presented, and the software interface is created using Visual Studio and C#. The conducted experiments highlight the system's precision in maintaining desired motor speeds under varying loads. The article concludes by suggesting potential areas for future research, such as mathematical modeling and the implementation of advanced control methods, including PID controllers and Fuzzy Logic Controllers [2].

The conference paper discusses the development of a control system for a syringe infusion pump. Infusion pumps are crucial in medical settings for accurately delivering fluids, like medications and nutrients, into a patient's bloodstream. This paper focuses on syringe infusion pumps, known for their precision in delivering low-flow rates, making them suitable for pediatric and intensive therapy applications [3]. The control system is divided into four key components: the electrical/mechanical unit of the infusion pump, a module with a stepping motor and its controller, an Arduino microcontroller module, and an LCD for visual feedback. A web interface facilitates the input of infusion parameters and provides a means to manage patient records. The Arduino microcontroller calculates and manages infusion parameters, triggers alarms, and communicates with the pump. Alarms can indicate the absence of a syringe, the end of an infusion, and battery voltage levels. All relevant data, including equipment parameters, user details, and patient records, are stored in a database. The paper suggests possible improvements, including better mechanical control and additional features like numerical keypads for enhanced functionality [3].

A team comprised of A. Andrew Silva, N. Chiranjeevi, V. Kaushikan, and R. Venkatesh, under the guidance of Prof. Muhammadu Sathik Raja, conducted research and developed a syringe pump system that is controlled by a Raspberry

Pi. This innovative system offers a cost-effective and adaptable solution for various industries, including healthcare and pharmaceuticals, where precise fluid delivery is essential [4]. The system's primary components consist of a Raspberry Pi, a stepper motor, a lead screw mechanism, and a motor controller driver. Operating on a Linux platform and programmed in Python, the Raspberry Pi provides a user-friendly interface. The stepper motor ensures precise control of flow rates, while the lead screw mechanism transforms rotary motion into linear motion to drive the syringe plunger. Through experimentation, the system achieved a flow rate of 0.07 ml/min, closely matching the theoretically predicted value of 0.1 ml/min. Thanks to its flexibility in controlling flow rates and its utilization of open-source technology, it has proven to be a valuable tool for a wide range of applications that demand accurate and controlled fluid delivery [5].

The article explores the role of surface micromachining in microfluidic systems and the application of syringe pumps for precise fluid delivery. It focuses on the design and testing of an 8051 microcontroller-based syringe pump system, particularly its performance with water at various flow rates [6]. This syringe pump is actuated by a meticulous stepper motor coupled with a lead screw mechanism, ensuring the seamless conversion of rotary to linear motion. The study takes a dual approach to validate the system: rigorous simulations using online software and real-world experiments, establishing the foundation for its precision. The precision of this syringe pump shines as it administers water at diverse flow rates, with each droplet's mass meticulously measured using a precision balance. The congruence of experimental results and numerical predictions underscores the system's reliability. An error function adds an extra layer of precision to maintain consistent and controlled flow rates, complemented by meticulous system alignment [7].

The versatility of the syringe pump is a notable feature; it is easily adaptable for various fluids through numerical simulations, making it suitable for specific microchannel applications. Additionally, the article delves into controlling stepper motors and optimizing kinematic design for seamless operation [8].

Execution Principle

The research focuses on the syringe, with the servo adjusting its degree of rotation based on the syringe's volume. The volume and speed of the dispensed medicine or fluid depend on the pressure applied and the motor's angle of rotation. To ensure accurate dispensing, the applied pressure must match that at the syringe's outlet. Therefore, Bernoulli's Equation is used to calculate the pressure.

Fig. (1), A1 represents the area of the syringe holder (plunger), and A2 represents the area of the syringe outlet. Additionally, V1 and V2 denote the velocities of the plunger and the liquid, respectively. According to Bernoulli's Equation:

Fig. (1). Basic Syringe.

$$\frac{P1}{\rho g} + \frac{V1^2}{2g} + \Delta 1 = \frac{P2}{\rho g} + \frac{V2^2}{2g} + \Delta 2 \tag{1}$$

To ensure precision during injection, we need to inject only when the syringe is in a flat position, so $\Delta 1h$ will be equal to $\Delta 2$zero, and $\frac{P2}{\rho g}$ P2 will be equal to 11.2 to account for atmospheric pressure. This leads to the equation

$$\frac{P1}{\rho g} = 11.2 + \frac{V2^2}{2g} - \frac{V1^2}{2g} \tag{2}$$

Next, we must calculate the required force to apply. By acknowledging that force is equal to the product of pressure and area, we can substitute the value of P1 to calculate the force. Following this, we will assess the speed and rotation of the servo motor.

Experimental Configuration

The diagram in Fig. (2) illustrates the experimental setup used in our research. It showcases the 8051 microcontroller, a syringe, and a 5V servo motor. Medical professionals can adjust the liquid volume for administration using either buttons or a keyboard interface. Once the desired flow rate is selected, the motor initiates rotation to control the movement of the rack and pinion, which are directly linked to the syringe. As the motor turns, the connected pinion also rotates, governing the movement of the rack. This reciprocal motion of the rack enables the suction and delivery of the liquid.

Fig. (3) shows us the stepper motor configuration. We use a motor driver ULN2003, which helps us interface the motor with a microcontroller.

Fig. (4) shows us the flowchart that represents the workflow and how the system operates.

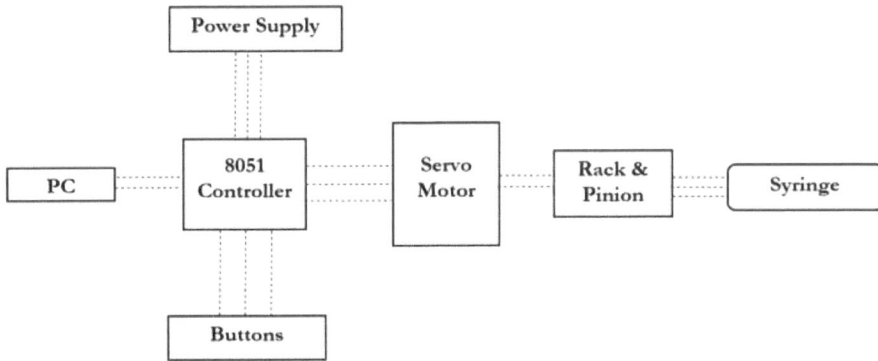

Fig. (2). Experimental set (Servo motor).

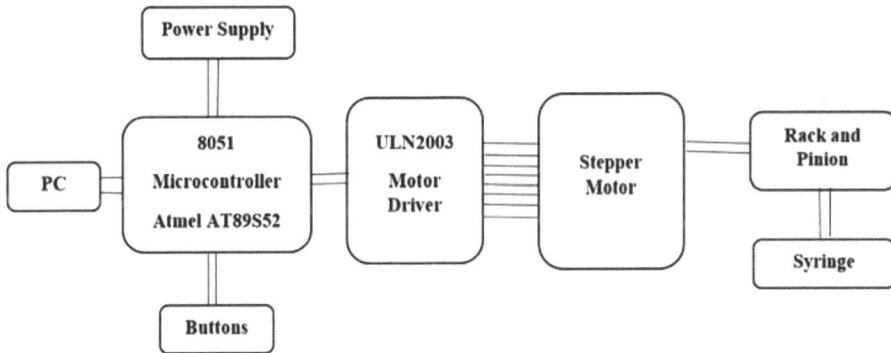

Fig. (3). Experimental setup (stepper).

RESULTS

In the realm of controlling a syringe using either a stepper motor or a servo motor in conjunction with an 8051 microcontroller, distinct approaches come into play, each with its unique advantages and drawbacks. These methodologies primarily differ in terms of precision and accuracy, complexity, and torque and power requirements, ultimately providing diverse solutions to the task at hand.

Let's first discuss the importance of precision and accuracy. Figs. (**5** and **6**) illustrate the hardware setup for controlling a syringe using a stepper and servo motor, along with a microcontroller. Stepper motors are known for their exceptional precision and accuracy in position control, achieved through their movement in discrete steps. This feature is particularly beneficial in applications requiring meticulous control, such as precise fluid dispensing. Stepper motors excel in accurately controlling the volume of dispensed fluid. On the other hand, servo motors also offer high precision but in a continuous manner. Their ability to provide consistent control over the syringe, especially when combined with feedback mechanisms like encoders, is invaluable, making them suitable for applications with dynamic load or environmental conditions where real-time accuracy is crucial.

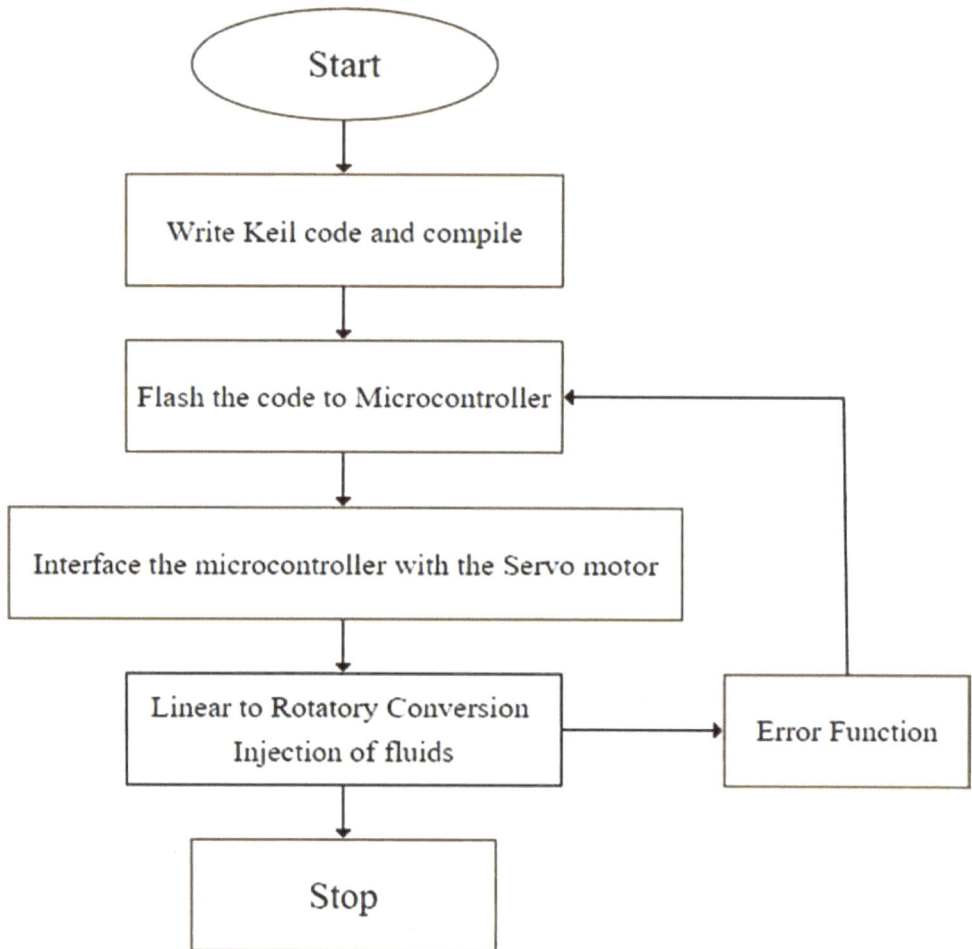

Fig. (4). Flowchart of system operations.

Both approaches offer unique benefits when considering torque and power requirements. Stepper motors are adept at providing high torque at low speeds, making them an excellent choice for precision applications like syringe control that demand precise positioning or gradual movement without a constant power source.

Fig. (5). Syringe control using a stepper.

Fig. (6). Syringe control using servo.

On the other hand, servo motors provide greater torque at high speeds, making them the preferred option for applications requiring fast and precise movements. It's important to note, however, that servo motors typically consume continuous power to maintain their position, which is a crucial factor in power-efficient applications.

Table 1. Comparison of stepper motor and servo motor.

Characteristic	Stepper Motor	Servo Motor
Power	Energy efficient	Less energy efficient
Position control	Less control	More accurate control
Response speed	Fast response	Slow response
Design	Simple to use	Complex
Speed	Slow	Fast
Motor weight	Light Weight	Heavy
Repeatability	Lower	Higher
Cost	Low Cost	High Cost

CONCLUSION

In summary, the decision between using a stepper motor or a servo motor for syringe control applications with an 8051 microcontroller depends on the specific requirements and priorities of the task at hand. Stepper motors are valued for their precision, simplicity, and energy-efficient operation, making them a wise choice when precise control is the main focus. On the other hand, servo motors are known for their continuous accuracy, adaptability to changing conditions, and responsiveness to rapid changes, especially in situations where speed and precision are crucial, even at the expense of constant power consumption. Ultimately, the suitability of each approach depends on the unique needs of the syringe control application being addressed.

REFERENCES

[1] Mathias A.M, Jayaraj S, Sivanandan KS. Driving Circuits for DC Motor Control Using 8051 Microcontroller Suitable for Applications Related to Prosthetic Legs. Int J Curr Eng Technol. [cited 2023 Dec 4].

[2] Adesina, Ganiyu, Shoewu, Olatinwo SO and omitola OO development of a microcontroller-based motor speed control system using intel 8051. Journal of Advancement In Engineering and Technology. 2014. [cited 2014 Mar 12]

[3] Assuncao R, Barbosa P. and Ruge R, *et al.* Developing the control system of a syringe infusion pump. In: 2014 11th International Conference on Remote Engineering and Virtual Instrumentation (REV), Porto, Portugal, 2014. p. 254-255.
[http://dx.doi.org/10.1109/REV.2014.6784270]

[4] Raja S, Silva A, Year CN. Advanced control system for syringe and infusion pump using IoT. IJIRAE - International Journal of Innovative Research in Advanced Engineering 2019.

[5] Appaji MSV, Shivakanth Reddy G, Arunkumar S, Venkatesan M. An 8051 Microcontroller based Syringe Pump Control System for Surface Micromachining. Procedia Materials Science 2014; Vol. 5.

[6] Dhone KA, Joshi S, Sonaskar S. Next-gen medical technology a smart approach to syringe injection pump automation. J Phys Conf Ser 2024; 2763(1): 012028.

[http://dx.doi.org/10.1088/1742-6596/2763/1/012028]

[7]   Mallick B, Mohanta HC. Development of a syringe infusion pump 2023; 6: 19-32.
       [http://dx.doi.org/10.5281/zenodo.10401078]

[8]   Lee Y, Zalewski D, Oleksy P, *et al.* Usability of the SB11 pre-filled syringe (PFS) in patients with
       retinal diseases. Adv Ther 2024; 41(8): 3426-36.
       [http://dx.doi.org/10.1007/s12325-024-02937-3] [PMID: 38976126]

CHAPTER 22

Real-time ECG Analysis and Classification Using Neural Networks in IoT Devices

Mohamed Osman Zaid K.B.[1], Aayush Singh[1], Sivaraj Chandrasekaran[1] and **Sridhar Raj S.[1,*]**

[1] *School of Computer Science and Engineering, Vellore Institute of Technology, Vellore, India*

Abstract: This paper presents the development of a simple, cost-efficient, and accessible ECG analyzer. The proposed prototype can acquire ECG signals from subjects and display them in real time through IoT devices such as mobile phones. It can diagnose conditions by classifying subjects as having arrhythmia, congestive heart failure, or normal sinus rhythm. The prototype includes a pre-trained deep learning model that helps with ECG signal categorization, delivering rapid insights about the subject's probable medical care needs.

Keywords: Convolutional neural network, ECG, IoT, Neural networks.

INTRODUCTION

Heart-related problems are among the most common health issues faced by humans today. A recent survey reported that nearly one in three people have heart disease. Heart disease encompasses various conditions, including heart valve disease, heart infection, disease of the heart muscle, congenital heart defects, heart rhythm issues, coronary artery disease, and more. Some heart problems are genetic and unavoidable, while others are preventable. Risk factors include age, sex, family history, poor diet, smoking, high blood pressure, obesity, stress, physical inactivity, and more. The most common and low-cost tool used in healthcare for analyzing the heart's electrical signals is the Electrocardiogram (ECG). ECG can detect different types of arrhythmias. Despite the availability of deep learning-based automated arrhythmia classification techniques with high accuracy, healthcare professionals have not widely adopted them. The primary problems involve the utilization of imbalanced data for categorization, as well as the classification methods [1]. Continuous wavelet transform (CWT) is used to decompose ECG signals to obtain different time-frequency components, and the

* **Corresponding author Sridhar Raj S.:** School of Computer Science and Engineering, Vellore Institute of Technology, Vellore, India; E-mail: sridharselva394@gmail.com

Sivakumar Rajagopal, Prakasam P., Konguvel E., Shamala Subramaniam, Ali Safaa Sadiq Al Shakarchi & B. Prabadevi (Eds.)

surrounding four RR interval features are extracted and combined with CNN features to input into a fully connected layer for ECG classification [2]. Ribeiro *et al.* proposed a lightweight solution using quantized one-dimensional deep convolutional neural networks, which are ideal for the real-time continuous monitoring of cardiac rhythm. Raw input data is used as input for the classifier, eliminating the need for complex data processing on low-powered wearable devices; data analysis can be carried out locally on edge devices, providing privacy and portability [3]. Granados *et al.* detailed the implementation of an IoT platform for real-time analysis and management of a network of bio-sensors and gateways, as well as a cloud-based deep neural network architecture for the classification of ECG data for multiple cardiovascular conditions [4]. In 2022, Kumar *et al.* proposed an approach where ECG signals collected using IoT nodes are processed to generate the QRS complex and the RR interval for establishing the feature vector for arrhythmia classification. The proposed Coy-Grey Wolf optimization-based deep convolution neural network (Coy-GWO-based Deep CNN) classifier detects anomalies in the ECG signal [5]. A paper presents a 1-dimensional convolutional neural network (CNN) for heartbeat classification from ECG signals obtained from an ambulatory device, classifying heartbeats into five classes as specified in the AAMI standard and tested using the Physionet MIT-BIH Arrhythmia database [6]. Another study proposed a compressed learning (CL) algorithm combined with a one-dimensional (1-D) convolutional neural network (CNN) that directly learns on ECG signals in the compression domain without expanded normalization, bypassing the reconstruction step and minimizing the raw input data dimension, significantly reducing processing power [7]. Li *et al.* conducted a study using convolutional neural networks to classify ECG image types using IoT to develop ECG signal measurement prototypes and simultaneously classify signal types through deep neural networks. The obtained signal is divided into QRS widening, sinus rhythm, ST depression, and ST elevation, with three models—ResNet, AlexNet, and SqueezeNet—developed with 50% of the training set and test set [8]. Another study used data from the MIT-BIH database for experiments, pre-processed the signals using different filters such as low pass filter and median filter, and utilized the discrete wavelet transform (DWT) technique to extract features, leading to the use of a deep neural network (DNN) for the classification model [9]. ECG machines are not typically set up at home, are not portable, and are expensive.

Related Works

The main concerns include imbalanced data for classification and the algorithm used for classification [12 - 15]. Continuous wavelet transform or CWT is used to decompose ECG signals to obtain different time-frequency components, and the surrounding four RR interval features are extracted and combined with CNN features to input into a fully connected layer for ECG classification [9]. Ribeiro *et al.* proposed a quantized one-dimensional deep convolutional neural network to monitor the cardiac rhythm. They have used the raw data to evaluate the classifier and eliminate the burden of data pre-processing on lighter devices with less power consumption. Data analysis is carried out on edge devices, ensuring mobility and privacy [10]. Granados *et al.* discussed the IoT platform implementation for real-time data management of bio-sensor networks. Also, they performed classification on multiple cardiovascular conditions using ECG data through deep neural networks [11].

In 2022, Kumar *et al.* [5] suggested a strategy in which IoT nodes gather ECG data and create the QRS complex and RR interval. Later, it was used to construct feature vectors to accomplish arrhythmia classification using the proposed Coy-Grey Wolf optimization-based deep convolution neural network (Coy-GW--based Deep CNN) classifier, which finds abnormalities in the ECG data. A paper discusses using a one-dimensional convolutional neural network (CNN) to classify heartbeats from an ambulatory device into various classes as specified by the AAMI standard, tested using the Physionet MIT-BIH Arrhythmia database [6]. Also, in another study, a compressed learning or CL algorithm combined with a CNN that directly learns on ECG signals in the compression domain is proposed, where such an approach bypasses the reconstruction step and minimizes the raw input data dimension, which significantly reduces the processing power [7]. Li *et al.* carried out a study that uses convolutional neural networks to classify ECG image types using the Internet of Things (IoT) to develop ECG signal measurement prototypes and simultaneously classify signal types through deep neural networks, divided into QRS widening, sinus rhythm, ST depression, and ST elevation, where three models, ResNet, AlexNet, and SqueezeNet, are developed with 50% of the training set and test set [8].

In an additional study, the MIT-BIH database was used, and the signals were pre-processed with a variety of filters, including low-pass and median filters. To improve efficiency, the discrete wavelet transform (DWT) approach was used for feature extraction and a deep neural network (DNN) for classification [9]. ECG machines cannot be set up at home, are awkward, and are not inexpensive.

Proposed Methodology

Input

Conductive pads, known as electrodes, are attached to the skin to record electrical currents. These electrodes connect to a signal conditioning block, AD8232, which gathers the electrical activity, collects the signal, and performs elementary pre-processing on the acquired signal.

Signal Pre-processing

The acquired signals are recorded and fed into the developed algorithm/model in MATLAB/Simulink. These models perform further signal pre-processing to cancel out noise and filter the signal for further use.

Deep Learning Neural Network

A dataset of 300+ ECG signals has been collected and classified into three categories: Normal, Arrhythmia, and Congestive Heart Failure. A pre-trained DL network is selected and trained through various techniques with various training and test data. An accurate model is chosen and made available to predict the manual signals obtained from a subject in real time.

Dataset used: https://www.kaggle.com/datasets/protobioengineering/mit-bih-arrhythmia-database-modern-2023

QRS Detection and Heart Rate Display

This model processes the signal and displays the heart rate of the subject along with the ECG, as shown in Fig. (**1**).

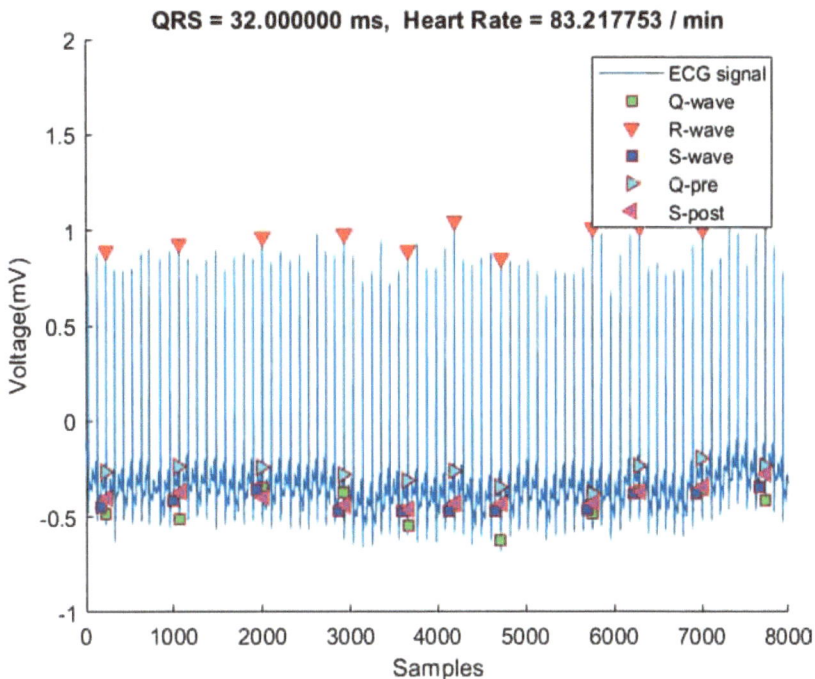

Fig. (1). QRS detection model.

Deployment of the Model to IoT Enterprises

This entire software-simulated model is made accessible through smartphones. After entering the signals from the AD8232 panel into their mobile device, the user may access a mobile application created using all of the previously discussed procedures. The MATLAB Android toolbox makes it feasible to create an easy-to-use, free application accessible to the general public.

Data Description

- 162 ECG Signals of 3 classes (Arrhythmia, Congestive Heart Failure, Normal Sinus Rhythm).
- Each Signal is 65536 samples long.
- Raw and almost noise-free.

Implementation and Results Analysis

The AD8232 ECG sensor works with jumper wires and skin electrodes. A system with Python/MATLAB (containing the Arduino hardware package) and an Arduino UNO board, as well as a connecting cable for 3 Lead ECG installation, is provided. Leads are located on the left arm (LA), right arm (RA), and left limb (LL), as shown in Fig. (**2**).

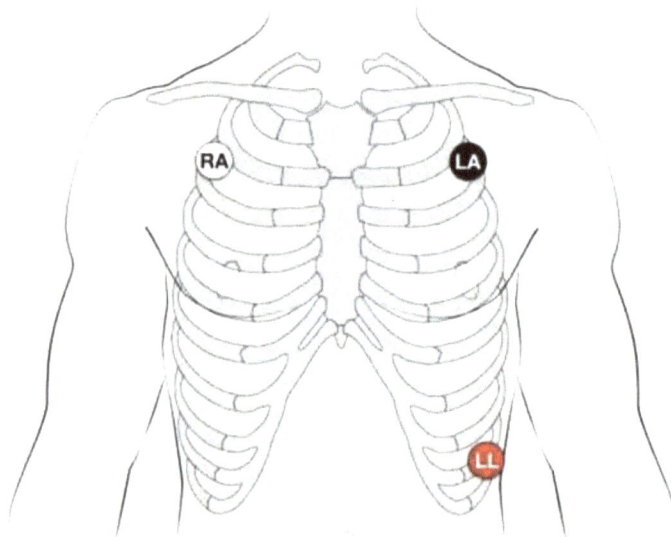

Fig. (2). Electrode placement.

LSTM Network

LSTM-based classification uses feature extraction wavelet scattering. The wavelet scattering network is used to extract low-variance features from real-valued time series. The wavelength scattering filter bank is reconstructed, reducing the feature set from 65536 samples long to 499 x 8. The total number of hidden units is 120, and the maximum epochs are 250. The training algorithm/solver used is adaptive moment estimation (Adam Solver), with an initial learning rate of 0.01. Data is divided randomly with weights of 80-20 for training and testing, as shown in Fig. (**3**).

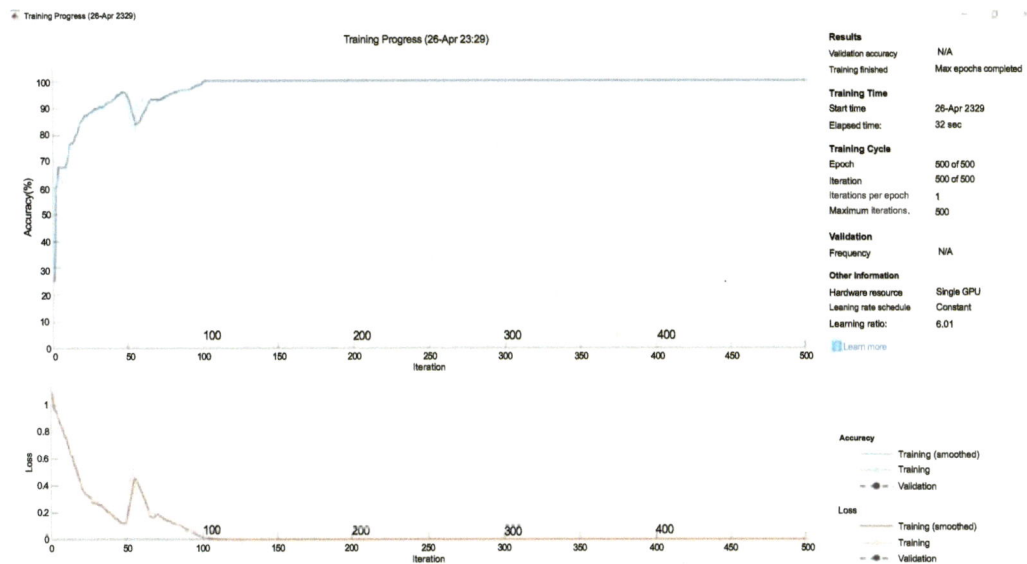

Fig. (3). LSTM training performance.

CNN Network Using Transfer Learning

The CNN-based architecture makes use of transfer learning to fine-tune pre-trained network models for the job of signal classification. There are other networks available, including AlexNet, with certain changes. CNN works well with visuals, as shown in Figs. (**4** and **5**).

The goal is to produce a time-frequency representation of the ECG signals, which represent how the signal content or the frequency content is involved as a function of time. This approach uses two techniques: the spectrogram technique, as shown in Fig. (**6**), and the scalogram technique, as shown in Fig. (**7**). The scalogram of all 162 signals is processed and stored. These scalogram images are saved in a local datastore to later feed them as input for CNN to train the model. Obtained images are of size 673x673x3 and are resized to 227x227x3 (Fig. **8**) to use in AlexNet with SDGM optimizer with the following equation.

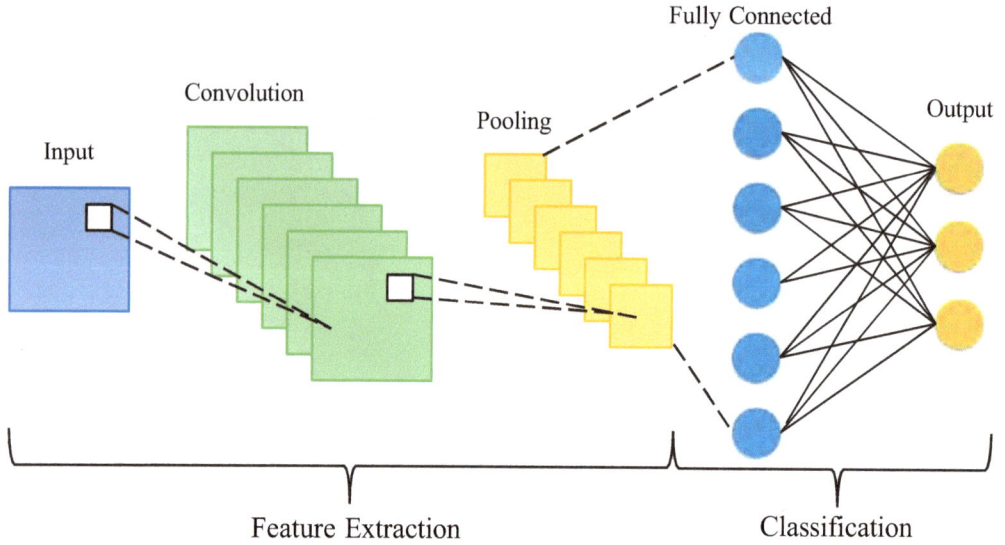

Fig. (4). CNN Architecture.

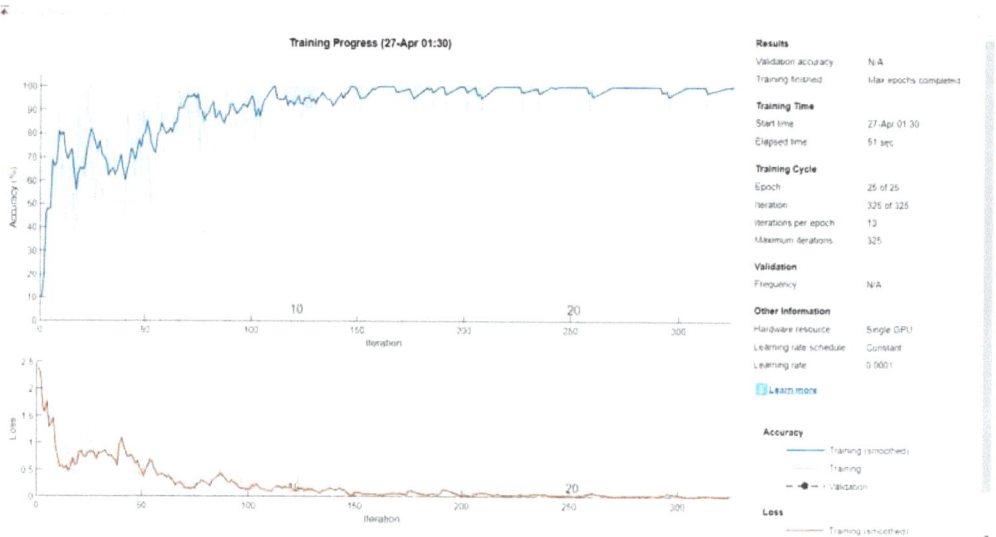

Fig. (5). CNN training performance.

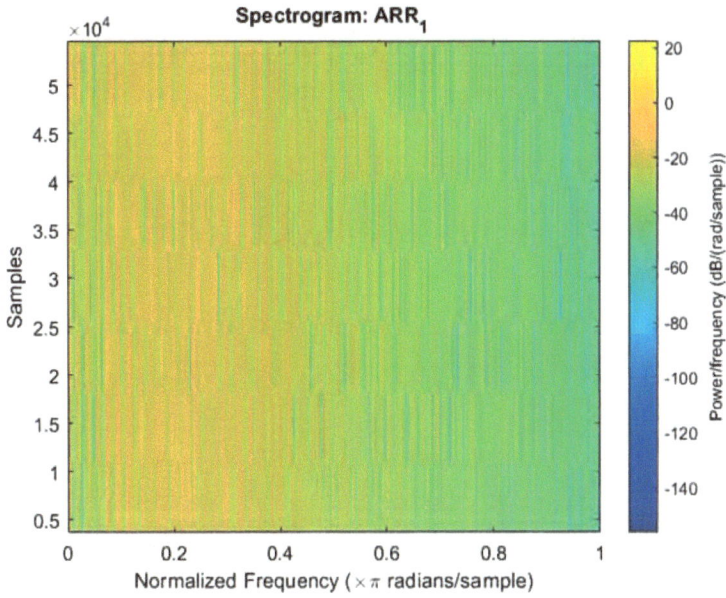

Fig. (6). Spectrogram ARR.

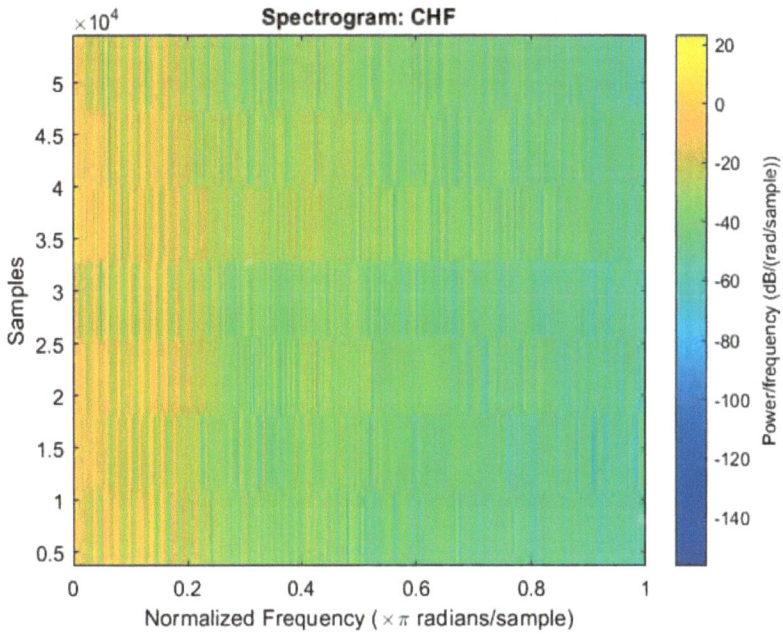

Fig. (7). Spectrogram CHF.

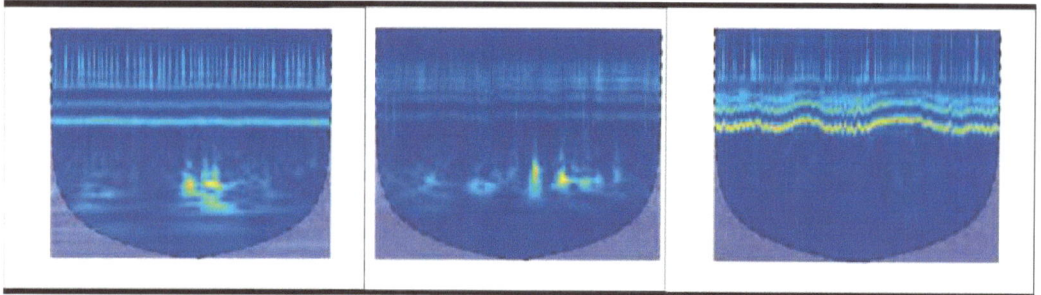

Fig. (8). Scalogram CHF.

Real-time Acquisition Using UBIDOTS

Ubidots is an Internet of Things (IoT) application builder with data analytics and visualization capabilities. It turns sensor data into information that matters for business decisions, machine-to-machine interactions, educational research, and increased economization of global resources. Ubidots is the cloud platform where the real-time ECG signal is uploaded. The cloud platform is accessible from anywhere in the world *via* an internet connection (Figs. **9** and **10**).

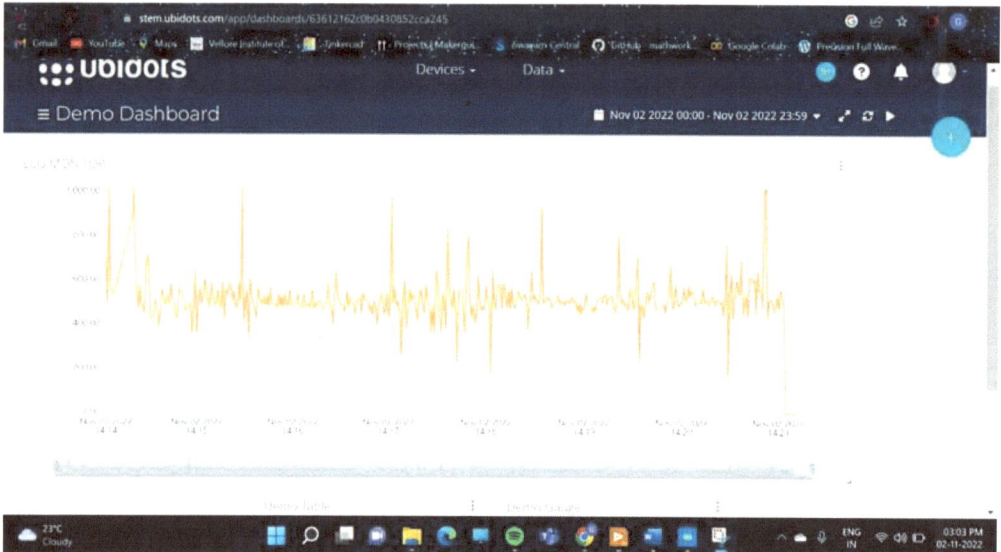

Fig. (9). Ubidots.

Fig. (10). Code to get real-time data.

Pseudo Code

- Initialize and clear workspace
- Load data
- Split data into training and testing sets
  - Use 'dividerand' to split indices into 80% training and 20% testing
- Extract signals and labels
  - Extract a specific signal for plotting
  - Extract labels
- Plot a single sample signal
  - Plot the data points of the signal
- Plot a sample from 'user'
  - Plot the signal 'aa' with x-axis limits from 200 to 800
- Spectrogram and Continuous Wavelet Transform (CWT)
  - Create a CWT filter bank
  - Check if 'data' directory exists; if not, create it
  - Loop through each signal to compute and save CWT images
- Train a model
  - Define the path to the image data
  - Create an image datastore with images from the specified path
  - Split the images into training and testing sets
- Load and modify AlexNet
  - Load the pre-trained AlexNet model
  - Modify the network layers to match the number of classes

- Set custom read function for training images
  - Define a custom function to preprocess images
- Train the network
  - Set training options (learning rate, epochs, mini-batch size, *etc.*)
  - Train the network using the training images and modified AlexNet layers
- Evaluate the model
  - Load the trained network
  - Set custom read function for test images
  - Classify test images using the trained network
  - Calculate the accuracy of the model
- Evaluate the model with a specific image
  - Define the path to a specific image
  - Create an image datastore for the specific image
  - Set custom read function for the specific image
  - Classify the specific image using the trained network
  - Define the true label for the specific image
- Generate confusion matrices
  - Compute and plot confusion matrix for test images
  - Compute and plot confusion matrix for the specific image
- Define custom read function
  - Read image from file
  - Resize image to 227x227 pixels

Table 1. Comparative analysis of the accuracy.

Researchers	Preprocessing Technique	Database	Modeling Technique	Accuracy (%)
M.Das and S. Ari (2014)	Feature extraction: ST, DWT level, Pan Tompkin's QRS detection algorithm	MIT-BIH arrhythmia	MLPNN classifier	97.5%
J.Nasiri *et al.* (2009)	Noise removal: DWT, Feature reduction: PCA, GA (meta-heuristic)	MIT-BIH arrhythmia	Genetic algorithm - SVM	93.46%
A.Muthuchudar and S. Baboo(2013)	Noise removal: Wavelet transform (UWT)	MIT-BIH arrhythmia	Feed-forward network with backpropagation algorithm as training algorithm	96%
Sadiq and N. Shukr (2013)	Denoised using DWT, Low pass and High pass filter, Wavelet selection: Harr filter, Daubechies wavelet	MIT-BIH arrhythmia	ID3 Decision tree Haar-ID3	94%

(Table 1) cont.....

Researchers	Preprocessing Technique	Database	Modeling Technique	Accuracy (%)
V. Srivastava and D. Prasad (2013)	Feature extraction (DWT)	MIT-BIH arrhythmia	Feedforward neuro fuzzy combination of Fuzzy logic and MLPNN	85%
Singh, M. Zaid, S. Chandrasekaran (2023)	Noise removal: Continuous Wavelet Transform (CWT); Feature extraction Wavelet Scattering	MIT-BIH arrhythmia	LSTM Network	98.83%

RESULTS AND DISCUSSIONS

Once the acquired signal is uploaded to the cloud service database, the CNN is trained to classify/predict whether the subject has arrhythmia, congestive heart failure, or normal sinus rhythm, as shown in Fig. (**11**). The model achieved an accuracy of 98.83%.

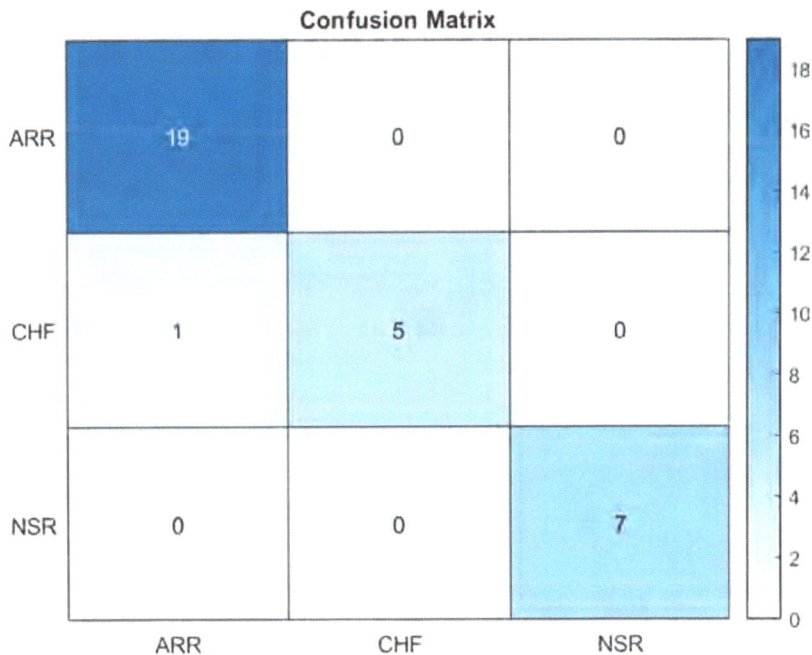

Fig. (11). Confusion matrix.

CONCLUSION

This project has developed an Easily available ECG kit that is affordable. Anyone with a smartphone can use it, hence reaching many people. Inexpensive diagnosis for pre-elementary conditions requires ECG tests.

No usage of sophisticated electronic parts and readily available replaceable spares is required. This kit would help in the preliminary diagnosis of a subject who is suspected to be a patient of cardiovascular disease. This cost-effective kit helps in narrowing down the possible conclusion a physician can get from a complete ECG device rather than an accurate one. Comparing the cost of 1 ECG recording session to the one proposed in this project is the takeaway and goal of this project.

REFERENCES

[1]   Byeon YH, Pan SB, Kwak KC. Intelligent deep models based on scalograms of electrocardiogram signals for biometrics. Sensors (Basel) 2019; 19(4): 935.
[http://dx.doi.org/10.3390/s19040935] [PMID: 30813332]

[2]   Liu Z, Yao G, Zhang Q, Zhang J, Zeng X. Wavelet scattering transform for ECG beat classification. Comput Math Methods Med 2020; 2020: 1-11.
[http://dx.doi.org/10.1155/2020/3215681] [PMID: 33133225]

[3]   Sepúlveda A, Castillo F, Palma C, Rodriguez-Fernandez M. Emotion recognition from ECG signals using wavelet scattering and machine learning. Appl Sci (Basel) 2021; 11(11): 4945.
[http://dx.doi.org/10.3390/app11114945]

[4]   Kleć M. Early detection of heart symptoms with convolutional neural network and scattering wavelet transformation. InFoundations of Intelligent Systems: 24th International Symposium, ISMIS 2018, Limassol, Cyprus, October 29–31, 2018. Proceedings 2018; 24: 24-31. [Springer International Publishing.].

[5]   Thakor NV, Zhu YS. Applications of adaptive filtering to ECG analysis: noise cancellation and arrhythmia detection. IEEE Trans Biomed Eng 1991; 38(8): 785-94.
[http://dx.doi.org/10.1109/10.83591] [PMID: 1937512]

[6]   Lyon A, Mincholé A, Martínez JP, Laguna P, Rodriguez B. Computational techniques for ECG analysis and interpretation in light of their contribution to medical advances. J R Soc Interface 2018; 15(138): 20170821.
[http://dx.doi.org/10.1098/rsif.2017.0821] [PMID: 29321268]

[7]   Gupta R, Bera JN, Mitra M. Development of an embedded system and MATLAB-based GUI for online acquisition and analysis of ECG signal. Measurement 2010; 43(9): 1119-26.
[http://dx.doi.org/10.1016/j.measurement.2010.05.003]

[8]   Silipo R, Marchesi C. Artificial neural networks for automatic ECG analysis. IEEE Trans Signal Process 1998; 46(5): 1417-25.
[http://dx.doi.org/10.1109/78.668803]

[9]   Tyagi PK, Rathore N, Agrawal D. A review on heartbeat classification for arrhythmia detection using ECG signal processing. Conference on Electrical, Electronics and Computer Science (SCEECS) 2023; 1-6.
[http://dx.doi.org/10.1109/SCEECS57921.2023.10063143]

[10]   Anbalagan T, Nath MK, Vijayalakshmi D, Anbalagan A. Analysis of various techniques for ECG Signal in Healthcare, Past, Present, and Future. Biomedical Engineering Advances. 2023 May 29:100089.

[11]   Aditya Y, Devi SS, Prasad BDCN. A real-time ECG CTG based ensemble feature extraction and unsupervised learning based classification framework for multi-class abnormality prediction. Int J Adv Comput Sci Appl 2023; 14(3).
[http://dx.doi.org/10.14569/IJACSA.2023.0140396]

[12]   Karri M, Annavarapu CSR. A real-time embedded system to detect QRS-complex and arrhythmia classification using LSTM through hybridized features. Expert Syst Appl 2023; 214: 119221.

[http://dx.doi.org/10.1016/j.eswa.2022.119221]

[13]   Kumar SS, Rinku DR, Kumar AP, Maddula R, Palagan CA. An IOT framework for detecting cardiac arrhythmias in real-time using deep learning resnet model. Measurement. Sensors (Basel) 2023; 29: 100866.

[14]   Duong LT, Doan TTH, Chu CQ, Nguyen PT. Fusion of edge detection and graph neural networks to classifying electrocardiogram signals. Expert Syst Appl 2023; 225: 120107.
[http://dx.doi.org/10.1016/j.eswa.2023.120107]

[15]   Karri M, Annavarapu CSR, Pedapenki KK. A real-time cardiac arrhythmia classification using hybrid combination of delta modulation, 1D-CNN and blended LSTM. Neural Process Lett 2023; 55(2): 1499-526.
[http://dx.doi.org/10.1007/s11063-022-10949-9]

Appendix

The neuromorphic computing application of spiking neural network-related datasets and Python codes are available in the following links.

A. DATASETS

i. https://archi ve.ics.uci.edu/ml/datas ets/Exase ns

ii. https://physionet.org/content/chbmit/1.0.0/.

iii. https://physionet.org/content/mitdb/1.0. 0/.

iv. http://ninaweb.hevs.ch/node/3.

v. https://figshare.com/s/d03a91081824536f12a8

vi. http://www2.imse-cnm.csic.es/caviar/MNISTDVS.html

vii. http://www.garrickorchard.com/datasets

B. PYTHON CODES FOR SNN IN NEUROMORPHIC COMPUTING

i. https://github.com/Pouya-SZ/Bioneuromorphics/blob/master/Analog2Binary_ Conversion. ipynb

ii. https://github.com/djlouie/snntorch

iii. https://github.com/nengo/nengo

iv. https://github.com/norse/norse

v. https://github.com/SynSense/rockpool

APPENDIX

The following table presents the Python source codes for the several transfer learning strategies used in this chapter.

Table 2. Python codes for various transfer learning techniques

Python Source Code
VGG Net 16

// Importing all necessary libraries
from keras.layers import Input,Lambda,Dense,Flatten
from keras.models import Model
from keras.applications.vgg16 import VGG16
from keras.applications.vgg16 import preprocess_input
from keras.preprocessing import image
from keras.models import Sequential
from keras.preprocessing.image import ImageDataGenerator
import numpy as np
from glob import glob
import matplotlib.pyplot as plt
// Taken image default size as a 224*224
image_size=[224,224]
// Load the dataset from the drive

(Table 2) cont.....

Python Source Code

```
train_path='/content/drive/MyDrive/Colab Notebooks/Alzheimer_s Dataset/train'
test_path='/content/drive/MyDrive/Colab Notebooks/Alzheimer_s Dataset/test'
// Create object for VGG 16 pretrained model
vgg=VGG16(input_shape=image_size+ [3],weights='imagenet',include_top=False)
for layer in vgg.layers:
layer.trainable=False
folders = glob('/content/drive/MyDrive/Colab Notebooks/Alzheimer_s Dataset/train/*')
x = Flatten()(vgg.output)
prediction = Dense(len(folders), activation='softmax')(x)
model = Model(inputs=vgg.input, outputs=prediction)
model.summary()
// Compile the model
model.compile(
loss='categorical_crossentropy',
optimizer='adam',
metrics=['accuracy']
)
// Generating more number of images using data augmentation
from keras.preprocessing.image import ImageDataGenerator
train_datagen = ImageDataGenerator(rescale = 1/255,
shear_range = 0.2,
zoom_range = 0.2,
horizontal_flip = True)
test_datagen = ImageDataGenerator(rescale = 1./255)
training_set = train_datagen.flow_from_directory('/content/drive/MyDrive/Colab Notebooks/Alzheimer_s Dataset/train',
target_size = (224, 224),
batch_size = 32,
class_mode = 'categorical')
test_set = test_datagen.flow_from_directory('/content/drive/MyDrive/Colab Notebooks/Alzheimer_s Dataset/test',
target_size = (224, 224),
batch_size = 32,
class_mode = 'categorical')
r = model.fit_generator(
training_set,
validation_data=test_set,
epochs=30,
)
// Plotting the graph for the following performance metrics training accuracy and validation accuracy.
plt.plot(r.history['accuracy'],label='train_accuracy')
plt.plot(r.history['val_accuracy'],label='validation_accuracy')
plt.xlabel('Number of epochs')
plt.ylabel('Accuracy')
plt.legend()
plt.show()
// Plotting the graph for the following performance metrics training loss and validation loss.
plt.plot(r.history['loss'],label='train_loss')
plt.plot(r.history['val_loss'],label='validation_loss')
plt.xlabel('Number of epochs')
plt.ylabel('Loss')
plt.legend()
plt.show()
```

Resnet 50

```
// Importing all necessary libraries
from keras.layers import Input,Lambda,Dense,Flatten
from keras.models import Model
from keras.applications.inception_resnet_v2 import InceptionResNetV2
from keras.applications.resnet_v2 import ResNet50V2
from google.colab import drive
drive.mount('/content/drive')
from keras.applications.vgg16 import preprocess_input
from keras.preprocessing import image
from keras.models import Sequential
from keras.preprocessing.image import ImageDataGenerator
```

(Table 2) cont.....

Python Source Code

```
import numpy as np
from glob import glob
import matplotlib.pyplot as plt
// Set image default size as 224 * 224
image_size=[224,224]
// Load the dataset from the drive
train_path='/content/drive/MyDrive/Colab Notebooks/Alzheimer_s Dataset/train'
test_path='/content/drive/MyDrive/Colab Notebooks/Alzheimer_s Dataset/test'
// Create object for ResNet50V2 pretrained model
res50=ResNet50V2(input_shape=image_size+ [3],weights='imagenet',include_top=False)
for layer in res50.layers:
layer.trainable=False
folders = glob('/content/drive/MyDrive/Colab Notebooks/Alzheimer_s Dataset/train/*')
x = Flatten()(res50.output)
prediction = Dense(len(folders), activation='sigmoid')(x)
model = Model(inputs=res50.input, outputs=prediction)
model.summary()
// Compile the model
model.compile(
loss='binary_crossentropy',
optimizer='adam',
metrics=['accuracy']
)
// generate more images using Data Augmentation
from keras.preprocessing.image import ImageDataGenerator
train_datagen = ImageDataGenerator(rescale = 1./255,
shear_range = 0.2,
zoom_range = 0.2,
horizontal_flip = True)
test_datagen = ImageDataGenerator(rescale = 1./255)
training_set = train_datagen.flow_from_directory('/content/drive/MyDrive/Colab Notebooks/Alzheimer_s Dataset/train',
target_size = (224, 224),
batch_size = 32,
class_mode = 'categorical')
test_set = test_datagen.flow_from_directory('/content/drive/MyDrive/Colab Notebooks/Alzheimer_s Dataset/test',
target_size = (224, 224),
batch_size = 32,
class_mode = 'categorical')
// Fit and prediction of model
r = model.fit_generator(training_set,
validation_data=test_set,
epochs=30,
steps_per_epoch=len(training_set),
validation_steps=len(test_set)
)
// Plotting the following metrics such training accuracy and validation accuracy
plt.plot(r.history['accuracy'],label='train_accuracy')
plt.plot(r.history['val_accuracy'],label='validation_accuracy')
plt.xlabel('Number of epochs')
plt.ylabel('accuracy')
plt.legend()
plt.show()
// Plotting the following metrics such training loss and validation loss
plt.plot(r.history['loss'],label='train_loss')
plt.plot(r.history['val_loss'],label='validation_loss')
plt.xlabel('Number of epochs')
plt.ylabel('Loss')
plt.legend()
plt.show()
```

ResNet 101

```
// Import all necessary libraries
from keras.layers import Input,Lambda,Dense,Flatten
from keras.models import Model
from keras.applications.resnet_v2 import ResNet50V2
```

Python Source Code

```
from keras.applications.resnet_v2 import ResNet101V2
from keras.applications.vgg16 import preprocess_input
from keras.preprocessing import image
from keras.models import Sequential
from keras.preprocessing.image import ImageDataGenerator
import numpy as np
from glob import glob
import matplotlib.pyplot as plt
// Set image sizes as 224*224
image_size=[224,224]
// Load the dataset from the file
train_path='/content/drive/MyDrive/Colab Notebooks/Alzheimer_s Dataset/train'
test_path='/content/drive/MyDrive/Colab Notebooks/Alzheimer_s Dataset/test'
// Create a ResNet V2101 pretrained Model
res101=ResNet101V2(input_shape=image_size+ [3],weights='imagenet',include_top=False)
for layer in res101.layers:
layer.trainable=False
folders = glob('/content/drive/MyDrive/Colab Notebooks/Alzheimer_s Dataset/train/*')
x = Flatten()(res101.output)
prediction = Dense(len(folders), activation='softmax')(x)
model = Model(inputs=res101.input, outputs=prediction)
model.summary()
// Compile the model
model.compile(
loss='categorical_crossentropy',
optimizer='adam',
metrics=['accuracy']
)
// Generating more images using Data Augmentation
from keras.preprocessing.image import ImageDataGenerator
train_datagen = ImageDataGenerator(rescale = 1./255,
shear_range = 0.2,
zoom_range = 0.2,
horizontal_flip = True)
test_datagen = ImageDataGenerator(rescale = 1./255)
training_set = train_datagen.flow_from_directory('/content/drive/MyDrive/Colab Notebooks/Alzheimer_s Dataset/train',
target_size = (224, 224),
batch_size = 32,
class_mode – 'categorical')
test_set = test_datagen.flow_from_directory('/content/drive/MyDrive/Colab Notebooks/Alzheimer_s Dataset/test',
target_size = (224, 224),
batch_size = 32,
class_mode = 'categorical')
r = model.fit_generator(
training_set,
validation_data=test_set,
epochs=30,
steps_per_epoch=len(training_set),
validation_steps=len(test_set)
)
// Plotting the graph for training accuracy and validation accurcy
plt.plot(r.history['accuracy'],label='train_accuracy')
plt.plot(r.history['val_accuracy'],label='val_accuracy')
plt.xlabel('Number of Epochs')
plt.ylabel('Accuracy vs Validation Accuracy')
plt.legend()
plt.show()
// Plotting the graph for training loss and validation loss
plt.plot(r.history['loss'],label='train_loss')
plt.plot(r.history['val_loss'],label='val_loss')
plt. xlabel('Number of Epochs')
plt.ylabel('Loss vs validation Loss')
plt.legend()
plt.show()
```

Resnet 152

(Table 2) cont.....

Python Source Code

```
// Import all necessary libraries
from keras.layers import Input,Lambda,Dense,Flatten
from keras.models import Model
from keras.applications.resnet_v2 import ResNet152V2
from keras.applications.resnet_v2 import ResNet50V2
from keras.applications.resnet_v2 import ResNet152V2
from keras.applications.vgg16 import preprocess_input
from keras.preprocessing import image
from keras.models import Sequential
from keras.preprocessing.image import ImageDataGenerator
import numpy as np
from glob import glob
import matplotlib.pyplot as plt
image_size=[224,224]
train_path='/content/drive/MyDrive/Colab Notebooks/Skin_Cancer_dataset/Train'
test_path='/content/drive/MyDrive/Colab Notebooks/Skin_Cancer_dataset/Test'
// Create an object for ResNet152V2 pretrained model
res152=ResNet152V2(input_shape=image_size+ [3],weights='imagenet',include_top=False)
for layer in res152.layers:
layer.trainable=False
folders = glob('/content/drive/MyDrive/Colab Notebooks/Skin_Cancer_dataset/Train/*')
x = Flatten()(res152.output)
prediction = Dense(len(folders), activation='softmax')(x)
model = Model(inputs=res152.input, outputs=prediction)
model.summary()
model.compile(
loss='categorical_crossentropy',
optimizer='adam',
metrics=['accuracy']
)
// Create more number of images using Data Augmentation
from keras.preprocessing.image import ImageDataGenerator
train_datagen = ImageDataGenerator(rescale = 1./255,
shear_range = 0.2,
zoom_range = 0.2,
horizontal_flip = True)
test_datagen = ImageDataGenerator(rescale = 1./255)
training_set = train_datagen.flow_from_directory('/content/drive/MyDrive/Colab Notebooks/Skin_Cancer_dataset/Train',
target_size = (224, 224),
batch_size = 32,
class_mode = 'categorical')
test_set = test_datagen.flow_from_directory('/content/drive/MyDrive/Colab Notebooks/Skin_Cancer_dataset/Test',
target_size = (224, 224),
batch_size = 32,
class_mode = 'categorical')
// Fit and predict the model
r = model.fit_generator(
training_set,
validation_data=test_set,
epochs=50,
steps_per_epoch=len(training_set),
validation_steps=len(test_set)
)
// Plot the graph for training accuracy and validation accuracy
plt.plot(r.history['accuracy'],label='train_accuracy')
plt.plot(r.history['val_accuracy'],label='val_accuracy')
plt.ylabel('Accuracy vs Validation Accuracy')
plt.xlabel('Number of Epochs')
plt.legend()
plt.show()
Plot the graph for training loss and validation loss
plt.plot(r.history['loss'],label='train_loss')
plt.plot(r.history['val_loss'],label='val_loss')
plt.xlabel('Number of Epochs')
plt.ylabel('Loss vs Validation Loss')
plt.legend()
plt.show()
```

SUBJECT INDEX

www.ingramcontent.com/pod-product-compliance
Lightning Source LLC
Chambersburg PA
CBHW050759220326
41598CB00006B/63